Springer Tracts in Advanced Robotics

Volume 19

Editors: Bruno Siciliano · Oussama Khatib · Frans Groen

Springer Tracts in Advanced Robotics

Edited by B. Siciliano, O. Khatib, and F. Groen

Vol. 18: Barbagli, F.; Prattichizzo, D.; Salisbury, K. (Eds.)
Multi-point Interaction with Real and Virtual Objects
281 p. 2005 [3-540-26036-6]

Vol. 17: Erdmann, M.; Hsu, D.; Overmars, M.; van der Stappen, F.A (Eds.)
Algorithmic Foundations of Robotics VI
472 p. 2005 [3-540-25728-4]

Vol. 16: Cuesta, F.; Ollero, A.
Intelligent Mobile Robot Navigation
224 p. 2005 [3-540-23956-1]

Vol. 15: Dario, P.; Chatila R. (Eds.)
Robotics Research – The Eleventh International Symposium
595 p. 2005 [3-540-23214-1]

Vol. 14: Prassler, E.; Lawitzky, G.; Stopp, A.; Grunwald, G.; Hägele, M.; Dillmann, R.; Iossifidis. I. (Eds.)
Advances in Human-Robot Interaction
414 p. 2005 [3-540-23211-7]

Vol. 13: Chung, W.
Nonholonomic Manipulators
115 p. 2004 [3-540-22108-5]

Vol. 12: Iagnemma K.; Dubowsky, S.
Mobile Robots in Rough Terrain – Estimation, Motion Planning, and Control with Application to Planetary Rovers
123 p. 2004 [3-540-21968-4]

Vol. 11: Kim, J.-H.; Kim, D.-H.; Kim, Y.-J.; Seow, K.-T.
Soccer Robotics
353 p. 2004 [3-540-21859-9]

Vol. 10: Siciliano, B.; De Luca, A.; Melchiorri, C.; Casalino, G. (Eds.)
Advances in Control of Articulated and Mobile Robots
259 p. 2004 [3-540-20783-X]

Vol. 9: Yamane, K.
Simulating and Generating Motions of Human Figures
176 p. 2004 [3-540-20317-6]

Vol. 8: Baeten, J.; De Schutter, J.
Integrated Visual Servoing and Force Control
198 p. 2004 [3-540-40475-9]

Vol. 7: Boissonnat, J.-D.; Burdick, J.; Goldberg, K.; Hutchinson, S. (Eds.)
Algorithmic Foundations of Robotics V
577 p. 2004 [3-540-40476-7]

Vol. 6: Jarvis, R.A.; Zelinsky, A. (Eds.)
Robotics Research – The Tenth International Symposium
580 p. 2003 [3-540-00550-1]

Vol. 5: Siciliano, B.; Dario, P. (Eds.)
Experimental Robotics VIII
685 p. 2003 [3-540-00305-3]

Vol. 4: Bicchi, A.; Christensen, H.I.; Prattichizzo, D. (Eds.)
Control Problems in Robotics
296 p. 2003 [3-540-00251-0]

Vol. 3: Natale, C.
Interaction Control of Robot Manipulators – Six-degrees-of-freedom Tasks
120 p. 2003 [3-540-00159-X]

Vol. 2: Antonelli, G.
Underwater Robots – Motion and Force Control of Vehicle-Manipulator Systems
209 p. 2003 [3-540-00054-2]

Vol. 1: Caccavale, F.; Villani, L. (Eds.)
Fault Diagnosis and Fault Tolerance for Mechatronic Systems – Recent Advances
191 p. 2002 [3-540-44159-X]

Tine Lefebvre · Herman Bruyninckx · Joris De Schutter

Nonlinear Kalman Filtering for Force-Controlled Robot Tasks

With 86 Figures and 10 Tables

Springer

Professor Bruno Siciliano, Dipartimento di Informatica e Sistemistica, Università degli Studi di Napoli Federico II, Via Claudio 21, 80125 Napoli, Italy, email: siciliano@unina.it

Professor Oussama Khatib, Robotics Laboratory, Department of Computer Science, Stanford University, Stanford, CA 94305-9010, USA, email: khatib@cs.stanford.edu

Professor Frans Groen, Department of Computer Science, Universiteit van Amsterdam, Kruislaan 403, 1098 SJ Amsterdam, The Netherlands, email: groen@science.uva.nl

STAR (Springer Tracts in Advanced Robotics) has been promoted under the auspices of EURON (European Robotics Research Network)

Authors

Dr. Tine Lefebvre
Prof. Herman Bruyninckx
Prof. Joris De Schutter

Katholieke Universiteit Leuven
Department of Mechanical Engineering
Ceelestijnenlaan 300 B
3001 Leuven (Heverlee)
Belgium

ISBN 978-3-642-06629-0 e-ISBN 978-3-540-31504-9

ISSN print edition: 1610-7438
ISSN electronic edition: 1610-742X

Springer is a part of Springer Science+Business Media
springeronline.com

Softcover reprint of the hardcover 1st edition 2005

Cover-Design: design & production GmbH, Heidelberg
Printed on acid-free paper 89/3141/Yu - 5 4 3 2 1 0

Foreword

At the dawn of the new millennium, robotics is undergoing a major transformation in scope and dimension. From a largely dominant industrial focus, robotics is rapidly expanding into the challenges of unstructured environments. Interacting with, assisting, serving, and exploring with humans, the emerging robots will increasingly touch people and their lives.

The goal of the new series of Springer Tracts in Advanced Robotics (STAR) is to bring, in a timely fashion, the latest advances and developments in robotics on the basis of their significance and quality. It is our hope that the wider dissemination of research developments will stimulate more exchanges and collaborations among the research community and contribute to further advancement of this rapidly growing field.

The monograph written by Tine Lefebvre, Herman Bruyninckx and Joris De Schutter is focused on how to achieve more robot autonomy by means of reliable processing skills. The latest developments in the areas of contact modelling, nonlinear parameter estimation and task plan optimisation for improved estimation accuracy are discussed. Kalman filtering techniques are applied to identify the contact state based on force sensing between a grasped object and the environment. The potential of the work is to be found not only for industrial robot operation in space, sub-sea or nuclear scenarios, but also for service robots operating in unstructured environments co-habited by humans where autonomous compliant tasks require active sensing.

Remarkably, the doctoral thesis at the basis of this monograph was a finalist for the Fourth EURON Georges Giralt PhD Award devoted to the best PhD thesis in Robotics in Europe. A fine addition to the series!

Naples, Italy,
June 2005

Bruno Siciliano
STAR Editor

Preface

During the last decades, a number of "smart machines" entered our homes: washing machines, vacuum cleaners, entertainment pet robots, ... This makes us dream about the next generation of service robots which take over household chores. However, several major difficulties have yet to be overcome before this dream can come true. One of these difficulties is to make contact in a safe and controlled way with inaccurately positioned objects.

More robot autonomy can only be achieved by reliable sensor processing skills. This book describes the latest developments in the areas of contact modelling, nonlinear parameter estimation and task plan optimisation for improved estimation accuracy. The obtained autonomy is indispensable for the execution of tasks with inaccurately known objects and for tasks in environments that do not allow a precise positioning of the objects. Application areas include industrial robot operation in space, sub-sea or nuclear environments. The same research is relevant for service robots operating in environments that have been designed for humans and where even simple tasks require a lot of sensing.

Meise, Leuven, Belgium,
March 2005

Tine Lefebvre
Herman Bruyninckx
Joris De Schutter

No book, however small, is produced without the help of others. This monograph is the result of four years of PhD research at the Division of Production Engineering, Machine Design & Automation (PMA), Department of Mechanical Engineering, Katholieke Universiteit Leuven, Belgium. I am grateful to all people who contributed to this work. I would like to give a special word of thanks to my supervisors Joris De Schutter and Herman Bruyninckx for offering me support and guidance and at the same time giving me the freedom to choose my research niche. Also many thanks to the Fund for Scientific Research Flanders for providing a fellowship for this research. Last but not least, I am indebted to my parents who always encouraged and supported me to study something in which I was interested.

Meise, Belgium,
March 2005

Tine Lefebvre

Contents

1

Introduction

Today's robots are very primitive, capable of understanding only a few simple instructions such as 'go left', 'go right', and 'build car'.

John Sladek, (1937–2000).

1.1 Introduction

Compliant motion tasks are tasks in which a robot moves a manipulated object while controlling the contact with object(s) in its environment. Still today, in industry, compliant motion tasks are often position-controlled. Hence, they require very structured environments, i.e., the work pieces or parts to assemble are accurately positioned and their dimensions are known. In these cases, the robot receives and executes a nominal task plan. Figure 1.1 shows an example of a compliant motion task: the robot has grasped the manipulated object (cube) and assembles it with an environment object (corner with three walls) by executing a sequence of contact formations, i.e., vertex-face contact, edge-face contact, etc.

The presented work increases the robot's autonomy when performing *autonomous compliant motion* tasks in *less structured environments*, reducing the time and money needs for exact positioning of a manufacturing robot's workpieces and tools. This autonomy is also a step forward in the development of service robots which need to operate in unstructured human environments.

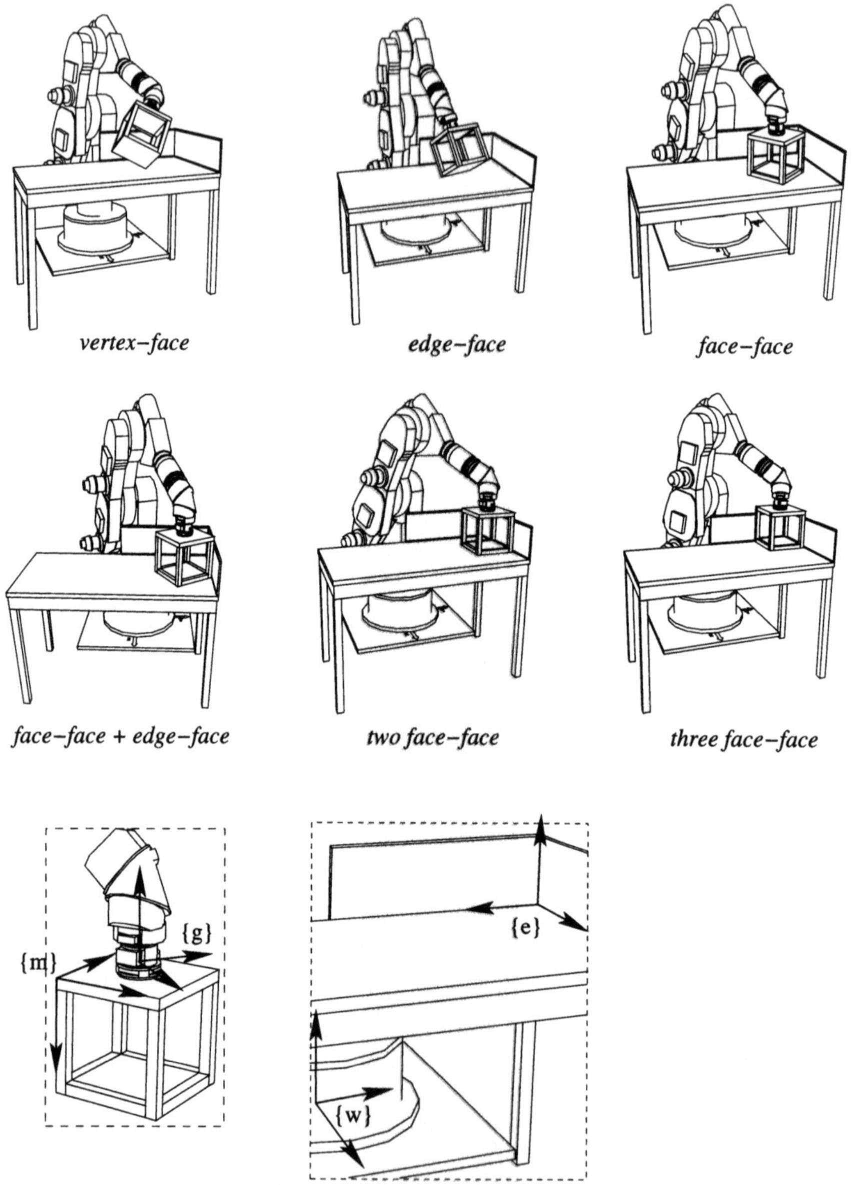

Fig. 1.1. Cube-in-corner assembly: the robot has grasped the manipulated object (cube) and assembles it with an environment object (corner with three walls) by executing a sequence of contact formations (CFs), i.e., vertex-face CF, edge-face CF, etc. The figures at the bottom show the reference frames used later on

Models

The task plan, i.e., the robot actions needed at each time to achieve the goal, is modelled by means of an *implicit* or an *explicit* model. The *implicit* models, e.g., neural networks, are a function of parameters which are not explicitly linked to a physical interpretation. The model parameters are learnt based on teaching data, obtained by human demonstration or by unsupervised learning. The trained models are well suited for task executions inside the "range" of the teaching data used to learn the models.

In the case of *explicit* models, the structure of the models is manually derived and the model parameters have a physical interpretation. The inaccurately known model parameters depend on the task at hand. E.g., the models are a function of the uncertain positions, orientations and dimensions of the contacting objects. A more uncertain system models objects by their description as an unknown number of features (such as vertices, edges and faces for polyhedral objects) with inaccurately known positions, orientations and dimensions.

The task plan corresponding to the explicit models is either manually derived (allowing *one-off* tasks) or is learnt based on teaching data. The advantage of learning an explicit model over learning an implicit model is that the estimated parameters of the explicit model can be *interpreted*. Hence, the learning of the task execution (the robot actions) is separated from the identification of the uncertain geometrical parameters (e.g., positions and orientations of the contacting objects). This makes that *one* demonstration is sufficient to re-execute the task under *different* values for the geometrical parameters (assuming that no unexpected contacts show up during this re-execution).

This work focuses on solutions based on *explicit geometrical contact models*. A common prejudice states that solutions based on explicit models need intrinsically more developers' time than the use of implicit models. However, in this work, the contact models are generated *automatically* based on a library of elementary contacts. The once-only modelling of each elementary contact is straightforward.

We restrict the applications to contacts between one manipulated object and several environment objects. The positions, orientations and dimensions of the contacting objects are uncertain. Their shapes, however, are known (up to the dimensions).

Sensors

The robot gains information about the real-world state by means of its *sensors*. Even though vision as a global sensor gains more importance lately, we here restrict ourselves to wrench (i.e., force and moment), twist (i.e., translational and rotational velocity) and pose (i.e., position and orientation) sensing. These measurements provide only *local information* about the contact region. Robots working in hazardous environments such as space, sub-sea or nuclear environments often cannot dispose of accurate vision information because no adequate illumination is possible or because the environment itself affects the proper operation of the camera and/or blurs the images. When vision sensing is available, its information can be fused with the information of the wrench, twist and pose sensing, leading to a better overall performance.

Chapter Overview

This chapter presents the components of an autonomous compliant motion system (Sect. 1.2), discusses the problems of online parameter estimation for nonlinear systems (Sect. 1.3) and describes the structure (Sect. 1.4) of this work.

1.2 Autonomous Compliant Motion

Controlling different contacts requires a *hybrid* control system. During one specific contact the system behaves as a *continuous* system. Transitions to other contacts are *discrete.*

Definitions: Contact Configuration (CC), Contact Formation (CF) and CF-Compliant Motion

Different poses of the manipulated object are referred to as *contact configurations* (CCs).
The different discrete contact states are called *contact formations* (CFs) [74]. CFs include all contact configurations for which the same geometrical features of the manipulated object are in contact with the same geometrical features of the objects in the environment.
CFs are *described* by the pairs of contacting features. The first mentioned feature refers to the contacting feature on the manipulated object, the second one to the contacting feature on an environment object. E.g., a vertex-face CF indicates that a vertex of the manipulated object contacts a face of the environment.
The motion of the manipulated object in a CF, i.e., the motion which keeps the same geometrical features in contact, is called *CF-compliant motion.*

Components of an Autonomous Compliant Motion System

A general autonomous compliant motion system consists of *three components, i.e., the task planner, the force controller and the estimator*, which exchange information:

1. The *task planner* (*fine-motion planner*) calculates pose, twist and/or wrench set-points for the force controller. The set-points are generated based on a high level task plan, which generally consists of a sequence

of several subtasks. The estimator component notifies when the end-conditions of each subtask are reached.
A more intelligent system is able to recover from execution errors: the estimator component recognises errors and the planner *re-plans* the task in order to assure good execution.

2. The *force controller* receives the desired set-points from the planner component. It converts these set-points and the measurements from the robot system into robot joint commands.
3. The *estimator* monitors the *end-conditions* of the planner's subtasks. The subtasks often correspond to motions in different CFs. In this case, monitoring the end-conditions means monitoring CF transitions.
Some estimators can detect *execution errors*, e.g., an estimator recognising CFs notices when a CF transition results in a different CF than desired. More sophisticated estimators additionally *reduce the uncertainty* by estimating the positions, orientations and dimensions of the contacting features. Reducing the uncertainty is advantageous because it (i) speeds up the task execution, and (ii) when online task adaptation is available, it enlarges the class of tasks that can be executed autonomously.

At present no overall autonomous compliant motion system has been developed yet. Most research focuses on one specific (sub)component of this scheme. In experiments, the human developer replaces the missing components.

The Autonomous Compliant Motion System in this Work

This work uses and investigates all three components of the autonomous compliant motion system. All three rely on the same *geometrical CF models.* The models specify the constraints, the motion degree of freedom directions and the ideal contact force directions for the CF. These are a function of the inaccurately known geometrical parameters, i.e., the positions, orientations and dimensions of the contacting objects.

Figure 1.2 shows the interactions between the system components in our framework:

1. the *task planner* uses a *CF graph* indicating all possible CFs and transitions between them. In this graph, the planner specifies the *nominal sequence of CFs* to be executed. For the example of Fig. 1.1, this sequence consists of the following CFs: no contact; one vertex-face CF; one edge-face CF; one face-face CF; one face-face plus one edge-face CF; two face-face CF; and finally three face-face CF.
The planner describes the *CF-compliant path* to be executed for every possible contact configuration (CC) to move from one CF to the next. This path is given in the context of the geometrical CF model.
The desired sequence of CFs and the CF-compliant path in each of them can be optimised with respect to some objective function such as time, energy, sensor information, etc. A specific application is the optimisation

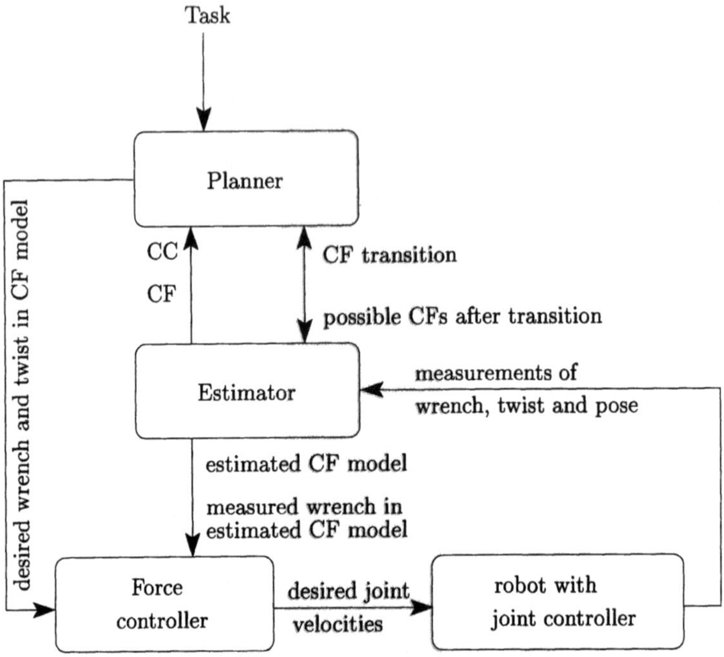

Fig. 1.2. Interaction between the planning, control and estimation component in our framework. The different components all use the same geometrical CF models. The autonomy is based on pose, twist and wrench measurements

of task plans in view of the accurate estimation of the geometrical parameters by the estimator. This is called *active sensing*.

2. the *hybrid force controller* receives the CF-compliant path defined by the task planner and interprets it in the geometrical CF model. The directions corresponding to the motion degrees of freedom are velocity-controlled, the others force-controlled.
3. during the task execution, the *estimator* estimates the inaccurately known *geometrical parameters* of the CF models used in the controller and the planner. To this aim the estimator interprets the wrench, twist and pose measurements in the geometrical CF model.
   CF transitions are detected because the measurements become inconsistent with the current model. After a CF transition, the estimator *recognises the next CF*.

Focus of this Work

The focus of this work is on the estimation of the geometrical parameters, i.e., on (i) the estimator component, and (ii) the optimisation of task plans to obtain accurate estimates of the geometrical parameters (active sensing).

A lot of other research groups omit the estimation of the geometrical parameters. The estimator components then only monitors subtask end-conditions. In a lot of applications, the estimation of the geometrical parameters is not indispensable because during the CF-compliant motion the uncertainty in the constraint directions is handled by the force controller. Estimating the geometrical parameters corresponds to estimating these same constraint directions. The advantage of the additional estimation is that the uncertainty reduction is permanent, i.e., the uncertainty stays small when the contact is lost. This increases the applications to tasks which cannot be fully written in terms of CF-compliant motions and their end-conditions, such as drilling holes at specific positions of the object.

The main reason why research groups omit the geometrical parameter estimation is the lack of off-the-shelve solutions. Both the required contact models and the estimation techniques are still under research. This work presents the latest developments in both areas. The described estimation techniques are general and also useful outside a robotics context.

1.3 Recursive State Estimation for Nonlinear Systems

Definitions: State, Parameter, Filter and Estimator

A system model is a function of a set of inaccurately known and/or changing variables. The *state* is the vector containing all these variables.
A *parameter* is a more specific term indicating a *static* state variable.
The algorithm to calculate a state estimate is called *filter* or *estimator*.

One key requirement of the estimation of the geometrical parameters during autonomous compliant motion is that the estimation needs to be performed *online* by the robot. The computational load of a batch processing of all past measurements increases with time, which is not a viable solution. Hence, the application imposes the use of a *recursive* estimator.

Recursive estimation is best performed by estimators which do not only calculate an estimated value for the state, but which additionally keep track of the *uncertainty* of this estimated value. Without this uncertainty, the estimator is not able to know which information was available in the previous measurements. The estimates of state variables which are not yet observed, measured by inaccurate measurements or with poorly conditioned measurement equations, have large uncertainty. The estimates of well-observed state variables, on the other hand, have small uncertainty.

There are different frameworks dealing with uncertainty. In this work *Bayesian techniques* are preferred over other techniques, such as interval or fuzzy theory, because Bayesian theory is well developed and consistent, i.e., calculating an estimate in different ways (e.g., by expressing the measurements or states in other frames or in other units) yields the same results.

Bayesian estimation corresponds to the calculation of a *probability density function* over the state variables. This probability density function represents the uncertainty on the state variables.

Although Bayesian theory is simple, the equations to calculate the probability density functions are often computationally intractable. An exception is the solution to *linear* systems subject to additive Gaussian uncertainty. For these systems, the probability density function over the state is *Gaussian* at all time. The Gaussian functions are characterised by their mean and covariance matrix. These are updated by simple linear algebraic formulas known as the *Kalman Filter algorithm*. Due to its low computational complexity, this algorithm is widely used in all kinds of applications. Unfortunately, the estimation of the geometrical parameters during autonomous compliant motion is a *nonlinear* estimation problem. For *small* uncertainties on the geometrical parameters, the models are linearized around the geometrical parameter estimates and the Kalman Filter algorithm is applied. We additionally present a new recursive Bayesian filter which can deal with the nonlinear contact models under *large* uncertainties of the geometrical parameters.

1.4 Overview of this Work

This book describes the latest research in the areas of fast recursive nonlinear estimation, contact modelling and fine-motion planning with active sensing.

Fast Recursive Estimation

Fast recursive filtering is usually obtained by the Kalman Filter (KF) algorithm. This filter is derived for linear system descriptions, i.e., systems with linear process and measurement models. For *nonlinear* systems, the models are linearized. Different ways of linearization result in different "KF variants". Chapter 4 presents an analysis of the best known variants, i.e., the Extended Kalman Filter, the Iterated Extended Kalman Filter and the Unscented Kalman Filter.

Chapter 5 describes a *new* Bayesian filter able to deal with nonlinear measurement models. This filter can be interpreted as a KF linearizing the process and measurement models in a higher-dimensional state space. Therefore, the filter is called the "Non-minimal State Kalman Filter".

This new filter solves the estimation problem of the geometrical parameters under large uncertainty (Chaps. 7 and 8). Previously, none of the Kalman Filter variants was able to deal with large uncertainties.

Contact Modelling

The latest achievements in contact modelling for geometrical parameter estimation (Chap. 6) are (i) the description of the contact models without the need for the "Virtual Contact Manipulator", which describes the relative degrees of freedom between the manipulated object and the environment objects;

and (ii) the automatic generation of the contact models for multiple-contact CFs based on a library of elementary CF models.

Fine-Motion Planning with Active Sensing

Chapter 9 presents the development of task plans in view of the accurate estimation of the geometrical parameters (active sensing). Active sensing is a new research area in fine-motion planning. This work decouples the optimisation of the task plan into an optimisation of the CF sequence and optimisation of the different CF-compliant paths for each of the CFs in the optimal sequence. The CF-compliant paths consist of (i) active sensing motions which allow the estimation of the geometrical parameters to the required accuracy, and (ii) transition motions which trigger the transition from one CF to the next.

Figure 1.3 illustrates the structure of the chapters. Chapters 2 and 3 contain literature surveys on autonomous compliant motion and Bayesian probability theory respectively. These chapters and their main section headings are indicated on the left and the right side of Fig. 1.3.

The main book topics are presented in six chapters, shown in the middle of Fig. 1.3. The arrows indicate the interdependence of the different text parts.

The chapters on online recursive estimation have their merits in other areas. Hence, they are treated outside the robotics context. Chapter 4 presents the analysis of the Kalman Filter variants commonly used for nonlinear systems [162, 159]. Chapter 5 describes the Non-minimal State Kalman Filter [166].

Chapters 6–9 focus on autonomous compliant motion: Chapter 6 describes contact modelling [160], Chapters 7 and 8 the geometrical parameter estimation and CF recognition [160, 161, 164, 163], and finally Chapter 9 task planning with active sensing [165].

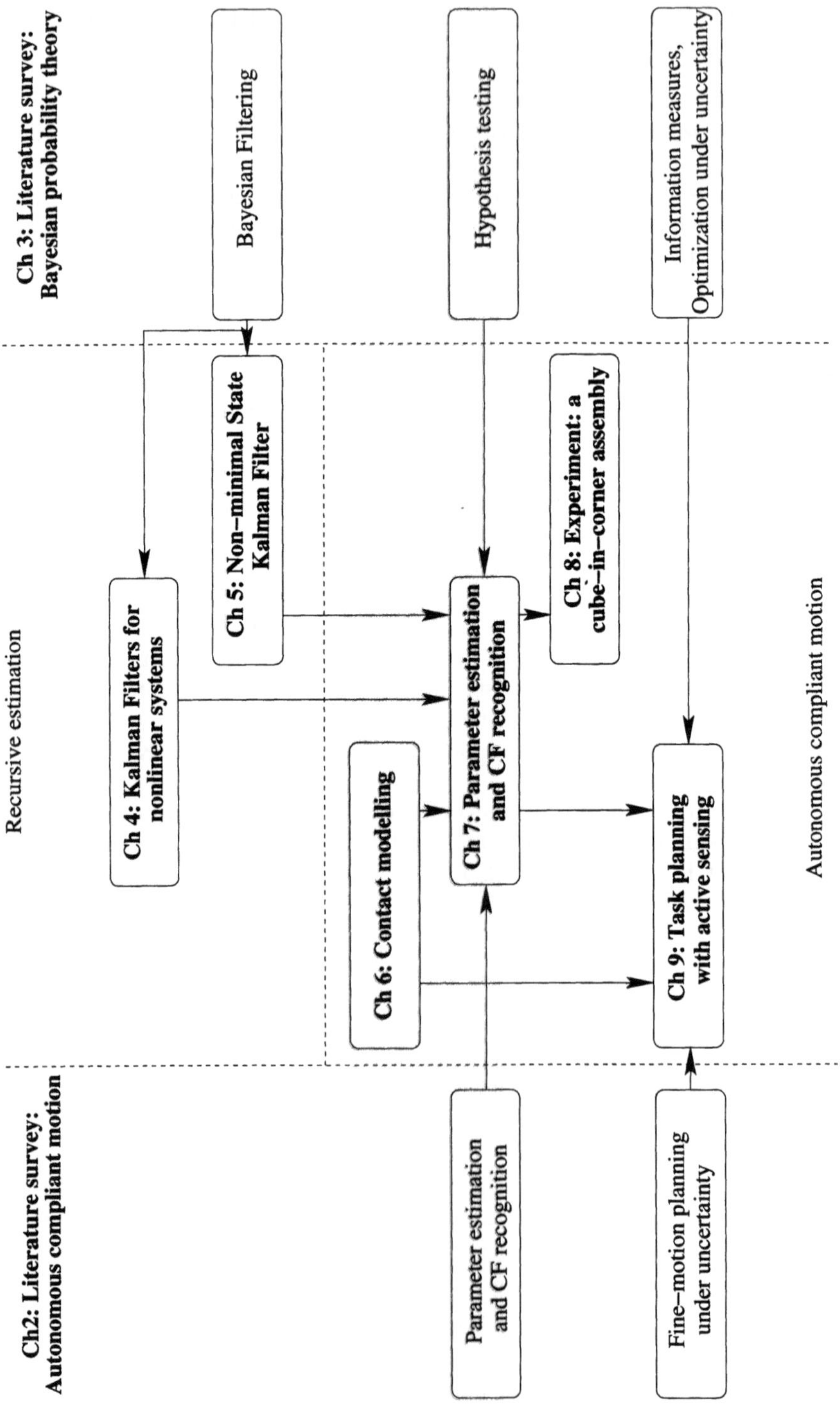

Fig. 1.3. Overview of the interdependence of the chapters

2

Literature Survey: Autonomous Compliant Motion

Nous devons donc envisager l'état présent de l'univers comme l'effet de son état antérieur et comme la cause de celui qui va suivre. (We must consider the present state of the universe as the effect of its past state and the cause of its future state.)

Pierre-Simon Laplace, *Essai Philosophique sur les Probabilités*, 1819.

2.1 Introduction

This chapter contains a survey of the research in autonomous compliant motion [167]. It presents the state-of-the art in the development of the different components of the autonomous compliant motion system described in the previous chapter, focusing on approaches using pose (i.e., positions and orientations), twist (i.e., translational and rotational velocities) and/or resultant[1] contact wrench measurements (i.e., contact forces and moments). A concise description of the force control component is presented in Sect. 2.2. Section 2.3 contains an overview of the research on the estimation component of the system. Previous research on fine-motion planning is summarised in Sect. 2.4. Finally Sect. 2.5 concludes.

[1] This work assumes the use of a force sensor measuring the resulting contact wrench, not the "tactile sensors" often used in grasping where the whole contact surface contains sensors measuring local contact pressure.

2.2 Force Control

Force control is the most mature component of the autonomous compliant motion system. This section provides a brief literature overview. The interested reader is referred to [266, 209, 280] for more detailed surveys.

Force control is divided into two branches:

1. *Passive* force control [267, 80] does not make use of a force sensor. The trajectory of the manipulated object is corrected for misalignments due to the compliance of the system. This way of force control needs special-purpose compliant end-effectors for each task and can only cope with small misalignments.
2. *Active* force control is based on force measurements which are fed back to the controller.
   The two most common active force control approaches are
   a) *impedance control* or *compliance control* [116, 7, 210, 222]. In this approach the manipulated object trajectory is modified due to contact wrenches. The deviations are modelled as the reaction to the contact wrenches of a system with given inertia, damping and stiffness matrices. These matrices have to be tuned or learnt for each particular task, hence this method implicitly requires knowledge of the task geometry.
   b) *hybrid force position/velocity control* [214, 142, 67]. In hybrid control, some of the degrees of freedom (usually in task space) are force-controlled while others are position/velocity-controlled. The control is based on an explicit geometrical contact model.

This book aims at one-off tasks, preferring active over passive control because it requires no task specific hardware. The experiments are performed with a hybrid controller.

The most common way to specify position/velocity-controlled and force-controlled directions for hybrid control is the *task frame formalism* [181, 69, 70, 42]. This formalism assumes that for every motion constraint an orthogonal reference frame exists (task frame) in which the force controlled and velocity controlled directions coincide with the frame's axes. However, the force controlled and velocity controlled directions are not orthogonal but *reciprocal* [168, 83]. The assumed orthogonality makes most of the above mentioned hybrid controllers non-invariant. References [265, 278, 1, 107, 43, 95] describe hybrid control correctly as acting on the *reciprocal* force controlled and velocity controlled directions.

2.3 Parameter Estimation and CF Recognition

The estimation and monitoring component of the autonomous compliant motion system improves the knowledge of the robot during task execution by (i) estimating the geometrical parameters, i.e., poses and dimensions of the contacting objects (Sect. 2.3.1), and/or (ii) recognising the CFs at runtime (Sect. 2.3.2). Section 2.3.3 describes the methods applied in this work.

2.3.1 Estimation of the Geometrical Parameters

Single Vertex-Surface Contacts

The identification of the contact location and orientation based on wrench, pose and/or twist measurements is often applied for *single vertex-surface contacts*. The main application area is 2D contour tracking. The estimation of the contact normal and/or contact point for frictionless contacts is described in [189, 141, 254, 69, 193, 206]. References [33, 13, 2, 143, 30, 277] include friction.

ElMaraghy and Payandeh [89], Liu [171], Muto and Shimokura [198] and Kang, Hwang, Kim, Lee and Lee [139] estimate the contact normal and/or contact point location between a workpiece and a known tool, e.g., with polyhedral or quadratic surfaces. Fedele, Fioretti, Manes and Ulivi [96] identify the local contour shape (approximated by a second order polynomial), the normal and tangent directions, and the friction coefficient. Demey, Bruyninckx and De Schutter [73] estimate the orientation of the contact normal and the surface curvature. Shape matching with a known contour model returns the position of the actual contact point on the contour. Finally, Charlebois, Gupta and Payandeh [53] estimate the shape of a surface from wrench measurements by fitting of a B-spline surface.

The research on single vertex-surface contacts with friction was extended to *soft-finger contacts* needed in grasping in the absence of tactile sensors on the finger surfaces, i.e., when only the resultant wrench on every finger is measured. Bicchi, Salisbury and Dario [29] identify the normal contact force and the friction wrench. Additional identification of the contact location is described in [30].

More General Contacts

Only few authors consider contact situations which are *more general than vertex-surface contacts*. Some first steps in this direction are the estimation of the instantaneous rotation centre of the manipulated object making multiple contacts with the environment by Ipri and Asada [122], the identification of the pose of the contacting features in vertex-face, edge-face and face-face contacts by Mimura and Funahashi [194], the estimation of the pose of a polyhedral environment object by touching it with a known manipulated object by Yoshikawa, Yu and Koike [279], the estimation of the contact normals at the different contacts of polyhedral contact configurations by McCarragher and Austin [186] and the estimation of the rotation of the constrained/free space of contacts for which the configuration space model is known by Eberman [88].

The most general contact cases between rigid objects with generally curved surfaces are considered by Bruyninckx, Dutré, De Schutter, De Geeter, Katupitiya, Demey and Lefebvre [41, 85, 68] and Debus, Dupont and Howe.

[71]. The former research group model contacts by a Virtual Contact Manipulator, describing the relative degrees of freedom between the contacting objects in function of the geometrical parameters. Identification of the parameters is based on wrench and twist measurements. Debus, Dupont and Howe estimate the parameters based on the penetration distance between the contacting objects calculated using pose measurements.

Estimation Algorithms

Most of the above mentioned references use deterministic estimation techniques and consider only the last measured data or the batch of all data. Exceptions are the work of Fedele et al. [96] who apply a recursive least-squares method with forgetting factor; McCarragher and Austin [186] adapt the estimates based on a user-defined weighting between the previous estimate and the measurement; Debus et al. [71] apply a *deterministic* non-linear least-squares estimation method based on a fixed-length moving data window; Blauer and Bélanger [33], Dutré et al. [86], and De Schutter et al. [68] use different Kalman Filter techniques; Demey et al. [73] compare non-recursive least-squares matching with the Extended Kalman Filter and the Page-Hinkley test for the contact point estimation; and Oussalah [206] applies a fuzzy linear regression estimation method.

2.3.2 CF Recognition

This section contains a survey of the literature on CF recognition. Techniques based on learnt contact models are described first. The learning data are obtained either in an unsupervised way or by human task demonstration. CF recognition based on learnt contact models can handle deviations of nominal parameter values for which learning data were available at training time. Next, techniques based on analytical contact models are discussed. Uncertainty is then described by probability density functions (pdfs) or uncertainty intervals. If the analytical models are expressed as a function of unknown geometrical parameters, the CF recognition is performed simultaneously with the parameter estimation.

CF Recognition Based on Learnt Contact Models

Hannaford and Lee [110] model assembly tasks by a Hidden Markov Model with "subtasks" (CFs) as states. Wrench measurement characteristics such as the mean and standard deviation for each CF and the transition probabilities are learnt.

Takahashi, Ogata and Muto [247] and Wang, De Schutter, Witvrouw and Graves [262] use qualitative measurements for CF recognition. The thresholds for clustering the measurements into "positive", "zero" and "negative" values are determined experimentally. Takahashi et al. [247] use twist measurements, Wang et al. [262] use wrench and relative distance measurements.

Cervera, Pobil, Marta and Serna [52] and Brignone and Howarth [38] recognise CFs by neural network structures based on wrench measurements. Nuttin, Rosell, Suárez, Van Brussel, Basañez and Hao [202] additionally consider pose measurements. Skubic et al. [236, 93, 237] compare the CF recognition by fuzzy and neural network classifiers based on pose and wrench measurements.

Sikka, Hovland and McCarragher recognise 2D CF *transitions* through wrench and change-in-wrench measurements. The recognition is either based on learnt discriminant functions [231], is rule-based using learnt measurement value thresholds [232], or is modelled by a trained Hidden Markov Model containing the time frequency model of the wrench measurements [118, 119]. The CF transition recognition is used in a Petri Net task model [183] to recognise CFs.

CF Recognition Based on Analytical Contact Models

Desai and Volz [75] recognise CFs based on pose and wrench measurements, considering uncertainty and friction. However, the contact model is based on *uniform* pressure distributions, an assumption rarely met (e.g., this means that a face-face contact cannot exert contact moments in the face).

Asada [10, 11] identifies 2D CFs based on wrench measurements given that the contact consists of a number of single vertex-edge contacts and that the relative pose between the contacting objects is known. The CF classification is implemented as the first two layers of a three-layered neural network. The network is designed to learn the nonlinear mapping of wrench measurements to twist commands, i.e., to learn the compliance for force feedback.

McCarragher and Asada [185] and Schulteis, Dupon, Millman and Howe [223] recognise 2D CFs based on qualitative template matching of every measurement component. The templates are designed by the programmer and indicate whether the measurement values are "positive", "zero" or "negative" for each CF. McCarragher and Asada use wrench and change-in-wrench measurements, Schulteis et al. use wrench and twist measurements. Note that in contrast with the other papers, the work of McCarragher and Asada does not only model CF recognition, but also CF *transition* recognition this way. The transition recognition is used in a Petri Net task structure to recognise CFs [183].

Xiao et al. [271, 275] recognise polyhedral CFs based on pose measurements by growing the objects with the pose uncertainty (defined as an interval) and by intersecting the grown regions. The method returns the set of possible valid principal contacts given the pose uncertainty. Xiao and Zhang [276] reduce the set to the geometrically valid CFs. Xiao and Liu [274] add wrench measurement information and interpret the CF recognition as the calculation of "fuzzy" membership values.

Hirai and Iwata [113], Hirai and Asada [112] and Hirukawa, Matsui and Takase [114] model polyhedral CFs with *polyhedral convex cones* indicating the ranges of possible wrenches and displacements (twists) measurable in a CF. The CF can be determined by applying linear discriminant functions (representing the face vectors of the cones) to the measurements. Farahat,

Graves and Trinkle [94] and Mosemann, Raue and Wahl [196] extend this work to include friction. Farahat, Graves and Trinkle additionally consider pose uncertainty by testing all of a discrete set of poses; the CFs considered for testing at each pose are given by [271].

Suárez, Basañez and Rosell [243, 245] describe 2D CF recognition by verifying if the pose and wrench measurements lie in predefined regions. These regions are calculated offline for every CF based on the configuration space description of the system given the uncertainties. Reference [244] compares this analytical approach with several learning methods for a 2D block-into-corner problem.

Spreng [241, 242] recognises polyhedral CFs based on the distance (with its uncertainty) between the objects, the history of the motion (i.e., its instantaneous approach direction), the compatibility of a hypothesised contact (one constraint lost or gained at each transition), and the distance (with its uncertainty) between the geometrical contacting elements of the objects. Discriminating between feasible CFs is based on movability tests: each test consists of a small twist which is feasible in case of the presence of one subset of the hypotheses but infeasible in case of its complement.

Simultaneous CF Recognition and Geometrical Parameter Estimation

If the *analytical models are a function of some unknown geometrical parameters*, simultaneous CF recognition and geometrical parameter estimation is to be performed.

Kitagaki, Ogasawara and Suehiro [143] estimate the location of the contact point based on wrench increments for a single vertex-face contact. A transition between the vertex-face and an edge-face contact is observed by monitoring the absolute difference between the expected and the observed moment. In [144] the estimation of a pseudo contact point is added. This pseudo contact point is the intersection between the force screw axis and the perpendicular to the force screw axis through the contact point. Large changes in pseudo contact point indicate a contact transition.

Mimura and Funahashi [194] recognise vertex-face, edge-face and face-face CFs based on wrench and pose measurements. The different CF models are tested from the least to the most constrained until a model is found which is consistent with the data, i.e., a model for which the geometrical parameters can be determined.

Based on pose and wrench measurements, Eberman [88] calculates a maximum likelihood estimate of the number of constraints and the rotation of the constrained/free space for systems for which the configuration space model is known. Changes in CF are detected by a statistical consistency test (SNIS, Summed Normalised Innovation Squared). Possible next CFs are determined by a CF network model. CF recognition is based on the likelihood of these possible next CFs.

Debus, Dupont and Howe [71] focus on general CF recognition under uncertainty of the poses and dimensions of the contacting objects. For every probable CF a deterministic estimator is run on the pose measurements. The penetration distances of the objects from the different estimators are used as measurement in a Hidden Markov Model which has CFs as states.

Research by De Geeter et al. [65] and De Schutter et al. [68] is similar but uses *statistical* Kalman Filters providing estimates for the geometrical parameters. Different possible CFs are derived from a Multiple Hypothesis Tree [66]. A statistical consistency test (SNIS test) on the data indicates the valid filter, i.e., the valid CF. The geometrical parameter estimates are those returned by this filter. A slightly different approach is taken in [191, 192] by implementing an Interacting Multiple Model Filter which returns the *weighted* estimate of the different models. The weighting factors are the model probabilities determined from a Hidden Markov Model and the measurement likelihoods for each of the models.

Gadeyne, Lefebvre and Bruyninckx [105] describe the simultaneous Bayesian estimation of the CF and the geometrical parameters. This general approach, however, is computationally expensive. Practical approximations for online measurement processing will be needed.

2.3.3 Estimation in this Work

This work primarily aims at *the estimation of the geometrical parameters* during complex multiple-contact CFs. The positions, orientations and dimensions of the contacting objects are estimated from pose, twist and wrench measurements. The estimation is based on the framework of [41, 85]. The book describes the latest successes in the generation of measurement models [160] and online recursive filtering techniques [162, 164, 161, 166, 159]. *The CF transition detection and the CF recognition* is performed with SNIS consistency tests [85, 68]. For small uncertainties on the geometrical parameters, this is sufficient. For larger uncertainties, more than one CF model stays probable and more elaborate techniques are needed [163, 105].

2.4 Fine-Motion Planning Under Uncertainty

This section describes research on *fine-motion planning*, i.e., the planning of motions which involve contacts. Fine-motion planning is complementary to *gross-motion planning* which considers collision-free motions and *grasp planning* which focuses on the stable grasping of objects.

Fine-motion planners often assume perfect positioning and geometry of the contacting parts. This section presents an overview of the literature which *explicitly* takes *uncertainty* into account. Four main different research approaches are planning with task skeletons (Sect. 2.4.1), task planning in configuration space (Sect. 2.4.2), task planning in contact space (Sect. 2.4.3) and task learning (Sect. 2.4.4). Section 2.4.5 describes the optimisation of task plans to maximise the accuracy of the model building, the geometrical parameter estimation and/or the CF recognition. Section 2.4.6 describes the approach to planning considered in this book.

2.4.1 Task Skeletons

A first idea to plan compliant-motion tasks under uncertainty was to specify the task in a number of manipulator routines, known as *task skeletons* (Taylor, Lozano-Pérez and Brooks [249, 174, 175, 39]). An assumption of this research is that only a relatively small number of basic operations (such as "grasp object", "insert the peg into the hole", "place the manipulated object such that ... is parallel to ...") are needed to perform fine-motion tasks in a particular domain.

The basic work on task skeletons is [249], in which Taylor describes how to propagate the effect of errors and uncertainties through a task skeleton. Brooks [39] extends Taylor's approach by making more complete use of symbolic constraints in the error computations. This results in constraints on the values of the skeleton parameters, such as the initial positions of the objects. E.g., a skeleton is not executed if an error estimate exceeds a threshold. When no good choice of parameters exists, the system chooses to execute sensing operations (other skeletons) which reduce the uncertainty enough to guarantee success.

Note that planning with task skeletons consists of *two phases*: for each skeleton first a nominal plan is defined, then the method calculates the accuracy of the parameters that is required to execute this nominal plan.

Currently, only few researchers proceed in this research direction. This is due to the observation that even simple variations in task geometry often require radical changes in motion strategy [178].

2.4.2 Planning in Configuration Space

This section describes planning in *configuration space* (Arnold and Lozano-Pérez [9, 176, 177]). The *configuration* of a rigid moving object is the set of parameters required to specify completely the position and orientation of the object, i.e., the object's degrees of freedom. Every configuration of the moving object corresponds to one point in its configuration space. Stationary object surfaces in real space define constraints on the manipulated object's degrees of freedom. These constraints correspond to surfaces (C-surfaces) in configuration space. Computing C-surfaces is difficult. Currently exact solutions only exists for 2D polygonal objects [18, 40, 76, 129]. Planning motions involving contacts between the objects in real space corresponds to planning motions of the point representing the moving object on the C-surfaces in configuration space.

The need for a rigorous treatment of uncertainty gave rise to the *LMT approach* for fine-motion planning (Lozano-Pérez, Mason, Taylor and Latombe [178, 154]). In this approach, the uncertainty is incorporated directly within *one planning phase.* In the simplest form of LMT planning, a starting region and a goal region are represented in configuration space. A command twist is chosen which guarantees to *reach* the desired goal *recognisably* despite control and sensing uncertainties (through termination conditions on the pose, wrench and time measurements). The *directional pre-image* is defined as the region from which the robot is guaranteed to recognisably reach the goal region via the *given motion command.* Once the goal's pre-image is calculated, it becomes a subgoal region and the process iterates until the starting region is enveloped by the recursive pre-images of subgoals. The fine-motion plan consists of the resulting sequence of command twists. Friction is taken into account in order to avoid sticking.

More generally, the LMT approach uses a *non-directional pre-image*, i.e., the region from which *some motion command* exists with which the goal can be recognisably reached [90]. The robot determines the necessary motion command online. The most general form of LMT uses an even larger non-directional pre-image, consisting of regions from which one of a *set of goals* may be reached. An important simplification is made by Erdmann [91] who approximates the pre-images by *back-projections* from which particular motions are guaranteed to reach the goal. This goal *does not need to be recognisable.*

The research on LMT planning considers uncertainty in sensing and control. Donald [77, 78] includes also *uncertainty on the geometrical model.* In this case the configuration space is augmented with additional dimensions for size and shape parameters varying in known tolerance zones. Because it is not always possible to construct a plan which is guaranteed to reach a goal region, Donald introduces the notion of an *error detecting and recovery strategy (EDR)*, consisting of a motion plan that is guaranteed to succeed if the goal is attainable (and recognisable), or to signal failure if the goal cannot be reached, e.g., this occurs when the dimensions of the objects prevent a desired insertion.

The LMT approach guarantees a plan which is robust against the largest possible errors. Hence, for a group of tasks the LMT approach does not find a solution, and the solutions found for other tasks are sub-optimal because the uncertainty during task execution can be(come) smaller. The LMT algorithm is based on the (difficult to obtain) configuration space representation of the held object and the motion planning requires exponential time in the number of planning steps [47]. This makes the algorithm computationally very expensive. The algorithm is often *approximated* and/or used in *restricted domains* such as 2D polygonal configuration space and pure translational motions [257, 45, 243, 20, 60]. Several of these implementations partition the C-space in discrete contact states and make a state graph in order to guide the search.

2.4.3 Planning in Contact Space

The planning approaches presented in this section do not rely on the configuration space representation, but use the topological CF representation[2]

[2] E.g., between two convex polyhedral objects six stable CFs can be established, i.e., a vertex-face CF, a face-vertex CF, an edge-edge CF, an edge-face CF, a face-edge CF and a face-face CF. A further simplification can be made by describing the contacts between edges and faces or between faces as combinations of vertex-face or face-vertex CFs.

instead. This keeps the computational cost of the algorithms small. The planning is decomposed into a *sequence of CFs* and the *CF-compliant motions* in each of these. The following works consider uncertainty in the planning and/or in the execution phase.

Laugier and Theveneau [156, 155] describe assembly planning based on the *disassembly* of two parts. Possible sequences of CFs are produced by successively relaxing each of the existing contacts. This leads to a CF graph. The next step consists of choosing a "good" CF sequence in the previous graph by computing the motion parameters. This step applies various heuristics such as the length of the path and the reliability of the selected motions under uncertainty.

Xiao and Volz [273] develop patch-plans to recover from unwanted CF transitions during task execution. They work under the assumption of convex objects moving with translational velocities. In [270] also rotational velocities are considered.

McCarragher [183, 184] models assembly tasks between polyhedral objects in a Petri Net structure representing the elementary contacts in which the CFs can be decomposed. The twist control command (used in a damping controller) is calculated based on contact disabling, enabling and maintaining conditions. Within the same representation, a wrench instead of twist control command can be calculated (Austin and McCarragher [15, 16, 17]).

Finally, Ji and Xiao [128] describe the planning of CF-compliant motions by extending the probabilistic road map paradigm for planning collision-free motion to the space of contact configurations. The algorithm builds a graph with (randomly generated) contact configurations as nodes. Two nodes are connected if there exists a straight-line path of configurations between them.

2.4.4 Learning Task Plans

A fourth approach to fine-motion planning is to *learn* the nonlinear relation between the sensed wrench and the desired twist *by human demonstration* or by *reinforcement learning*. When the learning is repeated for different geometrical parameter values, the learnt task plan is valid for execution under the corresponding geometrical uncertainty.

Literature on task learning is large, we describe some representative examples. Simons, Van Brussel, De Schutter and Verhaert [234] describe the learning of a wrench to twist mapping modelled with a stochastic automaton for a peg-in-hole task; Hirzinger and Landzettel [115] teach a compliant motion path or corrections to an existing path by human demonstration; Ahn, Cho, Ide, Miyazaki and Arimoto [3] describe the learning of a binary tree-type database of measurement-action pairs for a peg-in-hole task; Park and Cho [207] model misalignment corrections for a 2D peg-in-hole task by fuzzy rules; Gullapalli, Grupen and Barto [108, 109] and Nuttin and Van Brussel [204] describe the learning of neural networks for a peg-in-hole task; Kaiser and Dillmann [136, 135] train Radial-Basis Function Networks which model task skills, the learning of the skill *sequencing* is described in [104]; Koeppe, Breidenbach and Hirzinger [147, 146] train a rule-based neural network pre-structured to implement a fuzzy control system; Nuttin and Van Brussel [203, 205] describe the learning of the insertion of a light switch into a fixture modelled by regression trees and neural networks; and López-Juárez and Howarth [172] design a neural network controller for insertion tasks.

Some researchers exploit the hybrid nature of the task and separate the learning of the task in different CFs, by *combining task learning with CF recognition.*

Dufay and Latombe [81, 82] describe the learning of rules of the form "if <conditions> then <plan>". The rules are entered by a human demonstrator during several task demonstrations under different values of the geometrical parameters. The rules are merged into a program flowchart. Different contact "states" in this work can indicate different CFs, different alignments, etc.

Asada and Izumi [12, 13] and Wang, De Schutter, Witvrouw and Graves [262] present the learning of 3D hybrid force/position control programs for translational velocities for tasks which can be expressed in a task frame fixed to the base frame. The assumption is that the natural constraints during the task execution are parallel to the task frame axes. Wang and De Schutter [261] extend this work to full 6D motions and real-time program generation.

Asada [10, 11] trains a neural network which models the nonlinear mapping of wrench to twist commands. Although it was not used as an output, the first two layers (of the three-layered network) are designed in order to identify the CF. This research assumes known geometrical parameters and the network is trained in order to work well around a given relative pose between the objects.

Delson and West [72] describe the learning of the stiffness matrix of the controller and a reference position trajectory. This work assumes that every CF subtask consists of one straight line translational motion.

Hovland, Sikka and McCarragher [120] present the mapping from a 2D CF (modelled as a Hidden Markov Model state) to the desired twist control command. They also teach the transition probabilities between the CFs. This research is combined with the CF transition recognition of Hovland and McCarragher [118, 119].

Skubic and Volz [235] present the learning of both the desired sequence of CFs and the twists necessary to enhance the CF transitions. These twist commands are supposed to be constant during one CF.

Koeppe and Hirzinger [148] train neural networks which present the mapping from sensor measurements to hybrid control commands.

Chen and McCarragher [54] teach 2D assemblies by storing the CF sequence and the compliant path in each CF for different demonstrations. Afterwards, an automatic task planner generates CF sequences based on user-defined criteria such as execution time, stability and the frequency of the CF transitions in the demonstrations. The execution of the desired CF transitions is performed by replaying the corresponding demonstrated CF-compliant paths.

2.4.5 Active Sensing

Active sensing is the adaptation of task plans to optimise model building, CF recognition and/or geometrical parameter estimation. A lot of task parameters, such as the choice of sensors, the sensor placements and the executed trajectory, influence the task execution because they are directly related to the accuracy of these estimation processes. The modification of these task parameters to improve the accuracy of the estimation process is called *active sensing*. This section describes different types of active sensing found in literature of fine-motion planning. The overview is restricted to the parameters which can be adapted *online* once the system is running. E.g., we do not consider modifications in the system's measuring equipment.

Probably the oldest form of active sensing is the deliberate *randomisation* (vibration) of trajectories. This randomisation is applied in order to reduce the strategy's knowledge requirements, to simplify the planning and execution process, to blur the significance of some unmodelled effects (such as friction) and to persistently excite the modelled parameters. Examples of randomisation in fine-motion tasks are [220, 263, 127, 124, 92, 225, 157, 122, 117, 158].

In the process of model building, active sensing means the deliberate introduction of *exploratory motions* during the repetitive task execution. Exploratory motions in unexplored (unmodelled) directions can lead to the discovery of better task plans. A recent example is the improvement of assembly planning by a form of genetic algorithm [264]. In *reinforcement learning* [246], the learning process switches between optimal state/parameter estimation of the already modelled parts and optimal model building steps. The latter are performed by exploratory motions in order to gain knowledge about a previously unexplored part of the model. Examples of reinforcement learning in fine-motion planning are [108, 109, 136, 135, 203, 204, 205].

Active sensing for the accurate estimation of the system's state and model parameters is also known as "*optimal experiment design*". Optimal experiment design is popular in *mobile robotics* [233, 50, 149, 26, 34, 145, 102, 150, 218, 97, 252, 180, 35, 190]. Although the mobile robot task planning is restricted to 2D space, the optimisation of the planning is computationally expensive. Hence, in applications, the active sensing is limited to the optimisation of only the next action or of a limited number of parameters representing a parameterised action trajectory.

Active sensing for optimal *pose estimation using laser range data* is described by Caglioti [46] and De Geeter et al. [64, 63]. Caglioti localises a 3D polyhedral object whose geometry is known. The next action is optimised to minimise the determinant of the posterior covariance matrix over the estimates of the location parameters. De Geeter et al. localise a plate and a hole in a plate in 2D. They compare task optimisation by minimising the determinant, respectively the weighted trace, of the posterior covariance matrix over the estimates.

Also the number of applications of optimal experiment design in *3D fine-motion trajectory planning* are limited. Desai and Volz [75] and Spreng [242] discriminate between feasible CFs based on movability tests. Each test consists of a small motion which is feasible in case of the presence of one subset of the hypotheses but infeasible in case of its complement. There are indistinguishable CFs. Desai and Volz indicate the design constraints on the contacting objects which make the different CFs recognisable. Spreng describes the optimal choice of movability test as the minimisation of a generalised entropy function.

Kitagaki, Ogasawara and Suehiro [143] perform wrench based estimation of the location of the contact point for single vertex-face contacts by changing the applied wrench without moving. Two wrench measurements are necessary to estimate the contact point based on a least-squares pseudo-inverse. The desired wrenches are chosen to optimise the condition number of the matrix to invert.

2.4.6 Planning in this Work

This work calculates a nominal task plan in contact space (Sect. 2.4.3) and monitors errors during task execution (Sect. 2.3.3). The planning starts from a given CF graph [272] based on which feasible CF sequences leading to the goal CF can be derived. The transition motion between CFs is determined by contact disabling and enabling in contact space, cf. [183, 184].

The innovative element in the described planning approach is the optimisation of the task plan to assure the accurate observation of the geometrical parameters, i.e., *active sensing* (Sect. 2.4.5, [165]). Both the CF sequence and the CF-compliant motions are optimised. Active sensing for the accurate estimation of the geometrical parameters during fine-motion is a new research domain.

2.5 Conclusions

This chapter contains a survey of the literature on autonomous compliant motion. The focus of this chapter is on methods for estimation and planning, domains in which this work presents the lastest developments:

1. The positions, orientations and dimensions of the contacting objects are *estimated* from pose, twist and wrench measurements (Chaps. 7 and 8). The estimation is based on the framework of Bruyninckx [41] and Dutré [85]. The focus of the book is on both the online recursive filtering techniques (Chaps. 4 and 5, [162, 164, 161, 166, 159]) and the modelling of the measurement equations (Chap. 6, [160]).
2. Chapter 9 describes the optimisation of task plans in view of the accurate estimation of the geometrical parameters (active sensing) [165]. Both the CF sequence and the compliant motion in each CF are optimised. Active sensing for the accurate estimation of the geometrical parameters during fine-motion is a previously unexplored research domain.

3

Literature Survey: Bayesian Probability Theory

Frustra fit per plura quod potest fieri per pauciora.
(It is vain to do with more what can be done with less.)

William of Ockham, (1285–1347).

3.1 Introduction

This chapter presents some basic aspects of Bayesian probability theory [21, 153]. First of all, the difference between Bayesian and classical statistics is discussed (Sect. 3.2). Section 3.3 presents Bayesian *state estimation (filtering)* based on data measured at discrete time steps. Section 3.4 describes Bayesian *hypothesis testing.* Sections 3.5 and 3.6 focus on *optimal experiment design* (*active sensing*), i.e., the optimisation of the experiment in order to provide "optimal" state estimates. Section 3.5 presents ways to measure the "information content" of data and estimates. The algorithms for optimisation under state uncertainty are surveyed in Sect. 3.6. Section 3.7 concludes.

The problem of autonomous compliant motion deals with each of the discussed problems: the geometrical parameters of the contacting objects are *estimated* based on pose, twist and wrench measurements. The measurement models are different for each CF. The recognition of the valid CF at each time is a *hypothesis testing problem. Active sensing* corresponds to the calculation of a task plan (a CF sequence and a compliant motion trajectory in each of the CFs) which observes the geometrical parameters to the required uncertainty.

3.2 Bayesian Versus Classical Statistics

Bayesian Versus Classical Statistics

The main difference between the Bayesian and the classical approach (also called *frequentist* approach) to statistics is the *interpretation* of the concept *probability*. From a Bayesian point of view, "probability" is not interpreted in terms of the limit of some frequency that an event occurs, i.e., probability does not mean "randomness" but more generally as a *subjective measure of uncertainty*. E.g., assume a coin tossing experiment. The possible outcome of the experiment belongs to the discrete set of values {head, tail}. Figure 3.1 shows the probability distribution over this outcome if the probability of tossing a head is 0.55. The sum of the probability values for all possible outcomes is one, so this makes the probability of tossing a tail 0.45. In classical statistics the probability of tossing a head being 0.55 means that out of every n tosses 55% will yield heads if n goes to infinity, in Bayesian statistics it means that we *believe* we have 55% chance that the coin will show a head if we toss it. We can state this belief without doing any tossing at all.

For continuous-valued variables, the probability density function (pdf) is a function over the continuous domain of possible values. E.g., assume that we measure the time laps t between consecutive traffic accidents. Figure 3.2 shows a (fictive) pdf $p(t)$ corresponding to this problem. The probability of measuring a time laps in the infinitesimal interval $[t, t + dt]$ equals $p(t)dt$. In classical statistics this means that out of every n time lapses measured, $100p(t)dt$ % will be in the interval $[t, t+dt]$ for n going to infinity. In Bayesian statistics it means that we *believe* to have $100p(t)dt$ % chance that a time laps will fall in the interval $[t, t + dt]$.

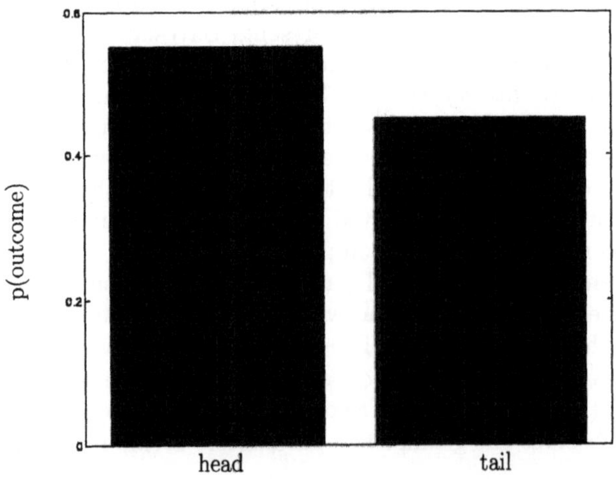

Fig. 3.1. Discrete probability distribution

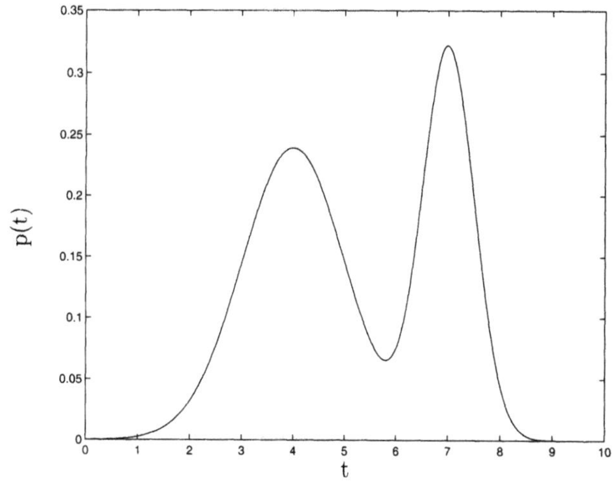

Fig. 3.2. Continuous probability density function

The *advantage* of Bayesian statistics is that it does not *need* the frequency related interpretation, e.g., Bayesian statisticians can talk about the probability distribution function (pdf) over a constant parameter. In Bayesian statistics this means that to every possible value corresponds a belief that the parameter equals this value. In classical statistics a pdf over a constant parameter does not make sense. The same interpretation can be made for the uncertainty on the models: in the frequentist approach uncertainty on the models is the representation of "noise" in the system, in a Bayesian approach, uncertainty on the models represents the belief of the validity of the model, due to noise, inaccurate modelling, lack of information, etc. Figure 3.3 shows an example of the state estimate based on a possibly biased measurement. The value $\hat{z}$ corresponds to a measurement of the value of interest x. The measurement is performed with a sensor with possible bias with value up to ± 2 (uniform distribution) and zero mean additive noise with standard deviation 0.1 (Gaussian distribution). The figure on the top shows the uncertainty over x given the measured value corresponding to the star on the horizontal axis. The figure on the bottom shows that more measurements (grouped in the symbol $\hat{\mathbf{Z}}$) make the estimate of x more accurate because the measurement noise is filtered out. The influence of the unknown bias, however, is still present in the uncertainty, i.e., the classical problem of "biased" estimates does not exist in the Bayesian framework. The use of *other sensors* can further reduce the uncertainty on x. In this case also a new pdf of the bias of the first sensor can be estimated, making the first sensor model more accurate.

Opponents of the Bayesian approach defend their choice by referring to the prior probabilities required by Bayes theorem as a *flaw* of the method. Indeed, there are no universally accepted ways of assigning prior probabilities. However, it is the belief of Bayesian statisticians that the classical approaches use prior probabilities (such as uniform distributions) *implicitly*. In Bayesian

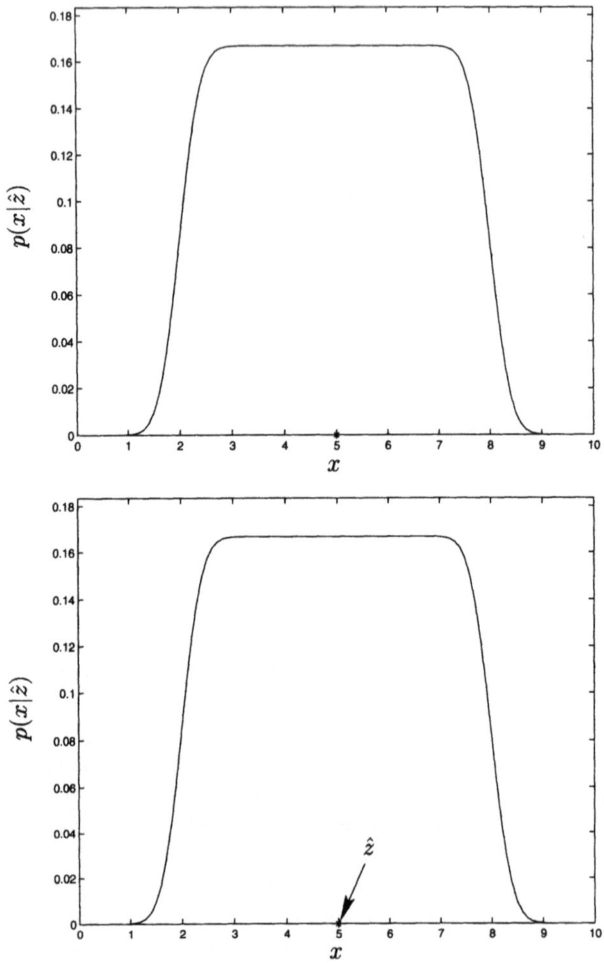

Fig. 3.3. (top) The uncertainty over x given a measurement $\hat{z}$ from a possibly biased sensor with additive Gaussian measurement noise. **(bottom)** Taking more measurements reduces the uncertainty on the estimate x caused by the Gaussian noise, the uncertainty due to the possible bias stays

probability theory the priors are mentioned *explicitly* in the formulas. In this point of view, the explicit use of priors is a *strength* of the method.

Bayesian Probability Rules

Bayesian statistics is based on two basic rules in probability theory, namely

1. the *marginalisation rule*

$$p(b) = \int_a p(a,b)\, da \quad \text{or} \quad p(b) = \sum_a p(a,b);$$

   for continuous and discrete distributions respectively;

2. the *product rule*

$$p(a, b) = p(a|b)\, p(b).$$

Bayes' theorem is derived from these rules as

$$p(a|b) = \frac{p(b|a)\, p(a)}{\int_a p(b|a)\, p(a)\, da} \quad \text{or} \quad p(a|b) = \frac{p(b|a)\, p(a)}{\sum_a p(b|a)\, p(a)}.$$

The importance of Bayes' theorem (also called the *inverse probability law*) becomes clear when replacing the general variable a with a variable of interest $\mathbf{x}$ (such as a state or parameter we want to know) and b with a variable $\mathbf{z}$ correlated to $\mathbf{x}$ (such as a measurement). Bayes' rule expresses that the probability of a certain value $\hat{\mathbf{x}}$ if we know[1] the value $\mathbf{z} = \hat{\mathbf{z}}$ equals

$$p(\mathbf{x} = \hat{\mathbf{x}}|\mathbf{z} = \hat{\mathbf{z}}) = \frac{p(\mathbf{z} = \hat{\mathbf{z}}|\mathbf{x} = \hat{\mathbf{x}})\, p(\mathbf{x} = \hat{\mathbf{x}})}{p(\mathbf{z} = \hat{\mathbf{z}})};$$

noted concisely as

$$p(\hat{\mathbf{x}}|\hat{\mathbf{z}}) = \frac{p(\hat{\mathbf{z}}|\hat{\mathbf{x}})\, p(\hat{\mathbf{x}})}{p(\hat{\mathbf{z}})}.$$

Writing this down for all possible values of $\mathbf{x}$, yields the formula to calculate the pdf over the variable $\mathbf{x}$ given the value $\hat{\mathbf{z}}$:

$$p(\mathbf{x}|\hat{\mathbf{z}}) = \frac{p(\hat{\mathbf{z}}|\mathbf{x})\, p(\mathbf{x})}{p(\hat{\mathbf{z}})}; \tag{3.1}$$

$$\textit{posterior} = \frac{\textit{likelihood} \times \textit{prior}}{\textit{evidence}};$$

where

- $p(\mathbf{x}|\hat{\mathbf{z}})$ is the *posterior* pdf over $\mathbf{x}$ given the value $\hat{\mathbf{z}}$;
- $p(\hat{\mathbf{z}}|\mathbf{x})$ is the "measurement model" representing the probability of $\hat{\mathbf{z}}$ *given* $\mathbf{x}$. $p(\hat{\mathbf{z}}|\mathbf{x})$ is called the *likelihood* of $\hat{\mathbf{z}}$.
- $p(\mathbf{x})$ is the *prior* pdf over $\mathbf{x}$, i.e., the belief function over $\mathbf{x}$ before $\hat{\mathbf{z}}$ was known.
- $p(\hat{\mathbf{z}})$ is the probability of $\mathbf{z}$ having the value $\hat{\mathbf{z}}$ when no knowledge about $\mathbf{x}$ is available, this is called the *evidence*. $p(\hat{\mathbf{z}})$ is independent of $\mathbf{x}$ and hence is only a scaling factor in (3.1) scaling $p(\hat{\mathbf{z}}|\mathbf{x})p(\mathbf{x})$ to a pdf. Therefore, Bayes rule is often denoted as:

$$p(\mathbf{x}|\hat{\mathbf{z}}) \sim p(\hat{\mathbf{z}}|\mathbf{x})\, p(\mathbf{x});$$
$$\textit{posterior} \sim \textit{likelihood} \times \textit{prior}.$$

Figure 3.4 illustrates this formula for a 1D continuous-valued x.

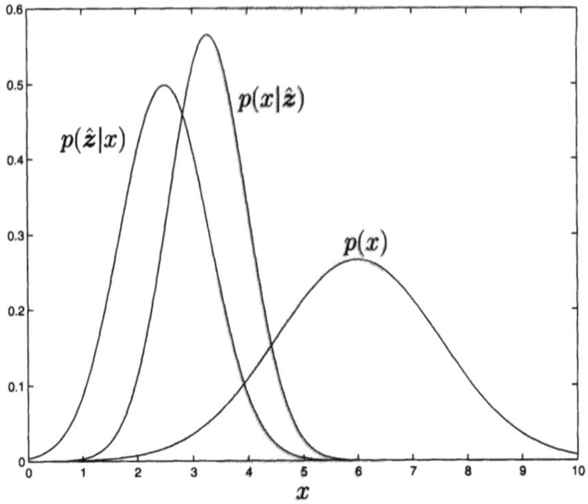

Fig. 3.4. Bayes rule for continuous-valued $\mathbf{x}$

The remainder of this chapter describes the application of the Bayesian framework to state estimation, hypothesis testing and optimal experiment design. Although Bayesian theory is only based on the aforementioned simple laws, it can be hard to calculate the pdfs in practical applications. This chapter additionally describes a number of approximate solutions. Knowing the optimal Bayesian framework, simplifications can and should be justified when applying them.

3.3 Bayesian Filtering

This section describes Bayesian *filtering* (or *estimation*), an application of Bayesian statistics where the pdf over the (possibly variable) states[2] $\mathbf{x}_0, \ldots, \mathbf{x}_k$ is calculated, given the measurements $\hat{\mathbf{z}}_1, \ldots, \hat{\mathbf{z}}_k$.

Notation

We adopt the following notation:

- $\mathbf{X}_k = \{\mathbf{x}_0, \ldots, \mathbf{x}_k\}$ denotes the state variables up to time k. The term *state* refers to the (static or dynamic) variables to be estimated. E.g., in a mobile robot localisation problem, the state is the position and orientation of the robot in its (known) environment. *Static* state variables are also

[1] Bold symbols denote vectors or matrices. Symbols with a "hat" refer to "realisations" (measured, estimated or known values) of the variable with the same name.

[2] The subscript of a variable denotes the time step for which it is valid.

called *parameters*. E.g., in the autonomous compliant motion problem of this work, the (static) state corresponds to the geometrical parameters (the inaccurately known position, orientation and dimensions) of the contacting objects. These parameters are expressed with respect to well-chosen frames on the robot such that they do not change over time.

- $\hat{\mathbf{U}}_{k-1} = \{\hat{\mathbf{u}}_0, \ldots, \hat{\mathbf{u}}_{k-1}\}$ are the system inputs. $\hat{\mathbf{u}}_i$ denotes the input to the system transforming $\mathbf{x}_i$ into $\mathbf{x}_{i+1}$. E.g., in the mobile robot example the inputs can be the robot velocities. In the autonomous compliant motion problem, the state is static, meaning that no inputs changing the state are to be considered in this system.
- $\hat{\mathbf{Z}}_k = \{\hat{\mathbf{z}}_1, \ldots, \hat{\mathbf{z}}_k\}$ represents the measured data. For instance for the mobile robot these can be distance measurements to known obstacles in the environment obtained by stereo vision. In this work, the pose and twist of the manipulated object and the contact wrenches are used as measurements for the estimation of the geometrical parameters.
- $\hat{\mathbf{S}}_k = \{\hat{\mathbf{s}}_1, \ldots, \hat{\mathbf{s}}_k\}$ are the measurement inputs, e.g., the focal lengths of the cameras on the mobile robot.

3.3.1 Bayesian Filtering Theory

The estimation problem (filtering problem) consists in calculating the posterior pdf $p(\mathbf{X}_k|\hat{\mathbf{Z}}_k, \hat{\mathbf{S}}_k, \hat{\mathbf{U}}_{k-1})$. Using Bayes' rule we rewrite this posterior pdf as the product of the system's *measurement model* $p(\hat{\mathbf{Z}}_k|\mathbf{X}_k, \hat{\mathbf{S}}_k, \hat{\mathbf{U}}_{k-1})$ and its *process model* $p(\mathbf{X}_k|\hat{\mathbf{S}}_k, \hat{\mathbf{U}}_{k-1})$:

$$p(\mathbf{X}_k|\hat{\mathbf{Z}}_k, \hat{\mathbf{S}}_k, \hat{\mathbf{U}}_{k-1}) \sim p(\hat{\mathbf{Z}}_k|\mathbf{X}_k, \hat{\mathbf{S}}_k, \hat{\mathbf{U}}_{k-1})\, p(\mathbf{X}_k|\hat{\mathbf{S}}_k, \hat{\mathbf{U}}_{k-1}).$$

The measurement model $p(\hat{\mathbf{Z}}_k|\mathbf{X}_k, \hat{\mathbf{S}}_k, \hat{\mathbf{U}}_{k-1})$ links the measurements to the true system state values, the process model $p(\mathbf{X}_k|\hat{\mathbf{S}}_k, \hat{\mathbf{U}}_{k-1})$ links the state values over different time steps. Both models are pdfs expressing that the models are subject to uncertainty.

Often we are interested in only the *last* state estimate $\mathbf{x}_k$ (e.g., the current robot position and orientation). The posterior pdf over this variable is obtained by marginalising over $\mathbf{X}_{k-1}$:

$$p(\mathbf{x}_k|\hat{\mathbf{Z}}_k, \hat{\mathbf{S}}_k, \hat{\mathbf{U}}_{k-1}) = \int_{\mathbf{X}_{k-1}} p(\mathbf{X}_k|\hat{\mathbf{Z}}_k, \hat{\mathbf{S}}_k, \hat{\mathbf{U}}_{k-1})\, d\mathbf{X}_{k-1}$$

or

$$p(\mathbf{x}_k|\hat{\mathbf{Z}}_k, \hat{\mathbf{S}}_k, \hat{\mathbf{U}}_{k-1}) = \sum_{\mathbf{X}_{k-1}} p(\mathbf{X}_k|\hat{\mathbf{Z}}_k, \hat{\mathbf{S}}_k, \hat{\mathbf{U}}_{k-1})$$

Recursive Filtering for Markov Models

Unfortunately, the complexity of the above formulas increases with time. Due to time constraints, *online* estimation requires a *recursive* formulation using at each time step only a fixed part of the past measurements. Recursive filtering is possible for systems modelled by *Markov models.* For instance, state space models of the form

$$\begin{aligned} \mathbf{x}_k &= \mathbf{f}_{k-1}(\mathbf{x}_{k-1}, \mathbf{u}_{k-1}, \boldsymbol{\rho}_{p,k-1}); \\ \mathbf{z}_k &= \mathbf{h}_k(\mathbf{x}_k, \mathbf{s}_k, \boldsymbol{\rho}_{m,k}); \end{aligned}$$

where the noise terms $\boldsymbol{\rho}_{p,k-1}$ and $\boldsymbol{\rho}_{m,k}$ are mutually uncorrelated and uncorrelated over time, are Markovian[3]. Formulated more generally, *a Markov model is a model in which, given the current state, both the future measurements and states do not depend on any past information (past measurements, states, system inputs and sensor inputs).* For these models, the Bayesian state estimation can be described recursively in two steps.

1. The *measurement update* step is

$$\begin{aligned} p(\mathbf{x}_k|\hat{\mathbf{Z}}_k, \hat{\mathbf{S}}_k, \hat{\mathbf{U}}_{k-1}) &\sim p(\hat{\mathbf{z}}_k|\mathbf{x}_k, \hat{\mathbf{Z}}_{k-1}, \hat{\mathbf{S}}_k, \hat{\mathbf{U}}_{k-1}) \\ &\qquad\qquad p(\mathbf{x}_k|\hat{\mathbf{Z}}_{k-1}, \hat{\mathbf{S}}_k, \hat{\mathbf{U}}_{k-1}); \\ &\sim p(\hat{\mathbf{z}}_k|\mathbf{x}_k, \hat{\mathbf{s}}_k) p(\mathbf{x}_k|\hat{\mathbf{Z}}_{k-1}, \hat{\mathbf{S}}_{k-1}, \hat{\mathbf{U}}_{k-1}). \end{aligned} \tag{3.2}$$

2. $p(\mathbf{x}_k|\hat{\mathbf{Z}}_{k-1}, \hat{\mathbf{S}}_{k-1}, \hat{\mathbf{U}}_{k-1})$ is calculated in the *process update* step as

[3] The *one-to-one mapping between the state-space model and the pdfs in the Bayesian formulas* is illustrated for the following linear state space description with additive Gaussian model uncertainty:

$$\begin{aligned} \mathbf{x}_k &= \mathbf{F}_{k-1}\mathbf{x}_{k-1} + \boldsymbol{\rho}_{p,k-1}; \\ \mathbf{z}_k &= \mathbf{H}_k\mathbf{x}_k + \boldsymbol{\rho}_{m,k}. \end{aligned}$$

$\boldsymbol{\rho}_{p,k-1}$ and $\boldsymbol{\rho}_{m,k}$ denote zero mean Gaussian uncertainties with covariance matrices $\mathbf{Q}_{k-1}$ and $\mathbf{R}_k$. The corresponding pdfs $p(\mathbf{x}_k|\mathbf{x}_{k-1})$ and $p(\hat{\mathbf{z}}_k|\mathbf{x}_k)$ are:

$$\begin{aligned} p(\mathbf{x}_k|\mathbf{x}_{k-1}) &= \frac{1}{\sqrt{(2\pi)^n|\mathbf{Q}_{k-1}|}} e^{-\frac{1}{2}(\mathbf{x}_k-\mathbf{F}_{k-1}\mathbf{x}_{k-1})^T\mathbf{Q}_{k-1}^{-1}(\mathbf{x}_k-\mathbf{F}_{k-1}\mathbf{x}_{k-1})}; \\ p(\hat{\mathbf{z}}_k|\mathbf{x}_k) &= \frac{1}{\sqrt{(2\pi)^m|\mathbf{R}_k|}} e^{-\frac{1}{2}(\hat{\mathbf{z}}_k-\mathbf{H}_k\mathbf{x}_k)^T\mathbf{R}_k^{-1}(\hat{\mathbf{z}}_k-\mathbf{H}_k\mathbf{x}_k)}; \end{aligned}$$

n and m denote the dimensions of the state $\mathbf{x}$ and the measurement $\mathbf{z}$. More information on the specific case of Bayesian filtering for linear systems is given in Chap. 4.

$$
\begin{aligned}
p(\mathbf{x}_k|\hat{\mathbf{Z}}_{k-1}, \hat{\mathbf{S}}_{k-1}, \hat{\mathbf{U}}_{k-1})
&= \int_{\mathbf{x}_{k-1}} p(\mathbf{x}_k|\mathbf{x}_{k-1}, \hat{\mathbf{Z}}_{k-1}, \hat{\mathbf{S}}_{k-1}, \hat{\mathbf{U}}_{k-1}) \\
&\qquad\qquad p(\mathbf{x}_{k-1}|\hat{\mathbf{Z}}_{k-1}, \hat{\mathbf{S}}_{k-1}, \hat{\mathbf{U}}_{k-1})\, d\mathbf{x}_{k-1}; \\
&= \int_{\mathbf{x}_{k-1}} p(\mathbf{x}_k|\mathbf{x}_{k-1}, \hat{\mathbf{u}}_{k-1})\, p(\mathbf{x}_{k-1}|\hat{\mathbf{Z}}_{k-1}, \hat{\mathbf{S}}_{k-1}, \hat{\mathbf{U}}_{k-2})\, d\mathbf{x}_{k-1}; \quad (3.3)
\end{aligned}
$$

or for discrete-valued states

$$
\begin{aligned}
p(\mathbf{x}_k|\hat{\mathbf{Z}}_{k-1}, \hat{\mathbf{S}}_{k-1}, \hat{\mathbf{U}}_{k-1})
&= \sum_{\mathbf{x}_{k-1}} p(\mathbf{x}_k|\mathbf{x}_{k-1}, \hat{\mathbf{Z}}_{k-1}, \hat{\mathbf{S}}_{k-1}, \hat{\mathbf{U}}_{k-1}) \\
&\qquad\qquad p(\mathbf{x}_{k-1}|\hat{\mathbf{Z}}_{k-1}, \hat{\mathbf{S}}_{k-1}, \hat{\mathbf{U}}_{k-1}); \\
&= \sum_{\mathbf{x}_{k-1}} p(\mathbf{x}_k|\mathbf{x}_{k-1}, \hat{\mathbf{u}}_{k-1})\, p(\mathbf{x}_{k-1}|\hat{\mathbf{Z}}_{k-1}, \hat{\mathbf{S}}_{k-1}, \hat{\mathbf{U}}_{k-2}).
\end{aligned}
$$

Further on, the notation of $p(\mathbf{x}_k|\hat{\mathbf{Z}}_{k-1}, \hat{\mathbf{S}}_{k-1}, \hat{\mathbf{U}}_{k-1})$ is simplified to $p(\mathbf{x}_k|\hat{\mathbf{Z}}_{k-1})$ and $p(\mathbf{x}_k|\hat{\mathbf{Z}}_k, \hat{\mathbf{S}}_k, \hat{\mathbf{U}}_{k-1})$ to $p(\mathbf{x}_k|\hat{\mathbf{Z}}_k)$. The general state space model is denoted as

$$
\mathbf{x}_k = \mathbf{f}_{k-1}(\mathbf{x}_{k-1}, \boldsymbol{\rho}_{p,k-1}); \quad (3.4)
$$

$$
\mathbf{z}_k = \mathbf{h}_k(\mathbf{x}_k, \boldsymbol{\rho}_{m,k}); \quad (3.5)
$$

Examples of Recursive Filtering

For some special priors, process and measurement models, the solution to (3.2)–(3.3) can be developed analytically by using pdfs with a fixed number of parameters. Examples are the Kalman Filter for linear systems subject to additive Gaussian model uncertainty (see Chap. 4) and some systems described by general exponential distributions [61]. However, for general priors and models the number of necessary parameters grows boundless as a function of time [125]. Hence, on-line implementations require *approximations*. Examples are Monte Carlo and grid-based Markov filters [101, 103], Kalman Filter variants linearizing the process and measurement models (see Chap. 4), higher-order filters [248, 106, 182] and Gaussian sum filters [123, 6].

3.3.2 Bayesian Filtering in this Work

This work presents the estimation of inaccurately known positions, orientations and dimensions of contacting objects during compliant motion. These positions, orientations and dimensions are *static* state variables, hence, the filters do not need a process update step. For small uncertainties the filtering is performed with *Iterated Extended Kalman Filters* (Chap. 4), an *approximation* of the Kalman filter for nonlinear systems. For larger uncertainties these filters yield inconsistent results. Chapter 5 presents a new *exact Bayesian* filter for static systems with additive Gaussian measurement model uncertainty. The filter can be interpreted as a Kalman Filter working in a higher-dimensional state space in which the measurement equations are linear. The filter is also valid for a class of dynamic systems.

3.4 Hypothesis Testing

This section describes how different hypotheses are compared in a Bayesian framework. Different hypotheses for instance represent different *system models*, or different *sequences of system models* in changing systems. An example of the latter are the CF sequences in assembly tasks.

Considering different hypotheses corresponds to adding a discrete valued parameter H to the system description. This reduces the hypothesis testing to *an augmented state estimation problem* (see Sect. 3.3) where the posterior pdf $p(\mathbf{x}_k, H|\hat{\mathbf{Z}}_k)$ is to be calculated. The marginal distribution over the state $\mathbf{x}_k$ is calculated as:

$$p(\mathbf{x}|\hat{\mathbf{Z}}_k) = \sum_H p(\mathbf{x}, H|\hat{\mathbf{Z}}_k) = \sum_H p(\mathbf{x}|H, \hat{\mathbf{Z}}_k)p(H|\hat{\mathbf{Z}}_k). \tag{3.6}$$

The Bayesian way of dealing with hypotheses as an extra parameter to be estimated becomes computationally intractable for large numbers of hypotheses. In problems where the hypotheses represent *sequences* of system models in changing systems the complexity of the problem even grows with time. Reducing the computational complexity is often done by determining the "valid" hypothesis based on the measurement data and discarding the other hypotheses. This section analyses this *approximation* into more detail, the selection of valid/invalid hypotheses is based on the posterior pdf

$$p(H|\hat{\mathbf{Z}}_k) \sim p(\hat{\mathbf{Z}}_k|H)p(H). \tag{3.7}$$

Note that for every specific value $\hat{H}$ this is a scalar value.

Section 3.4.1 describes the hypothesis testing problem when the hypotheses represent whatever discrete set (system models, sequences of system models, etc). Section 3.4.2 presents specific approximations for testing hypotheses corresponding to *model sequences.* In both sections, the problem is described independently of whether the different system models are expressed in an unknown state vector or not, by describing the likelihood of (possibly part of) the data $\hat{\mathbf{Z}}_k$ with respect to a specific model $\hat{M}^i$ as $p(\hat{\mathbf{Z}}_k|\hat{M}^i)$. However, if the models have an unknown state vector $\mathbf{x}^i$ this pdf corresponds to:

$$p(\hat{\mathbf{Z}}_k|\hat{M}^i) = \int_{\mathbf{x}^i} p(\hat{\mathbf{Z}}_k, \mathbf{x}^i|\hat{M}^i)\, d\mathbf{x}^i = \int_{\mathbf{x}^i} p(\hat{\mathbf{Z}}_k|\mathbf{x}^i, \hat{M}^i)\, p(\mathbf{x}^i|\hat{M}^i)\, d\mathbf{x}^i; \quad (3.8)$$

or

$$p(\hat{\mathbf{Z}}_k|\hat{M}^i) = \sum_{\mathbf{x}^i} p(\hat{\mathbf{Z}}_k, \mathbf{x}^i|\hat{M}^i) = \sum_{\mathbf{x}^i} p(\hat{\mathbf{Z}}_k|\mathbf{x}^i, \hat{M}^i)\, p(\mathbf{x}^i|\hat{M}^i).$$

Equation (3.8) describes a high-dimensional integral which in general does not have an analytical solution. Therefore, $p(\hat{\mathbf{Z}}_k|\hat{M}^i)$ is often *approximated.* Section 3.4.3 describes common approximations and their validity for use in model selection.

3.4.1 General Hypothesis Testing

Comparing different hypotheses in a Bayesian way means comparing the posterior pdf $p(H|\hat{\mathbf{Z}}_k)$ for the different hypotheses H. A specific hypothesis $\hat{H}^i$ explains the data better than a hypothesis $\hat{H}^j$ if $p(\hat{H}^i|\hat{\mathbf{Z}}_k) > p(\hat{H}^j|\hat{\mathbf{Z}}_k)$, i.e., if

$$\frac{p(\hat{H}^i|\hat{\mathbf{Z}}_k)}{p(\hat{H}^j|\hat{\mathbf{Z}}_k)} = \frac{p(\hat{H}^i)\, p(\hat{\mathbf{Z}}_k|\hat{H}^i)}{p(\hat{H}^j)\, p(\hat{\mathbf{Z}}_k|\hat{H}^j)} > 1.$$

$\frac{p(\hat{H}^i|\hat{\mathbf{Z}}_k)}{p(\hat{H}^j|\hat{\mathbf{Z}}_k)}$ is called the *posterior odds*, $\frac{p(\hat{H}^i)}{p(\hat{H}^j)}$ the *prior odds* and $\frac{p(\hat{\mathbf{Z}}_k|\hat{H}^i)}{p(\hat{\mathbf{Z}}_k|\hat{H}^j)}$ the *Bayes factor* or *likelihood ratio.*

Bayesian decision theory [126, 211] rejects a hypothesis $\hat{H}^j$ in favour of a hypothesis $\hat{H}^i$ if the posterior odds exceeds a user-defined threshold. This threshold is application dependent and is a trade-off between quick recognition and possible false result.

The prior odds is supposed to be known (e.g., the process of model sequencing can be described by a HMM). If no prior preference of one of both hypotheses is assumed, then $p(\hat{H}^i)$ is taken equal to $p(\hat{H}^j)$ and the decision is made on the likelihood ratio only.

3.4.2 Model Selection for Switching Models

This section presents the recognition of sequences of models for systems with switching models. This corresponds to applying the hypothesis testing described above with *different hypotheses representing different possible model sequences.* The number of hypotheses to be considered grows exponentially in time [268]. For computational reasons several *approximations* are proposed:

1. *Sequential hypothesis testing* [260, 230] assumes a Markovian system for which the time for the algorithm to "uniquely" recognise the current model is short compared to the holding time before a model transition. Therefore, (i) at any time only one model needs to be considered true, and (ii) only one transition from this model (to possibly several other models) is possible. For every possible model transition, i.e., every possible next model and transition time, a hypothesis $\hat{H}^i(\hat{r}, \hat{M}^i)$ is constructed which

assumes the transition from the current model $\hat{M}^c$ to the model $\hat{M}^i$ at transition time $\hat{r}$. The likelihood of every transition hypothesis $\hat{H}^i$ versus the no-change hypothesis $\hat{H}^c$ (the hypothesis that all data comes from model $\hat{M}^c$) is

$$\frac{p(\hat{\mathbf{Z}}_k|\hat{H}^i)}{p(\hat{\mathbf{Z}}_k|\hat{H}^c)} = \frac{p(\hat{\mathbf{z}}_1, \dots \hat{\mathbf{z}}_{r-1}|\hat{M}^c)p(\hat{\mathbf{z}}_r, \dots \hat{\mathbf{z}}_k|\hat{M}^i)}{p(\hat{\mathbf{z}}_1, \dots, \hat{\mathbf{z}}_k|\hat{M}^c)};$$
$$= \frac{p(\hat{\mathbf{z}}_r, \dots \hat{\mathbf{z}}_k|\hat{M}^i)}{p(\hat{\mathbf{z}}_r, \dots \hat{\mathbf{z}}_k|\hat{M}^c)}.$$

The most likely hypothesis is

$$\hat{H}^i = \arg\max_{H^i} \left(\log \frac{p(\hat{\mathbf{Z}}_k|H^i)}{p(\hat{\mathbf{Z}}_k|\hat{H}^c)} \right);$$
$$= \arg\max_{M^i} \max_{0 \le r \le k} \left(\log \frac{p(\hat{\mathbf{Z}}_k|H^i)}{p(\hat{\mathbf{Z}}_k|\hat{H}^c)} \right).$$

If this most likely ratio exceeds a user-defined threshold, the current model becomes $\hat{M}^i$ and the change time is considered to be $\hat{r}$, otherwise hypothesis $\hat{H}^c$ is kept (with current model $\hat{M}^c$).
To limit the computation time for sequential likelihood testing, the search can be constrained to a data window of fixed size M [269]:

$$\hat{H}^i = \arg\max_{M^i} \max_{k-M+1 \le r \le k} \left(\log \frac{p(\hat{\mathbf{Z}}_k|H^i)}{p(\hat{\mathbf{Z}}_k|\hat{H}^c)} \right).$$

2. Another frequently applied *approximation* is to reduce the increasingly growing number of hypotheses by a fixed number M by discarding the less likely ones [256, 255].
3. Finally, a third often applied approximation is the use of *consistency tests* instead of likelihood ratios. This approximation assumes that a consistency test for a model $\hat{M}^i$ exists which detects inconsistency as soon as the data originates from one of the other models $\hat{M}^j$, $j \neq i$. Examples of such consistency tests for linear systems subject to Gaussian model uncertainty are the testing of the innovation[4] process for zero mean, whiteness and a given covariance [188]; and the testing of the sum of a number of Normalised Innovation Squared values[4, p. 36] (SNIS) [19, 268] to lie in a confidence interval.

[4] An innovation $\boldsymbol{\eta}$ is the difference between the measurement value and the (by the model) expected one. The Normalised Innovation Squared (NIS) is the square of this vector normalised with the inverse of its covariance $\mathbf{S}$: $NIS = \boldsymbol{\eta}^T \mathbf{S}^{-1} \boldsymbol{\eta}$. The NIS value for a linear system subject to Gaussian model uncertainty is χ^2 distributed.

3.4.3 Likelihood for Parameterised Models

The previous sections described hypothesis testing independent of the fact that the models M are expressed in unknown state vectors or not. If however a model $\hat{M}^i$ is expressed in a state $\mathbf{x}^i$, the data likelihood is to be calculated as:

$$\begin{aligned} p(\hat{\mathbf{Z}}_k|\hat{M}^i) &= \int_{\mathbf{x}^i} p(\hat{\mathbf{Z}}_k, \mathbf{x}^i|\hat{M}^i)\, d\mathbf{x}^i; \\ &= \int_{\mathbf{x}^i} p(\hat{\mathbf{Z}}_k|\mathbf{x}^i, \hat{M}^i)\, p(\mathbf{x}^i|\hat{M}^i)\, d\mathbf{x}^i; \end{aligned} \tag{3.9}$$

or

$$\begin{aligned} p(\hat{\mathbf{Z}}_k|\hat{M}^i) &= \sum_{\mathbf{x}^i} p(\hat{\mathbf{Z}}_k, \mathbf{x}^i|\hat{M}^i); \\ &= \sum_{\mathbf{x}^i} p(\hat{\mathbf{Z}}_k|\mathbf{x}^i, \hat{M}^i)\, p(\mathbf{x}^i|\hat{M}^i). \end{aligned}$$

In general the high-dimensional integrals of (3.9) have no closed-form analytical solution. This section describes *approximations* to calculate $p(\hat{\mathbf{Z}}_k|\hat{M}^i)$ and discusses the effect of the approximations on the hypothesis testing problem.

Ockham's Razor and the Likelihood Ratio

Assume a data set and two models: a simple model $\hat{M}^1$ from which the data originates and a more difficult (more flexible) model $\hat{M}^2$. With the likelihood ratio test, *the simpler model is preferred* over the more flexible one if the data can be explained by this more simple model. This principle is generally accepted in science and is known as *Ockham's razor* (Occam's razor). Figure 3.5 illustrates this. The indicated data set $\hat{\mathbf{Z}}_k$ (stars on the horizontal axis) originates from a model of the form $\hat{M}^1$ (for a specific value of $\mathbf{x}^1$). The more flexible model $\hat{M}^2$ is able to predict a greater variety of data sets than the simpler model $\hat{M}^1$, hence the likelihood of this particular data set $\hat{\mathbf{Z}}_k$ given $\hat{M}^2$ is smaller.

Approximations of the Likelihood (Ratio)

The likelihood ratios of Sects. 3.4.1 and 3.4.2 for parameterised models are often *approximated* by the likelihood ratio over the models with their most likely state values:

$$\frac{p(\hat{\mathbf{Z}}_k|\hat{M}^i)}{p(\hat{\mathbf{Z}}_k|\hat{M}^j)} \approx \frac{p(\hat{\mathbf{Z}}_k|\hat{\mathbf{x}}^i, \hat{M}^i)}{p(\hat{\mathbf{Z}}_k|\hat{\mathbf{x}}^j, \hat{M}^j)};$$

$$\hat{\mathbf{x}}^i = \arg\max_{\mathbf{x}^i} p(\hat{\mathbf{Z}}_k|\mathbf{x}^i, \hat{M}^i);$$

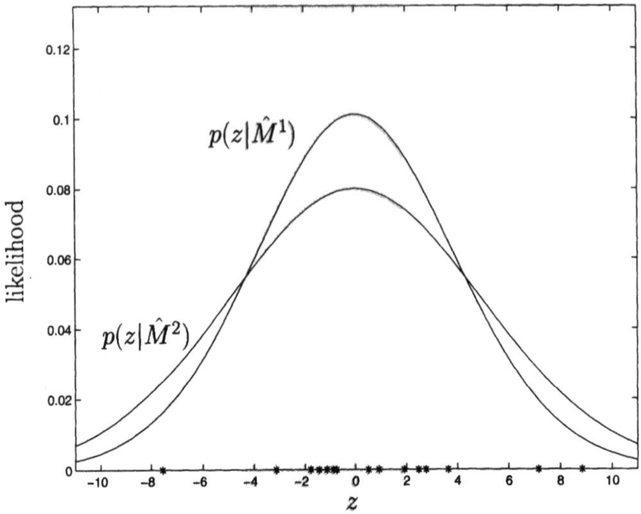

Fig. 3.5. Likelihood of the data given two models $\hat{M}^1$ and $\hat{M}^2$. The data $\hat{\mathbf{Z}}_k$ (stars on the horizontal axis) originate from a model of the form $\hat{M}^1$ (for a specific value of $\mathbf{x}^1$). The more flexible model $\hat{M}^2$ can predict a larger domain of measurements. The likelihood of the data $\hat{\mathbf{Z}}_k$ for the simpler model $\hat{M}^1$ is larger than for the model $\hat{M}^2$

$$\hat{\mathbf{x}}^j = \arg\max_{\mathbf{x}^j} p(\hat{\mathbf{Z}}_k|\mathbf{x}^j, \hat{M}^j);$$

$\frac{p(\hat{\mathbf{Z}}_k|\hat{\mathbf{x}}^i,\hat{M}^i)}{p(\hat{\mathbf{Z}}_k|\hat{\mathbf{x}}^j,\hat{M}^j)}$ is called the *generalised likelihood ratio.*

However, replacing the likelihood ratio with the generalised likelihood ratio for the purpose of model testing, *introduces the phenomenon of over-fitting.* E.g., for the previous example it means that the most likely flexible model $\hat{M}^2$ (for a specific state value $\hat{\mathbf{x}}^2$) fits the data better than the most likely model $\hat{M}^1$ (for a specific state value $\hat{\mathbf{x}}^1$). Figure 3.6 illustrates this.

Different *correction factors* are proposed to obey Ockham's rule when using the generalised likelihood ratio:

- after a number of measurements, the factor $p(\hat{\mathbf{Z}}_k|\mathbf{x}^i, \hat{M}^i)p(\mathbf{x}^i|\hat{M}^i)$ of (3.9) usually turns out to be a sharply peaked Gaussian-like function around a value $\hat{\mathbf{x}}^i$ with covariance matrix $(\mathbf{I}^i_{FI}(\hat{\mathbf{x}}^i))^{-1}$ where

$$\begin{aligned}\mathbf{I}^i_{FI}(\mathbf{x}^i) &= -\nabla_{\mathbf{x}^i}\nabla^T_{\mathbf{x}^i} \ln\left(p(\hat{\mathbf{Z}}_k|\mathbf{x}^i, \hat{M}^i)p(\mathbf{x}^i|\hat{M}^i)\right);\\ &= -\nabla_{\mathbf{x}^i}\nabla^T_{\mathbf{x}^i} \ln\left(p(\mathbf{x}^i|\hat{\mathbf{Z}}_k, \hat{M}^i)\right);\end{aligned}$$

  where $\nabla_{\mathbf{x}^i}\nabla^T_{\mathbf{x}^i}$ denotes the Hessian matrix with respect to $\mathbf{x}^i$ (Fisher [99, 100]).
  By applying the Laplace methods for integrals [62], (3.9) is approximated as

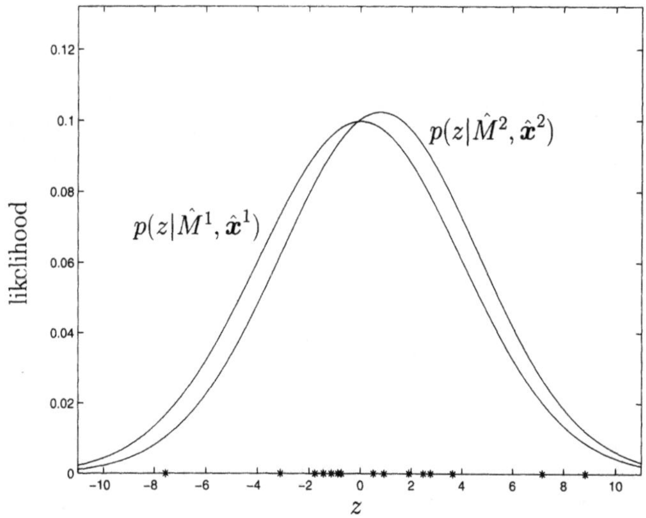

Fig. 3.6. Likelihood of the data given two models $\hat{M}^1, \hat{\mathbf{x}}^1$ and $\hat{M}^2, \hat{\mathbf{x}}^2$. The data $\hat{\mathbf{Z}}_k$ (stars on the horizontal axis) originate from a model of the form $\hat{M}^1$ (for a specific value of $\mathbf{x}^1$). $\hat{\mathbf{x}}^1$ and $\hat{\mathbf{x}}^2$ are the state values for which the likelihood of this data set $\hat{\mathbf{Z}}_k$ is maximal. The specific model $\hat{M}^2, \hat{\mathbf{x}}^2$ of the more flexible form $\hat{M}^2$ fits the data $\hat{\mathbf{Z}}_k$ better than the most likely model $\hat{M}^1, \hat{\mathbf{x}}^1$

$$p(\hat{\mathbf{Z}}_k|\hat{M}^i) \approx \frac{p(\hat{\mathbf{Z}}_k|\hat{\mathbf{x}^i}, \hat{M}^i)p(\hat{\mathbf{x}}^i|\hat{M}^i)}{\sqrt{|\mathbf{I}^i_{FI}(\hat{\mathbf{x}}^i)/(2\pi)|}}; \tag{3.10}$$

or equivalently

$$\log p(\hat{\mathbf{Z}}_k|\hat{M}^i) \approx \log p(\hat{\mathbf{Z}}_k|\hat{\mathbf{x}^i}, \hat{M}^i) + \log \frac{p(\hat{\mathbf{x}}^i|\hat{M}^i)}{\sqrt{|\mathbf{I}^i_{FI}(\hat{\mathbf{x}}^i)/(2\pi)|}}. \tag{3.11}$$

The correction factor $\frac{p(\hat{\mathbf{x}}^i|\hat{M}^i)}{\sqrt{|\mathbf{I}^i_{FI}(\hat{\mathbf{x}}^i)/(2\pi)|}}$ is called the *Ockham factor* [179].

- Akaike [4, 5] derives a correction factor by maximising the predictive accuracy of the models when the number of measurements k goes to infinity. This leads to the Akaike Information Criterion (AIC):

$$\log p(\hat{\mathbf{Z}}_k|\hat{M}^i) \approx \log p(\hat{\mathbf{Z}}_k|\hat{\mathbf{x}^i}, \hat{M}^i) - n^i;$$

  where n^i is the dimension of the state vector $\mathbf{x}^i$. References [229] and [140] show that, even asymptotically, the AIC favours the more flexible models, i.e., the models with larger state dimension.
- The Bayesian Information Criterion (BIC, Schwartz [224]) favours the model which maximises the likelihood of the measured data when the number of measurements k goes to infinity:

$$\log p(\hat{\mathbf{Z}}_k|\hat{M}^i) \approx \log p(\hat{\mathbf{Z}}_k|\hat{\mathbf{x}}^i, \hat{M}^i) - \frac{n^i \log k}{2}.$$

- The minimum description length principle (MDL; Rissanen [216, 217]) selects the model that provides the most compact data description of the model itself and the data given the model, when the number of measurements k goes to infinity. The MDL principle turns out to be the same as the BIC:

$$\log p(\hat{\mathbf{Z}}_k|\hat{M}^i) \approx \log p(\hat{\mathbf{Z}}_k|\hat{\mathbf{x}}^i, \hat{M}^i) - \frac{n^i \log k}{2}.$$

3.4.4 Hypothesis Testing in this Work

During the execution of compliant motion tasks, the system passes through different CFs. Each CF corresponds to a different measurement model. Our goal, in addition to estimating the geometrical parameters, is to recognise the CF at each time step. This corresponds to the recognition of a model *sequence*. Contact transitions are detected by running a SNIS *consistency test* within the Kalman Filter corresponding to the current model. The next valid CF is based on the knowledge of which transitions are possible (described in the CF graph). For every possible next model a filter with corresponding SNIS consistency test is run. For systems with small uncertainty on the geometrical parameters, the invalid models are ruled out very quickly. However, when the uncertainty on the geometrical parameters is larger, several models stay possible. In the latter case a multiple model approach calculating the discrete probability distribution over the different possible CFs is needed [163, 105].

3.5 Information Measures

The previous sections presented Bayesian parameter estimation and model selection. The described filters are *passive* in the sense that the estimates are calculated based on the data *received* from some other system component. *Active sensing* focuses on the *active* search for system and measurement inputs such that the measurement data are "informative" and the posterior estimate will be "accurate". E.g., the task plan of the autonomous compliant motion, i.e., the executed CF sequence and the compliant motion trajectories in each CF, influence the accuracy of the geometrical parameter estimates. Active sensing consists in optimising this task plan in order to obtain accurate estimates.

The formulation of the active sensing problem needs a *measure of the accuracy/uncertainty* of an estimate (also called "*information measure*"). Measuring the amount of uncertainty requires the choice of a *scalar function* of the pdf. Different scalar information measures exist, all with their particular properties and (dis)advantages. This section presents two well-known ways of measuring uncertainty, namely the *entropy* as a measure of uncertainty of a pdf $p(\mathbf{x})$ (Sect. 3.5.1) and *loss functions of the covariance matrix* as measures of uncertainty of a state value $\hat{\mathbf{x}}$ (Sect. 3.5.2). Section 3.5.3 presents an illustration of the difference between these two measures.

In this section $p(\mathbf{x})$ denotes *a known pdf* over a variable $\mathbf{x}$, e.g. this pdf could be $p(\mathbf{x}_k|\hat{\mathbf{Z}}_k)$. The *expected posterior distribution* over the variable $\mathbf{x}$ after some *future* measurements $\mathbf{Z}$ (measurements of which the specific measured value is not known yet) is denoted as $p(\mathbf{x}|\mathbf{Z})$, e.g, this could represent $p(\mathbf{x}_k|\hat{\mathbf{Z}}_k, \mathbf{z}_{k+1}, \ldots, \mathbf{z}_l)$. Similarly, $p(\mathbf{Z})$ represents the pdf over the future measurements $\mathbf{Z}$ and $p(\mathbf{x}, \mathbf{Z})$ the joint pdf of the variable and the future measurements.

3.5.1 Entropy

An often applied measure of *uncertainty of a pdf* is its *entropy*: the larger the entropy of the pdf, the more uncertain the state is. This section presents the definition of entropy, the comparison between two pdfs based on the entropy measure and an expression for the expected information of future measurements.

Accuracy of the Estimate

Shannon [227, 228] defined the *informational entropy*, also called *Shannon entropy*, as a measure of the "amount of uncertainty" represented by a probability distribution[5]:

$$H_{\mathbf{x}} = E_{p(\mathbf{x})}[-\log p(\mathbf{x})].$$

Any choice of "log" is possible; this changes only the units of the entropy result ($\log_2$: [bits], ln: [nats]).

E.g., for a Gaussian distribution over the n-dimensional state $\mathbf{x}$ with mean $\hat{\mathbf{x}}$ and covariance matrix $\mathbf{P}$:

$$p(\mathbf{x}) = \frac{1}{\sqrt{(2\pi)^n|\mathbf{P}|}} e^{-\frac{1}{2}(\mathbf{x}-\hat{\mathbf{x}})^T\mathbf{P}^{-1}(\mathbf{x}-\hat{\mathbf{x}})};$$

$$H_{\mathbf{x}} = \log\left(\sqrt{(2\pi e)^n|\mathbf{P}|}\right) = \frac{1}{2}\log\left((2\pi e)^n|\mathbf{P}|\right);$$

where $|\mathbf{P}|$ denotes the determinant of the covariance matrix.

For discrete distributions the entropy measure is *absolute*, in the continuous case the measurement *depends* on the coordinate system. The difference between the entropies of two pdfs expressed in the same coordinate system, however, is *independent* of the choice of the coordinate system.

[5] $E_{p(\mathbf{x})}[\cdot]$ denotes the expected value calculated with respect to $p(\mathbf{x})$. For continuous pdfs $E_{p(\mathbf{x})}[\mathbf{f}(\mathbf{x})] = \int \mathbf{f}(\mathbf{x})\ p(\mathbf{x})\ d\mathbf{x}$, while for discrete probability distributions $E_{p(\mathbf{x})}[\mathbf{f}(\mathbf{x})] = \sum_x \mathbf{f}(\mathbf{x})\ p(\mathbf{x})$.

Comparing Two Pdfs

Comparing the accuracy of two estimates, represented by their pdfs $p_1(\mathbf{x})$ and $p_2(\mathbf{x})$ is performed through calculation of

- the *change in entropy* between the distributions:

$$H_1(\mathbf{x}) - H_2(\mathbf{x}) = E_{p_1(\mathbf{x})}[-\log p_1(\mathbf{x})] - E_{p_2(\mathbf{x})}[-\log p_2(\mathbf{x})].$$

- their *Kullback-Leibler distance* (Kullback and Leibler [151, 152]) or *relative entropy*:

$$D(p_2(\mathbf{x})||p_1(\mathbf{x})) = E_{p_2(\mathbf{x})}[\log \frac{p_2(\mathbf{x})}{p_1(\mathbf{x})}].$$

  Note that the relative entropy is *not a metric* on the space of all pdfs, since it is not a symmetric function of its arguments:

$$D(p_2(\mathbf{x})||p_1(\mathbf{x})) \neq D(p_1(\mathbf{x})||p_2(\mathbf{x})).$$

The change in entropy and the relative entropy are *different* measures. The change in entropy quantifies how much the *form* of the pdfs differs; the relative entropy additionally incorporates a measure of how much the pdfs are *translated*: if $p_1(\mathbf{x})$ and $p_2(\mathbf{x})$ represent the same pdf translated to another mean value, their change in entropy is zero while their relative entropy is not.

Expected Information in the Measurement

The expected information in the measurements is called the *mutual information* [57] and is defined as the expected change in entropy due to future measurements $\mathbf{Z}$:

$$\begin{aligned} I_{MI}(\mathbf{x}, \mathbf{Z}) &= E_{p(\mathbf{Z})}\left[\, E_{p(\mathbf{x})}[-\log p(\mathbf{x})] - E_{p(\mathbf{x}|\mathbf{Z})}[-\log p(\mathbf{x}|\mathbf{Z})]\,\right]; \\ &= H_{\mathbf{x}} - H_{\mathbf{x}|\mathbf{Z}}. \end{aligned}$$

In this equation $H_{\mathbf{x}|\mathbf{Z}}$ represents the *conditional entropy*, defined as the expected value of the posterior conditional distribution:

$$\begin{aligned} H_{\mathbf{x}|\mathbf{Z}} &= E_{p(\mathbf{Z})}\left[\, E_{p(\mathbf{x}|\mathbf{Z})}\left[-\log p(\mathbf{x}|\mathbf{Z})\right]\,\right]; \\ &= E_{p(\mathbf{x},\mathbf{Z})}\left[-\log p(\mathbf{x}|\mathbf{Z})\right]; \\ &= E_{p(\mathbf{x},\mathbf{Z})}\left[-\log p(\mathbf{x},\mathbf{Z})\right] - E_{p(\mathbf{Z})}\left[-\log p(\mathbf{Z})\right]; \\ &= H_{\mathbf{Z},\mathbf{x}} - H_{\mathbf{Z}}. \end{aligned}$$

3.5.2 Covariance Matrix

In applications where a *single estimated value* $\hat{\mathbf{x}}$ for the true state $\mathbf{x}^*$ is needed, one often chooses $\hat{\mathbf{x}}$ to minimise the expected mean square error

$$\hat{\mathbf{x}} = \arg\min_{\tilde{\mathbf{x}}} E_{p(\mathbf{x})}\left[(\tilde{\mathbf{x}} - \mathbf{x})^T (\tilde{\mathbf{x}} - \mathbf{x})\right];$$

i.e., $\hat{\mathbf{x}}$ equals the expected value of the probability distribution $p(\mathbf{x})$ [240]:

$$\hat{\mathbf{x}} = E_{p(\mathbf{x})}[\mathbf{x}].$$

This section presents the definition of the *covariance matrix* on an estimate $\hat{\mathbf{x}}$. It discusses how two estimates are compared based on *loss functions* of their covariance matrices and it presents an expression for the expected information of future measurements.

Accuracy of the Estimate

Loss functions of the *covariance matrix* $\mathbf{P}$ are often used as a measure for the uncertainty of the state estimate $\hat{\mathbf{x}}$. The covariance matrix $\mathbf{P}$ is defined as:

$$\mathbf{P} = E_{p(\mathbf{x})}\left[(\mathbf{x} - \hat{\mathbf{x}})(\mathbf{x} - \hat{\mathbf{x}})^T\right].$$

The inverse of the covariance matrix is called *information matrix* $\mathbf{I} = \mathbf{P}^{-1}$ and is a measure of the accuracy of the estimate.

Comparing Two Estimates

Two estimates $\hat{\mathbf{x}}_1$ and $\hat{\mathbf{x}}_2$ with covariance or information matrices $\mathbf{P}_1$ and $\mathbf{P}_2$ or $\mathbf{I}_1$ and $\mathbf{I}_2$ are often compared through:

- a *matrix inequality*. Estimate $\hat{\mathbf{x}}_1$ is more accurate than estimate $\hat{\mathbf{x}}_2$ if

  $$\mathbf{P}_2 > \mathbf{P}_1 \qquad \text{or} \qquad \mathbf{I}_2 < \mathbf{I}_1;$$

  meaning that $\mathbf{P}_2 - \mathbf{P}_1$ and $\mathbf{I}_1 - \mathbf{I}_2$ are positive semi-definite. If this is the case, estimate $\hat{\mathbf{x}}_1$ is more accurate than estimate $\hat{\mathbf{x}}_2$ in any direction of the state space. This classification is not always possible since an estimate $\hat{\mathbf{x}}_1$ can be more accurate in some directions of state space and an estimate $\hat{\mathbf{x}}_2$ in others.
- the values of a *scalar loss function* of their covariance/information matrices. Examples of such loss functions are the determinant of the matrices, their eigenvalues, their (weighted) trace, etc. Minimisation of these criteria (called D-, A-, E- and L-optimal design) is extensively described in the literature of optimal experiment design [98, 212, 63].

Expected Information in the Measurement

Fisher [99, 100] defined the *Fisher information matrix* as

$$\mathbf{I}_{FI}(\mathbf{x}) = E_{p(\mathbf{Z}|\mathbf{x})}\left[-\nabla_{\mathbf{x}}\nabla_{\mathbf{x}}^T \ln p(\mathbf{Z}|\mathbf{x})\right];$$

where $\nabla_{\mathbf{x}} \nabla_{\mathbf{x}}^T$ denotes the Hessian matrix with respect to $\mathbf{x}$. When evaluated in the *true state value*, $\mathbf{I}_{FI}(\mathbf{x}^*)$ describes the *information in the measurements* $\mathbf{Z}$ about the state $\mathbf{x}$. Its inverse $\mathbf{I}_{FI}^{-1}(\mathbf{x}^*)$ is the *Cramér-Rao lower bound* and defines a lower bound on the posterior covariance matrix of the state estimate (if no prior information is available) (Dugué [84], Rao [215] and Cramér [58]).

Van Trees [259] redefined the Fisher information matrix based on the *prior information* $p(\mathbf{x})$ instead of the true value $\mathbf{x}^*$:

$$\begin{aligned}\mathbf{I}_{VT} &= E_{p(\mathbf{x},\mathbf{Z})}\left[-\nabla_{\mathbf{x}} \nabla_{\mathbf{x}}^T \ln p(\mathbf{x},\mathbf{Z})\right];\\ &= E_{p(\mathbf{x})}\left[-\nabla_{\mathbf{x}} \nabla_{\mathbf{x}}^T \ln p(\mathbf{x})\right] + E_{p(\mathbf{x})}\left[E_{p(\mathbf{Z}|\mathbf{x})}\left[-\nabla_{\mathbf{x}} \nabla_{\mathbf{x}}^T \ln p(\mathbf{Z}|\mathbf{x})\right]\right].\end{aligned}$$

The first term $\mathbf{I}_{VT,prior} = E_{p(\mathbf{x})}\left[-\nabla_{\mathbf{x}} \nabla_{\mathbf{x}}^T \ln p(\mathbf{x})\right]$ reflects the information in the prior distribution, the second term

$$\mathbf{I}_{VT,meas} = E_{p(\mathbf{x})}\left[E_{p(\mathbf{Z}|\mathbf{x})}\left[-\nabla_{\mathbf{x}} \nabla_{\mathbf{x}}^T \ln p(\mathbf{Z}|\mathbf{x})\right]\right];$$

is the *expected* information in the measurements and is the expectation of the original Fisher information matrix over the a priori distribution of $\mathbf{x}$. The inverse of the matrix $\mathbf{I}_{VT}$ is called the *Van Trees version of Cramér-Rao lower bound* or the *posterior Cramér-Rao bound* [253] and defines the expected lower bound on the posterior covariance matrix of the state estimate (taking prior information into account).

3.5.3 Entropy Versus Covariance Matrix

The covariance matrix is a measure of the probable *deviation between the true state value and a single state estimate (the mean value of the distribution).* The entropy is a measure for the *"area" of the domain of probable state values.* Figure 3.7 illustrates the difference: the pdf on the left is a unimodal pdf, the pdf in the middle is a pdf with two separate peaks with the same

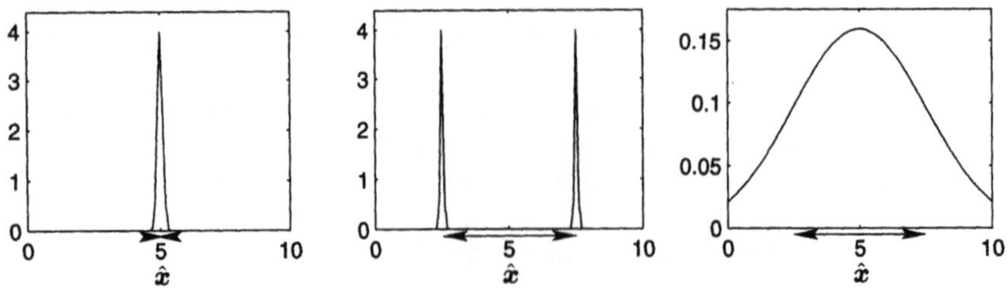

Fig. 3.7. Three pdfs. **(left)** A unimodal pdf. **(middle)** A bi-modal pdf with the same entropy but a larger covariance than the first. **(right)** A pdf with the same covariance but a larger entropy than the second. The arrows indicate the mean value $\hat{\mathbf{x}} \pm$ one standard deviation

entropy as the pdf on the left. The entropy of this bi-modal pdf is fully determined by the "area" of the two peaks, it is independent of the distance between them:

$$\begin{aligned} H &= E_{p(\mathbf{x})}[-\log p(\mathbf{x})] = -\int_x (\log p(\mathbf{x}))\ p(\mathbf{x})\ d\mathbf{x} \\ &= -\int_{peak1} (\log p(\mathbf{x}))\ p(\mathbf{x})\ d\mathbf{x} - \int_{peak2} (\log p(\mathbf{x}))\ p(\mathbf{x})\ d\mathbf{x} \\ &= H_{peak1} + H_{peak2} \end{aligned}$$

The covariance, however, is clearly influenced by this distance. This bi-modal pdf has a larger covariance than the pdf on the left (the arrows on the horizontal axis indicate the mean value $\hat{\mathbf{x}} \pm$ one standard deviation. The pdf on the right is unimodal with the same covariance as the second pdf, however, its entropy is larger.

3.5.4 Information Measures in this Work

In mobile robot localisation the entropy is an often applied measure of uncertainty on the robot pose [50, 102, 150, 218, 252, 180, 35]. This means that the task developers prefer *different* small uncertainty regions (with whatever distance between them) to a single but larger uncertainty region. In some papers the covariance matrix is used as a measure [26, 34, 97, 190]. This results in minimising the size of *one single* uncertainty region in space enclosing all possible robot locations.

In our application of autonomous compliant motion, we want to optimise task plans to get an accurate estimate $\hat{\mathbf{x}}$ of the geometrical parameters. The accurate task execution is based on this estimated value which needs to be determined unambiguously. Therefore this work uses the *covariance matrix* as a measure for the accuracy of the estimate.

In contrast to most mobile robotics problems where the goal of running the robot is to have pose estimates which are "as accurate as possible", our application is characterised by a *tolerance bound* on the estimates. More accuracy than this desired tolerance does not bring any benefit, in general there is even an associated cost (time, energy, etc). Therefore, this work uses the *matrix inequality* as comparison method between the covariance of the estimate and the desired covariance: an active sensing task plan is accepted if the expected covariance matrix of the estimate is more accurate than the desired tolerance covariance matrix. The matrix inequality is a constraint on the task plan, the plan itself can be optimised to minimise execution time, energy cost, etc.

3.6 Optimisation Under Uncertainty

This section describes the optimisation algorithms used in active sensing when the task execution is subject to uncertainty. The goal of the optimisation problem is to find the *sequence* of *discrete actions* $\mathbf{a}_i$ *(system inputs and sensor inputs)* that maximises an *expected value function* at execution time. The active sensing problem for autonomous compliant motion for instance aims at optimising task plans in order to limit the execution time (function to optimise) while assuring the estimation of the geometrical parameters to the desired accuracy (constraint expressing that the final covariance matrix is smaller than the tolerance matrix).

A specific sequence of actions is called a *policy* $\boldsymbol{\pi}_0 = \left[\mathbf{a}_0 \ldots \mathbf{a}_{N-1}\right]$. A policy is finite-horizon (over a fixed, finite number of time steps) or infinite-horizon ($N = \infty$). With every action, state and/or change in state a *value* V (a reward, or minus a cost) can be associated. Hence, with every initial state $\mathbf{x}_0$ and policy $\boldsymbol{\pi}_0$, an *expected* value $V^{\boldsymbol{\pi}_0}(\mathbf{x}_0)$ is associated. The goal is to find the policy $\boldsymbol{\pi}_0^*$ that maximises the expected value function: $V^{\boldsymbol{\pi}_0^*}(\mathbf{x}_0) \geq V^{\boldsymbol{\pi}_0}(\mathbf{x}_0), \forall \boldsymbol{\pi}_0, \mathbf{x}_0$.

Process model uncertainty makes the outcome of a system input on a state $\mathbf{x}_i$ uncertain, i.e., after a system input, the next state is described by a pdf over the state space $\mathbf{x}_{i+1}$. If the state is *fully observable* (no measurement model uncertainty), the optimal policy $\boldsymbol{\pi}^*$ is calculated for every possible state at every possible time step. This optimisation problem is called a *Markov Decision Process* (MDP, Sect. 3.6.1). If, however, the state is *not fully observable*, the optimal policy needs to be calculated for every possible *pdf* over the state space at every possible step. The latter optimisation problem is called a *Partially Observable Markov Decision Process* (POMDP, Sect. 3.6.2). Defining the optimal task plan for autonomous compliant motion is a POMDP: the executed CF sequence is uncertain (process model uncertainty) and the current CF is not always uniquely identifiable (measurement model uncertainty).

This section focuses on decision problems under uncertainty with *discrete* state and action spaces. Discrete (PO)MDP problems can be solved exactly, whereas the solutions to continuous (PO)MDPs can generally only be approximated. There are two basic strategies for approximating the solution to a continuous (PO)MDP [219]: (i) discretize the problem (grid-based, Monte Carlo, ...) or (ii) approximate the value function V and/or the policy $\boldsymbol{\pi}$. This section presents an overview of the optimisation algorithms without going into mathematical details. The interested reader is referred to the mentioned literature for more detailed descriptions.

3.6.1 Markov Decision Process

Markov Decision Processes (MDPs) are optimisation problems under process model uncertainty for which, at execution time, the state of the system is known, i.e., the measurements undoubtedly identify the state. MDPs are originally described by Bellman [22, 23] and Howard [121]. More recent works on MDPs are [27, 28, 170, 213]. The optimal policy, which maximises the *expected* future reward, can be calculated before the task execution for every possible state at every possible time step. Standard methods to solve Markov Decision Processes are:

- *Value iteration* [22] for both finite and infinite horizon problems. This optimisation converts the problem of optimising the N actions into a succession of N *sequential optimisation problems* to be solved for only one action. This way of optimising is called *dynamic programming.*
  The optimisation is based on *Bellman's principle*: "An optimal policy $\boldsymbol{\pi}_{k-1}^* = \left[\mathbf{a}_{k-1}\ \mathbf{a}_k\ \ldots\ \mathbf{a}_{N-1}\right]$ at time $k-1$ has the property that whatever the state $\mathbf{x}_{k-1}$ and the action $\mathbf{a}_{k-1}$ are, the remaining actions $\boldsymbol{\pi}_k^* = \left[\mathbf{a}_k\ \ldots\ \mathbf{a}_{N-1}\right]$ at time k must constitute an optimal policy with regard to the state $\mathbf{x}_k$ resulting from $\mathbf{x}_{k-1}$ and $\mathbf{a}_{k-1}$." This makes a recursive calculation of the optimal policy possible: finding an *optimal* policy for the system when $N-i$ time steps remain, is obtained by using the *optimal* policy for the next time step (when $N-i-1$ steps remain).
- The *policy iteration* algorithm [121] and the *modified policy algorithm* [213] for infinite horizon problems. Policy iteration is an iterative technique similar to dynamic programming, the modified policy algorithm is a combination of policy iteration and value iteration.
- *Linear programming* for infinite horizon problems [213, 59, 226]. The MDP problem is converted into a linear program, solved by the simplex method or the interior point method [199]. Because linear programming does not exploit the dynamic programming structure of the problem, it is generally less efficient than the previously mentioned techniques.

The previously mentioned methods are optimal algorithms. However, for larger MDPs, *approximate solutions* are needed [36, 219], e.g., by compacting the state, action and/or reward representation; by truncating the time horizon; by building special problem-specific structures such as graphs or sets of clauses; etc.

3.6.2 Partially Observable Markov Decision Process

If next to the process model uncertainty, the system is also subject to measurement model uncertainty, the optimisation problem is called a *Partially Observable Markov Decision Process* (POMDP) [79, 14, 173, 51]. Hence, at execution time only a *pdf* over the states is available. This complicates the solution a lot: an optimal policy for every possible pdf at every possible time step is needed.

Exact methods to solve the POMDP convert the POMDP in a belief-state MDP, and compute the exact value function $V(p(\mathbf{x}))$ [197]. Several exact algorithms analogous to the MDP value iteration algorithm are developed [239, 195, 87, 55, 173, 48, 169, 281]. However, for general problems these algorithms are computationally intractable [49].

Practical algorithms (i) approximate the value function $V(\ldots)$, the belief state $p(\mathbf{x})$, or both [111, 173, 251]; (ii) exploit the *structure* of the POMDP and compute exact tree-structured value functions and policies [37]; or (iii) solve the underlying MDP and use it as the basis of various *heuristics* [49].

3.6.3 Task Planning for Compliant Motion

As mentioned before, the task planning for autonomous compliant motion is a POMDP. Unfortunately, solving a POMDP is impossible for all but the smallest problems [49]. Therefore, we approximate the problem by calculating the optimal task plan assuming that the current state estimate is the true state value. The idea is to re-plan the task during the execution, i.e., when more accurate estimates are available.

3.7 Conclusions

This chapter contains a survey of Bayesian techniques for state estimation, hypothesis testing and active sensing. In this work Bayesian techniques are preferred over other techniques, such as interval or fuzzy theory, because the Bayesian theory is well developed and consistent, i.e., calculating an estimate in different ways yields the same results. However, for practical applications the Bayesian formulas are often computationally too complex. This chapter additionally describes approximations to the theory and presents the assumptions under which they are derived.

Bayesian Probability in this Work

This work presents the following improvements in estimation theory:

1. Chapter 4 makes a comparison between the well-known Kalman Filter variants, i.e., the Extended Kalman Filter, the Iterated Extended Kalman Filter, the Central Difference Filter, the first-order Divided Difference Filter and the Unscented Kalman Filter, which are commonly used as *approximate* estimators for nonlinear systems.
2. Chapter 5 describes a new *exact* Bayesian filter (called the "Non-minimal State Kalman Filter") for static systems with any kind of nonlinear measurement model subject to additive Gaussian uncertainty. The filter is also valid for a class of dynamic systems.

The following techniques are applied to the problem of autonomous compliant motion:

1. *Estimation:* Chaps. 7 and 8 describe the estimation of the geometrical parameters of the contacting objects based on pose, twist and wrench measurements. The measurement equations are nonlinear. This work applies the Iterated Extended Kalman Filter when the initial uncertainties on the geometrical parameter estimates are small and the Non-minimal State Kalman Filter when the initial uncertainties are large.
2. *Hypothesis testing:* the valid CF is recognised by SNIS consistency testing of the different possible models (Chaps. 7 and 8).

3. *Active sensing:* a task plan consists of both the CF sequence and the compliant motion trajectory in each CF. Chapter 9 optimises both to minimise the execution time under the constraint of observing the geometrical parameters to the required uncertainty. The covariance matrix of the estimate is used as a measure of uncertainty.

4

Kalman Filters for Nonlinear Systems

Although this may seem a paradox, all exact science is dominated by the idea of approximation.

Bertrand Russell, (1872–1970).

4.1 Introduction

The (Linear) Kalman Filter

The Kalman Filter (KF) [137, 240] is a special case of Bayesian filtering theory. It applies to the estimation of a continuous-valued state $\mathbf{x}$ if the state space description of the estimation problem has *linear process and measurement equations* subject to *additive Gaussian uncertainty.* I.e., if (3.4)–(3.5) reduce to

$$\mathbf{x}_k = \mathbf{F}_{k-1}\mathbf{x}_{k-1} + \mathbf{b}_{k-1} + \mathbf{C}_{k-1}\boldsymbol{\rho}_{p,k-1}; \tag{4.1}$$

$$\mathbf{z}_k = \mathbf{H}_k\mathbf{x}_k + \mathbf{d}_k + \mathbf{E}_k\boldsymbol{\rho}_{m,k}; \tag{4.2}$$

where $\mathbf{F}_{k-1}$, $\mathbf{b}_{k-1}$, $\mathbf{C}_{k-1}$, $\mathbf{H}_k$, $\mathbf{d}_k$ and $\mathbf{E}_k$ are (possibly nonlinear) functions of the system input $\mathbf{u}_{k-1}$ and the measurement input $\mathbf{s}_k$. $\mathbf{z}_k$ is the measurement vector at a time step k. $\boldsymbol{\rho}_{p,k-1}$ denotes the process model uncertainty, being a random vector sequence with zero mean and known covariance matrix $\mathbf{Q}_{k-1}$. $\boldsymbol{\rho}_{m,k}$ is the measurement model uncertainty and is a random vector sequence with zero mean and known covariance matrix $\mathbf{R}_k$; $\boldsymbol{\rho}_p$ and $\boldsymbol{\rho}_m$ are mutually uncorrelated and uncorrelated between sampling times[1]. Furthermore, assume a Gaussian prior pdf with mean $\hat{\mathbf{x}}_{0|0}$ and covariance matrix $\mathbf{P}_{0|0}$.

In this case, the pdfs $p(\mathbf{x}_k|\hat{\mathbf{Z}}_{k-1})$ and $p(\mathbf{x}_k|\hat{\mathbf{Z}}_k)$, obtained by (3.2)–(3.3), are also Gaussian distributions. The filtering formulas can be expressed as *analytical functions* calculating the mean vector and covariance matrix of the pdfs:

$$\begin{aligned}
\hat{\mathbf{x}}_{k|k-1} &= E_{p(\mathbf{x}_k|\hat{\mathbf{Z}}_{k-1})}\left[\mathbf{x}_k\right]; \\
&= \mathbf{F}_{k-1}\hat{\mathbf{x}}_{k-1|k-1} + \mathbf{b}_{k-1}; \qquad (4.3) \\
\mathbf{P}_{k|k-1} &= E_{p(\mathbf{x}_k|\hat{\mathbf{Z}}_{k-1})}\left[\left(\mathbf{x}_k - \hat{\mathbf{x}}_{k|k-1}\right)\left(\mathbf{x}_k - \hat{\mathbf{x}}_{k|k-1}\right)^T\right]; \\
&= \mathbf{F}_{k-1}\mathbf{P}_{k-1|k-1}\mathbf{F}_{k-1}^T + \mathbf{C}_{k-1}\mathbf{Q}_{k-1}\mathbf{C}_{k-1}^T; \qquad (4.4) \\
\hat{\mathbf{x}}_{k|k} &= E_{p(\mathbf{x}_k|\hat{\mathbf{Z}}_k)}\left[\mathbf{x}_k\right]; \\
&= \hat{\mathbf{x}}_{k|k-1} + \mathbf{K}_k\boldsymbol{\eta}_k; \qquad (4.5) \\
\mathbf{P}_{k|k} &= E_{p(\mathbf{x}_k|\hat{\mathbf{Z}}_k)}\left[\left(\mathbf{x}_k - \hat{\mathbf{x}}_{k|k}\right)\left(\mathbf{x}_k - \hat{\mathbf{x}}_{k|k}\right)^T\right]; \\
&= \left(\mathbf{I}_{n\times n} - \mathbf{K}_k\mathbf{H}_k\right)\mathbf{P}_{k|k-1}; \qquad (4.6)
\end{aligned}$$

where

$$\begin{aligned}
\boldsymbol{\eta}_k &= \hat{\mathbf{z}}_k - \mathbf{H}_k\hat{\mathbf{x}}_{k|k-1} - \mathbf{d}_k; \qquad (4.7) \\
\mathbf{S}_k &= \mathbf{E}_k\mathbf{R}_k\mathbf{E}_k^T + \mathbf{H}_k\mathbf{P}_{k|k-1}\mathbf{H}_k^T; \qquad (4.8) \\
\mathbf{K}_k &= \mathbf{P}_{k|k-1}\mathbf{H}_k^T\mathbf{S}_k^{-1}. \qquad (4.9)
\end{aligned}$$

$\boldsymbol{\eta}_k$ is called the *innovation*, its covariance is $\mathbf{S}_k$. $\mathbf{K}_k$ is the *Kalman gain*. Equations (4.3)–(4.4) are referred to as the *process update*, (4.5)–(4.9) as the *measurement update*. $\hat{\mathbf{x}}_{k|k-1}$ is called the *predicted* state estimate and $\hat{\mathbf{x}}_{k|k}$ the *updated* state estimate. If no measurement $\hat{\mathbf{z}}_k$ is available at a certain time step k, then (4.5)–(4.9) reduce to $\hat{\mathbf{x}}_{k|k} = \hat{\mathbf{x}}_{k|k-1}$ and $\mathbf{P}_{k|k} = \mathbf{P}_{k|k-1}$.

Equation (4.6) is sensitive to round-off errors which can lead to the loss of symmetry and positive definiteness of the covariance matrix. Replacing the update by

$$\mathbf{P}_{k|k} = \mathbf{P}_{k|k-1} - \mathbf{K}_k\mathbf{S}_k\mathbf{K}_k^T;$$

avoids the loss of symmetry. Replacing the update by

$$\mathbf{P}_{k|k} = \left(\mathbf{I}_{n\times n} - \mathbf{K}_k\mathbf{H}_k\right)\mathbf{P}_{k|k-1}\left(\mathbf{I}_{n\times n} - \mathbf{K}_k\mathbf{H}_k\right)^T + \mathbf{K}_k\mathbf{E}_k\mathbf{R}_k\mathbf{E}_k^T\mathbf{K}_k^T;$$

preserves symmetry and positive definiteness.

[1] Correlated uncertainties can be dealt with by augmenting the state vector, this is the original formulation of the KF [137]. Expressed in this new state vector, the process and measurement models are of the form (4.1) and (4.2) with uncorrelated uncertainties.

Some filter implementations do not update the covariance matrix $\mathbf{P}$ itself, but (i) one of its square roots $\mathbf{S}$ ($\mathbf{P} = \mathbf{S}\mathbf{S}^T$ with a square matrix $\mathbf{S}$ [138, 31]); or (ii) its U-D factorisation ($\mathbf{P} = \mathbf{U}\mathbf{D}\mathbf{U}^T$ with a unit upper triangular matrix $\mathbf{U}$ and a diagonal matrix $\mathbf{D}$, [250]).

The reduced sensitivity to round-off errors generally comes at the price of computational complexity. Reference [32] illustrates the numerical accuracy and the computational efficiency of the different implementations with an orbit determination case study.

An alternative description of the KF algorithm is the *Information Filter* which calculates the inverse of the covariance matrix, i.e., (4.4) and (4.6) are replaced by

$$\mathbf{P}^{-1}_{k|k-1} = \left(\mathbf{F}_{k-1}\left(\mathbf{P}^{-1}_{k-1|k-1}\right)^{-1}\mathbf{F}^T_{k-1} + \mathbf{C}_{k-1}\mathbf{Q}_{k-1}\mathbf{C}^T_{k-1}\right)^{-1};$$
$$\mathbf{P}^{-1}_{k|k} = \mathbf{P}^{-1}_{k|k-1} + \mathbf{H}^T_k\left(\mathbf{E}_k\mathbf{R}_k\mathbf{E}^T_k\right)^{-1}\mathbf{H}_k.$$

This filter needs the inversion of an $n \times n$ matrix, where n is the dimension of the state vector. The standard Kalman Filter inverts an $m \times m$ matrix, where m is the dimension of the measurement vector. The Information Filter can be started up without prior information, i.e., $\mathbf{P}^{-1}_{0|0} = \mathbf{0}_{n\times n}$.

Kalman Filters for Nonlinear Systems

In practise, because the KF is easy to use and because it has a low computational cost, the algorithm is often applied to systems with *nonlinear* process and measurement models[2],

$$\mathbf{x}_k = \mathbf{f}_{k-1}(\mathbf{x}_{k-1}) + \mathbf{C}_{k-1}\boldsymbol{\rho}_{p,k-1}; \tag{4.10}$$
$$\mathbf{z}_k = \mathbf{h}_k(\mathbf{x}_k) + \mathbf{E}_k\boldsymbol{\rho}_{m,k}; \tag{4.11}$$

by *linearizing* the models:

$$\mathbf{x}_k = \mathbf{F}_{k-1}\mathbf{x}_{k-1} + \mathbf{b}_{k-1} + \boldsymbol{\rho}^*_{p,k-1} + \mathbf{C}_{k-1}\boldsymbol{\rho}_{p,k-1}; \tag{4.12}$$
$$\mathbf{z}_k = \mathbf{H}_k\mathbf{x}_k + \mathbf{d}_k + \boldsymbol{\rho}^*_{m,k} + \mathbf{E}_k\boldsymbol{\rho}_{m,k}. \tag{4.13}$$

The difference between these models and models (4.1)–(4.2) is the presence of the terms $\boldsymbol{\rho}^*_{p,k-1}$ and $\boldsymbol{\rho}^*_{m,k}$ representing the linearization errors. The additional uncertainty on the linearized models due to these linearization errors is $\mathbf{Q}^*_{k-1}$ and $\mathbf{R}^*_k$. Unfortunately, applying the KF[3] (4.3)–(4.9) to systems with

[2] Models which are nonlinear functions of the uncertainties $\boldsymbol{\rho}_p$ and $\boldsymbol{\rho}_m$, can be dealt with by augmenting the state vector with the uncertainties. Expressed in this new state vector, the process and measurement models are of the form (4.10) and (4.11).

[3] $\mathbf{C}_{k-1}\boldsymbol{\rho}_{p,k-1}$ of (4.1) corresponds to $\boldsymbol{\rho}^*_{p,k-1} + \mathbf{C}_{k-1}\boldsymbol{\rho}_{p,k-1}$ of (4.12); hence instead of $\mathbf{C}_{k-1}\mathbf{Q}_{k-1}\mathbf{C}^T_{k-1}$ in (4.4), $\mathbf{Q}^*_{k-1} + \mathbf{C}_{k-1}\mathbf{Q}_{k-1}\mathbf{C}^T_{k-1}$ is used.
$\mathbf{E}_k\boldsymbol{\rho}_{m,k}$ of (4.2) corresponds to $\boldsymbol{\rho}^*_{m,k} + \mathbf{E}_k\boldsymbol{\rho}_{m,k}$ of (4.13); hence instead of $\mathbf{E}_k\mathbf{R}_k\mathbf{E}^T_k$ in (4.9), $\mathbf{R}^*_k + \mathbf{E}_k\mathbf{R}_k\mathbf{E}^T_k$ is used.

nonlinear process and/or measurement models leads to *non-optimal* estimates and covariance matrices. Different ways of linearizing the process and measurement models, i.e., different choices for $\mathbf{F}_{k-1}$, $\mathbf{b}_{k-1}$, $\mathbf{Q}^*_{k-1}$, $\mathbf{H}_k$, $\mathbf{d}_k$ and $\mathbf{R}^*_k$, yield other results. This chapter aims at making an objective comparison of the best known "KF variants".

Overview of this Chapter

This chapter describes (i) how the best known KF variants differ in their linearization of the process and measurement models; (ii) how they take the linearization errors into account; and (iii) how the quality of their state estimates depends on the previous two choices. The studied algorithms are:

1. The *Extended Kalman Filter* (EKF) [19, 248, 106, 182];
2. The *Iterated Extended Kalman Filter* (IEKF) [19, 248, 106, 182];
3. The *Central Difference Filter* (CDF) [221], the *first-order Divided Difference Filter* (DD1) [200, 201] and the *Unscented Kalman Filter* (UKF) [258, 133, 134, 132]. We describe the CDF, DD1 and UKF in a unified way as the *Linear Regression Kalman Filter* (LRKF).

The chapter gives the following insights:

1. The UKF is described as a filter which *does* linearize the process and measurement models. The developers of this filter claimed that the UKF did *not* perform this linearization.
   The UKF is shown to be a filter which linearizes the models by statistical linear regression. This allows a better understanding of the filter performance and creates the possibility to compare the filter to other KF variants. Also the less known CDF and DD1 fit in the framework of this "Linear Regression Kalman Filter" (LRKF, Sect. 4.2).
2. The quality of the estimates of the KF variants can be expressed by two criteria, i.e., the *consistency* and the *information content* of the estimates (defined in Sect. 4.3). This chapter relates the consistency and information content of the estimates to (i) how the linearization is performed and (ii) how the linearization errors are taken into account.
3. Although the filters use similar linearization techniques for the linearization of the process and measurement models, there can be a substantial difference in their performance for both updates:
   a) for the linearization of the *process* update, which describes the evolution of the state, the state estimate and its uncertainty are the only available information;
   b) the *measurement* update, on the other hand, describes the fusion of the information in the state estimate with the information in the new measurement. Hence, in this update also the new measurement is available and can be used to linearize the measurement model.

Therefore, it can be interesting to use different filters for both updates. This chapter compares the filter performances separately for process updates (Sect. 4.4) and measurement updates (Sect. 4.5) instead of their overall performance when they are both combined.
4. Two new insights on the performance of specific KF variants are: (i) the IEKF measurement update outperforms the EKF and LRKF updates if the state —or at least the part of it that causes the non-linearity in the measurement model— is instantaneously fully observable (Sect. 4.5.2); and (ii) for large uncertainties on the state estimate, the LRKF measurement update yields consistent but non-informative state estimates (Sect. 4.5.3).

4.2 The Linear Regression Kalman Filter

This section introduces the Linear Regression Kalman Filter (LRKF). The LRKF (i) linearizes the process and measurement functions by statistical linear regression of the functions through a number of regression points; and (ii) represents the extra uncertainty on a linearized function due to linearization errors by the sample covariance of the deviations between the nonlinear and the linearized function in the regression points.

Sections 4.2.1 and 4.2.2 present the process and measurement update formulas. The detailed derivations are given in Appendix A. The update formulas are equal to those of the Central Difference Filter (CDF), the first-order Divided Difference Filter (DD1) and the Unscented Kalman Filter (UKF). These filters are special cases of the LRKF, imposing specific choices of the regression points (Sect. 4.2.3).

4.2.1 Process Update

The LRKF uses the function values of r regression points $\boldsymbol{\mathcal{X}}^j_{k-1|k-1}$, $j = 1, \ldots, r$, in state space to model the behaviour of the process function in the "uncertainty region" around the updated state estimate $\hat{\mathbf{x}}_{k-1|k-1}$. The regression points are chosen such that their mean and covariance matrix equal the state estimate $\hat{\mathbf{x}}_{k-1|k-1}$ and its covariance matrix $\mathbf{P}_{k-1|k-1}$:

$$\hat{\mathbf{x}}_{k-1|k-1} = \frac{1}{r}\sum_{j=1}^{r} \boldsymbol{\mathcal{X}}^j_{k-1|k-1};$$

$$\mathbf{P}_{k-1|k-1} = \frac{1}{r}\sum_{j=1}^{r} \left(\boldsymbol{\mathcal{X}}^j_{k-1|k-1} - \hat{\mathbf{x}}_{k-1|k-1}\right)\left(\boldsymbol{\mathcal{X}}^j_{k-1|k-1} - \hat{\mathbf{x}}_{k-1|k-1}\right)^T.$$

This leaves some freedom in the choice of the points $\boldsymbol{\mathcal{X}}^j_{k-1|k-1}$, Sect. 4.2.3 describes some specific choices. The function values of the regression points are

$$\boldsymbol{\mathcal{X}}^j_{k|k-1} = \mathbf{f}_{k-1}(\boldsymbol{\mathcal{X}}^j_{k-1|k-1}).$$

The LRKF linearizes the process model by *statistical linear regression* through the $(\boldsymbol{\mathcal{X}}^j_{k-1|k-1}, \boldsymbol{\mathcal{X}}^j_{k|k-1})$ points; i.e., the deviations $\mathbf{e}_j$ between the function values in $\boldsymbol{\mathcal{X}}^j_{k-1|k-1}$ for the nonlinear and the linearized function are minimised in least-squares sense:

$$\mathbf{e}_j = \boldsymbol{\mathcal{X}}^j_{k|k-1} - \left(\mathbf{F}\boldsymbol{\mathcal{X}}^j_{k-1|k-1} + \mathbf{b}\right); \tag{4.14}$$

$$(\mathbf{F}_{k-1}, \mathbf{b}_{k-1}) = \arg\min_{(\mathbf{F},\mathbf{b})} \sum_{j=1}^{r} \mathbf{e}_j^T \mathbf{e}_j. \tag{4.15}$$

The sample covariance of the deviations $\mathbf{e}_j$

$$\mathbf{Q}^*_{k-1} = \frac{1}{r}\sum_{j=1}^{r} \mathbf{e}_j \mathbf{e}_j^T \tag{4.16}$$

gives an idea of the magnitude of the linearization errors in the "uncertainty region" around $\hat{\mathbf{x}}_{k-1|k-1}$.

The LRKF *process update equations* follow from the linear Kalman Filter equations[3,p. 53] (4.3)–(4.4). In Appendix A, the update equations are shown to be:

$$\hat{\mathbf{x}}_{k|k-1} = \frac{1}{r}\sum_{j=1}^{r} \boldsymbol{\mathcal{X}}^j_{k|k-1};$$

$$\mathbf{P}_{k|k-1} = \frac{1}{r}\sum_{j=1}^{r} \left(\boldsymbol{\mathcal{X}}^j_{k|k-1} - \hat{\mathbf{x}}_{k|k-1}\right)\left(\boldsymbol{\mathcal{X}}^j_{k|k-1} - \hat{\mathbf{x}}_{k|k-1}\right)^T + \mathbf{C}_{k-1}\mathbf{Q}_{k-1}\mathbf{C}^T_{k-1}.$$

4.2.2 Measurement Update

The LRKF evaluates the measurement function in r regression points $\boldsymbol{\mathcal{X}}^j_{k|k-1}$, $j = 1, \ldots, r$, in the "uncertainty region" around the predicted state estimate $\hat{\mathbf{x}}_{k|k-1}$. The $\boldsymbol{\mathcal{X}}^j_{k|k-1}$ are chosen such that their mean and covariance matrix are equal to the predicted state estimate and its covariance:

$$\hat{\mathbf{x}}_{k|k-1} = \frac{1}{r}\sum_{j=1}^{r} \boldsymbol{\mathcal{X}}^j_{k|k-1};$$

$$\mathbf{P}_{k|k-1} = \frac{1}{r}\sum_{j=1}^{r} \left(\boldsymbol{\mathcal{X}}^j_{k|k-1} - \hat{\mathbf{x}}_{k|k-1}\right)\left(\boldsymbol{\mathcal{X}}^j_{k|k-1} - \hat{\mathbf{x}}_{k|k-1}\right)^T.$$

This leaves some freedom in the choice of the points $\boldsymbol{\mathcal{X}}^j_{k|k-1}$, Sect. 4.2.3 describes some specific choices.

The function values of the regression points through the nonlinear function are

$$\mathcal{Z}_k^j = \mathbf{h}_k(\mathcal{X}_{k|k-1}^j).$$

The LRKF algorithm uses a linearized measurement model obtained by *statistical linear regression* through the $(\mathcal{X}_{k|k-1}^j, \mathcal{Z}_k^j)$ points. The statistical linear regression is such that the deviations $\mathbf{e}_j$ between the nonlinear and the linearized function in the regression points are minimised in least-squares sense:

$$\mathbf{e}_j = \mathcal{Z}_k^j - \left(\mathbf{H}\mathcal{X}_{k|k-1}^j + \mathbf{d}\right);$$

$$(\mathbf{H}_k, \mathbf{d}_k) = \arg\min_{(\mathbf{H},\mathbf{d})} \sum_{j=1}^{r} \mathbf{e}_j^T \mathbf{e}_j.$$

The sample covariance matrix of the deviations $\mathbf{e}_j$ gives an idea of the magnitude of the linearization errors:

$$\mathbf{R}_k^* = \frac{1}{r}\sum_{j=1}^{r} \mathbf{e}_j \mathbf{e}_j^T. \tag{4.17}$$

The LRKF *measurement update equations* follow from the linear Kalman Filter equations[3,p. 53] (4.5)–(4.9). In Appendix A, the update equations are shown to be:

$$\hat{\mathbf{x}}_{k|k} = \hat{\mathbf{x}}_{k|k-1} + \mathbf{K}_k \boldsymbol{\eta}_k;$$
$$\mathbf{P}_{k|k} = \left(\mathbf{I}_{n\times n} - \mathbf{K}_k\mathbf{H}_k\right)\mathbf{P}_{k|k-1};$$

where

$$\bar{\mathcal{Z}}_k = \frac{1}{r}\sum_{j=1}^{r} \mathcal{Z}_k^j;$$

$$\boldsymbol{\eta}_k = \hat{\mathbf{z}}_k - \bar{\mathcal{Z}}_k;$$

$$\mathbf{S}_k = \frac{1}{r}\sum_{j=1}^{r}\left(\mathcal{Z}_k^j - \bar{\mathcal{Z}}_k\right)\left(\mathcal{Z}_k^j - \bar{\mathcal{Z}}_k\right)^T + \mathbf{E}_k\mathbf{R}_k\mathbf{E}_k^T;$$

$$\mathbf{K}_k = \left(\frac{1}{r}\sum_{j=1}^{r}\left(\mathbf{X}_{k|k-1}^j - \hat{\mathbf{x}}_{k|k-1}\right)\left(\mathcal{Z}_k^j - \bar{\mathcal{Z}}_k\right)^T\right)\mathbf{S}_k^{-1}.$$

4.2.3 The UKF, CDF and DD1 Filters

The Central Difference Filter (CDF), the first-order Divided Difference Filter (DD1) and the Unscented Kalman Filter (UKF) are LRKFs. The general LRKF leaves the choice of the regression points open, it only states that their mean and covariance correspond to the mean and covariance of the previous state estimate. The CDF, DD1 and UKF propose specific choices of the regression points which assure that — for *continuous* process and measurement functions — the results are equally or less inconsistent than the results of the Extended Kalman Filter.

- The UKF [258, 133, 134, 132] is the best known LRKF. The UKF is derived as a filter transforming approximations of the pdfs through the nonlinear process and measurement functions. To this purpose, it it chooses $2n + 2\kappa$ points in state space. κ is a degree of freedom in the choice of the points. The optimal transformation of the mean and covariance of the pdf with these points as "samples" leads to the same update formulas as the LRKF (Sects. 4.2.1 and 4.2.2). This means that the UKF performs a statistical linear regression for the linearization of the process and measurement functions.
  Choosing the same regression point several times, corresponds to associating a non-unit weight to the point. The UKF then performs a weighted linear regression. Also real numbers can be chosen as weights.
- Sometimes, the UKF is used with a $\kappa < 0$, resulting in negative weights [133, 132]. In this case, the calculated covariance matrices can be non-positive semi-definite. To overcome this problem, the covariances are artificially increased.
- The Scaled Unscented Transformation [131] introduces one more degree of freedom in the choice of the regression points and their weights (parameter α). In this case, the calculated covariance matrices lie between those of the original formulation and the above modified form ($\kappa < 0$). The covariances can be increased even more by introducing another parameter (β).
- The Reduced Sigma Point Filters [258, 130] reduce the number of regression points to $n + 1$ (the so-called "Simplex Sigma Points") for an n-dimensional state space. This means that the linear regression is exact, i.e., the linearized function is a hyperplane through the regression points. Hence, $\mathbf{Q}^*$ and $\mathbf{R}^*$ are zero: the linearization errors are not taken into account. In this case, the calculated covariances are too small and need to be increased artificially.
- The Central Difference Filter (CDF) [221] and the first-order Divided Difference Filter (DD1) [200, 201] choose $2n$ regression points for an n-dimensional state vector. These filters calculate the *covariance matrices* by linear regression. The state update $\hat{\mathbf{x}}_{k|k-1}$ and the innovation $\boldsymbol{\eta}_k$ are obtained by the propagation of $\hat{\mathbf{x}}_{k-1|k-1}$ and $\hat{\mathbf{x}}_{k|k-1}$ through the nonlinear functions (as in the case of the Extended Kalman Filter).

4.3 Consistency and Information Content of the Estimates

The KF variants for nonlinear systems calculate an estimate $\hat{\mathbf{x}}_{k|i}$ and covariance matrix $\mathbf{P}_{k|i}$ for a pdf which is non-Gaussian. The performance of these KFs depends on how representative the Gaussian pdf with mean $\hat{\mathbf{x}}_{k|i}$ and covariance $\mathbf{P}_{k|i}$ is for the (unknown) pdf $p(x_k|\hat{\mathbf{Z}}_i)$. Figure 4.1 shows a non-Gaussian pdf $p(x_k|\hat{\mathbf{Z}}_i)$ and three possible Gaussian "approximations" $p_1(x_k|\hat{\mathbf{Z}}_i)$, $p_2(x_k|\hat{\mathbf{Z}}_i)$ and $p_3(x_k|\hat{\mathbf{Z}}_i)$. Intuitively we feel that $p_1(x_k|\hat{\mathbf{Z}}_i)$ is a good approximation because "the same" values of x are probable. Similarly $p_3(x_k|\hat{\mathbf{Z}}_i)$ is not a good approximation because a lot of probable values for x of the original distribution have a probability density of approximately zero in $p_3(x_k|\hat{\mathbf{Z}}_i)$. Finally pdf $p_2(x_k|\hat{\mathbf{Z}}_i)$ is "more uncertain" than $p(x_k|\hat{\mathbf{Z}}_i)$ because a larger domain of x values is uncertain.

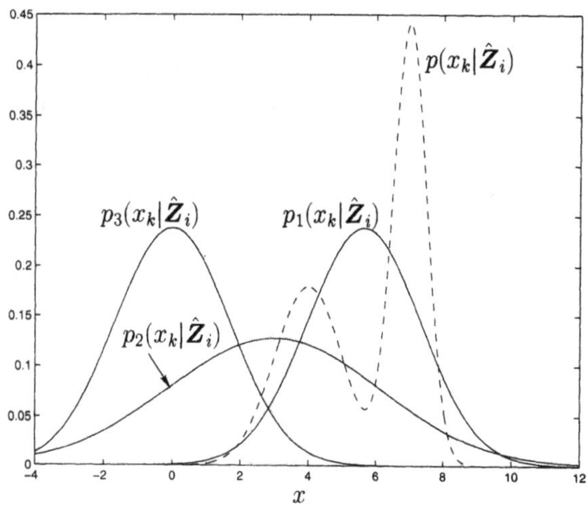

Fig. 4.1. Non-Gaussian pdf $p(x_k|\hat{\mathbf{Z}}_i)$ with three Gaussian approximations $p_1(x_k|\hat{\mathbf{Z}}_i)$, $p_2(x_k|\hat{\mathbf{Z}}_i)$ and $p_3(x_k|\hat{\mathbf{Z}}_i)$. $p_1(x_k|\hat{\mathbf{Z}}_i)$ and $p_2(x_k|\hat{\mathbf{Z}}_i)$ are consistent, $p_3(x_k|\hat{\mathbf{Z}}_i)$ is inconsistent. $p_1(x_k|\hat{\mathbf{Z}}_i)$ is more informative than $p_2(x_k|\hat{\mathbf{Z}}_i)$

These intuitive reflexions are formulated in two criteria: the *consistency* and the *information content* of the state estimate. The consistency of the state estimate is a necessary condition for a filter to be acceptable. The information content of the state estimates defines an ordering between all consistent filters.

1. the *consistency* of the state estimate.
   A state estimate $\hat{\mathbf{x}}_{k|i}$ with covariance matrix $\mathbf{P}_{k|i}$ is called consistent if

$$E_{p(\mathbf{x}_k|\hat{\mathbf{Z}}_i)}\left[\left(\mathbf{x}_k-\hat{\mathbf{x}}_{k|i}\right)\left(\mathbf{x}_k-\hat{\mathbf{x}}_{k|i}\right)^T\right] \leq \mathbf{P}_{k|i}. \tag{4.18}$$

For consistent results, matrix $\mathbf{P}_{k|i}$ is equal to or larger than the expected squared deviation with respect to the estimate $\hat{\mathbf{x}}_{k|i}$ under the (unknown) distribution $p(\mathbf{x}_k|\hat{\mathbf{Z}}_i)$. The mean and covariance of pdfs $p_1(x_k|\hat{\mathbf{Z}}_i)$ and $p_2(x_k|\hat{\mathbf{Z}}_i)$ in Fig. 4.1 obey (4.18). Pdf $p_3(x_k|\hat{\mathbf{Z}}_i)$, on the other hand, is inconsistent.
Inconsistency of the calculated state estimate $\hat{\mathbf{x}}_{k|i}$ and covariance matrix $\mathbf{P}_{k|i}$ is the most encountered problem with the KF variants ("divergence" of the filter). In this case, $\mathbf{P}_{k|i}$ is too small and does no longer represent a reliable measure for the uncertainty on $\hat{\mathbf{x}}_{k|i}$. Even more, once an inconsistent state estimate is met, the subsequent state estimates are also inconsistent. This is because the filter believes the inconsistent state estimate to be more accurate than it is in reality and hence, it attaches too much weight to this state estimate when processing new measurements.
Testing for inconsistency is done by *consistency tests* such as tests on the sum of a number of Normalised Innovation Squared values[4, p. 36] (SNIS) [19, 268].
2. the *information content* of the state estimate.
The calculated covariance matrix $\mathbf{P}_{k|i}$ indicates how uncertain the state estimate $\hat{\mathbf{x}}_{k|i}$ is: a large covariance matrix indicates an inaccurate (and little useful) state estimate; the smaller the covariance matrix, the larger the information content of the state estimate.
E.g., both pdfs $p_1(x_k|\hat{\mathbf{Z}}_i)$ and $p_2(x_k|\hat{\mathbf{Z}}_i)$ of Fig. 4.1 are consistent with $p(x_k|\hat{\mathbf{Z}}_i)$, however, $p_1(x_k|\hat{\mathbf{Z}}_i)$ has a smaller variance, hence it is more informative than $p_2(x_k|\hat{\mathbf{Z}}_i)$ (a smaller domain of x values is probable). The most informative, consistent approximation is the Gaussian with the same mean and covariance as the original distribution, i.e., $p_1(x_k|\hat{\mathbf{Z}}_i)$ for the example.
There is a *trade-off* between consistent and informative state estimates: inconsistency can be avoided by making $\mathbf{P}_{k|i}$ artificially larger, see (4.18). Making $\mathbf{P}_{k|i}$ too large, i.e., larger than necessary for consistency, corresponds to loosing information about the actual accuracy of the state estimate.

The different KF variants *linearize* the process and the measurement models in the uncertainty region around the state estimate. *Consistent* estimates are obtained by adding process and measurement model uncertainty on the linearized models to compensate for the linearization errors. In order for the estimates to be *informative* (i) the linearization errors need to be as small as possible; and (ii) the extra uncertainty on the linearized models should not be larger than necessary to compensate for the linearization errors.

The following sections describe how the Extended Kalman Filter, the Iterated Extended Kalman Filter and the Linear Regression Kalman Filter differ in their linearization of the process and measurement models; how they take

Table 4.1. Summary of the linearization of the process model by the Extended Kalman Filter (EKF), the Iterated Extended Kalman Filter (IEKF) and the Linear Regression Kalman Filter (LRKF)

	$\mathbf{F}_{k-1}$	$\mathbf{b}_{k-1}$	$\mathbf{Q}^*_{k-1}$
EKF	$\frac{\partial \mathbf{f}_{k-1}}{\partial \mathbf{x}}\Big\vert_{\hat{\mathbf{x}}_{k-1\vert k-1}}$	$\mathbf{f}_{k-1}(\hat{\mathbf{x}}_{k-1\vert k-1}) - \mathbf{F}_{k-1}\hat{\mathbf{x}}_{k-1\vert k-1}$	$\mathbf{0}_{n\times n}$
IEKF	$\frac{\partial \mathbf{f}_{k-1}}{\partial \mathbf{x}}\Big\vert_{\hat{\mathbf{x}}_{k-1\vert k-1}}$	$\mathbf{f}_{k-1}(\hat{\mathbf{x}}_{k-1\vert k-1}) - \mathbf{F}_{k-1}\hat{\mathbf{x}}_{k-1\vert k-1}$	$\mathbf{0}_{n\times n}$
LRKF	$\arg\min_{(\mathbf{F},\mathbf{b})} \sum_{j=1}^{r} \mathbf{e}_j^T \mathbf{e}_j$	$\arg\min_{(\mathbf{F},\mathbf{b})} \sum_{j=1}^{r} \mathbf{e}_j^T \mathbf{e}_j$	$\frac{1}{r}\sum_{j=1}^{r} \mathbf{e}_j \mathbf{e}_j^T$

the linearization errors into account; and how the quality[4] of the state estimates, expressed in terms of consistency and information content, depends on these two choices.

4.4 Nonlinear Process Models

This section contains a comparison between the *process* updates of the different KF variants when dealing with a nonlinear process model (4.10) with linearization (4.12). The KF variants differ by their choice of $\mathbf{F}_{k-1}$, $\mathbf{b}_{k-1}$ and $\mathbf{Q}^*_{k-1}$. After linearization, they all use process update equations[3,p. 53] (4.3)–(4.4) to update the state estimate and its uncertainty.

Section 4.4.1 describes the linearization of the process model by the EKF and IEKF, Sect. 4.4.2 by the LRKF. The formulas are summarised in Table 4.1. Section 4.4.4 presents some examples.

4.4.1 The (Iterated) Extended Kalman Filter

The EKF and the IEKF linearize[5] the process model by a first-order Taylor series around the *updated state estimate* $\hat{\mathbf{x}}_{k-1|k-1}$:

$$\mathbf{F}_{k-1} = \frac{\partial \mathbf{f}_{k-1}}{\partial \mathbf{x}}\Big|_{\hat{\mathbf{x}}_{k-1|k-1}};$$

$$\mathbf{b}_{k-1} = \mathbf{f}_{k-1}(\hat{\mathbf{x}}_{k-1|k-1}) - \mathbf{F}_{k-1}\hat{\mathbf{x}}_{k-1|k-1}.$$

The basic (I)EKF algorithms *do not take the linearization errors into account*:

$$\mathbf{Q}^*_{k-1} \equiv \mathbf{0}_{n\times n}.$$

This leads to *inconsistent* state estimates when these errors can not be neglected.

[4] The more nonlinear the behaviour of the process or measurement model in the "uncertainty region" around the state estimate, the more pronounced the difference in quality performance (consistency and information content of the state estimates) between the KF variants.

[5] The EKF and IEKF only differ in their measurement update (Sects. 4.5.1–4.5.2).

4.4.2 The Linear Regression Kalman Filter

The LRKF algorithm uses a linearized process function (4.1) where $\mathbf{F}_{k-1}$, $\mathbf{b}_{k-1}$ and $\mathbf{Q}^*_{k-1}$ are obtained by statistical linear regression through a number of regression points $\left(\mathcal{X}^j_{k-1|k-1}, \mathcal{X}^j_{k|k-1}\right)$, $j = 1, \ldots, r$ (see Sect. 4.2.1).

Of course, the state estimates of the LRKF are only consistent and informative if the $(\mathcal{X}^j_{k-1|k-1}, \mathcal{X}^j_{k|k-1})$ points are representative for the behaviour of the process model in the "uncertainty region" around $\hat{\mathbf{x}}_{k-1|k-1}$. The literature on the CDF, DD1 and UKF filters propose choices of the regression points which assure that —for *continuous* process functions— the calculated estimate $\hat{\mathbf{x}}_{k|k-1}$ with covariance matrix $\mathbf{P}_{k|k-1}$ is equally or less inconsistent than the result calculated by the EKF.

Intuitively we feel that *when enough*[6] *regression points are taken*, the state estimates of the LRKF process update are *consistent* and *informative*. They are consistent because $\mathbf{Q}^*_{k-1}$ gives a well founded approximation of the linearization errors, see (4.16). They are informative because the linearized model is a good approximation of the process model in the "uncertainty region" around $\hat{\mathbf{x}}_{k-1|k-1}$, see (4.14)–(4.15).

4.4.3 Extra Process Model Uncertainty

In all of the presented filters, the user can decide to *add extra process model uncertainty* $\mathbf{Q}^*_{k-1}$ (or to multiply the calculated covariance matrix $\mathbf{P}_{k|k-1}$ by a fading factor larger than 1 [19]. This is useful if the basic filter algorithm is not consistent. E.g., this is the case for the (I)EKF or for an LRKF with a number of regression points too limited to capture the nonlinear behaviour of the process model in the uncertainty region around $\hat{\mathbf{x}}_{k-1|k-1}$.

For a particular problem, values for $\mathbf{Q}^*_{k-1}$ that result in consistent and informative state estimates are obtained by *off-line tuning or on-line parameter learning* (adaptive filtering [187]). In many practical cases consistency is assured by taking the added uncertainty too large, e.g., by taking a constant $\mathbf{Q}^*$ over time which compensates for decreasing linearization errors. This, however, results in less informative estimates.

4.4.4 Examples

The different process updates are illustrated by a simple 2D nonlinear process model ($\mathbf{x}(i)$ denotes the ith element of $\mathbf{x}$):

$$\begin{cases} \mathbf{x}_k(1) = (\mathbf{x}_{k-1}(1))^2\,; \\ \mathbf{x}_k(2) = \mathbf{x}_{k-1}(1) + 3\,\mathbf{x}_{k-1}(2); \end{cases}$$

[6] This depends on the nonlinearity of the model in the "uncertainty region" around the state estimate. A possible approach is to increase the number of regression points until the resulting linearization (with error covariance) does not change any more. Of course, because the true pdf is unknown, it is not possible to *guarantee* that the iteration has converged to a set of regression points representative for the model behaviour in the "uncertainty region".

with no process model uncertainty: $\boldsymbol{\rho}_{p,k-1} \equiv \mathbf{0}_{2\times 1}$. $\mathbf{x}_k(1)$ depends nonlinearly on $\mathbf{x}_{k-1}$. The process update of $\mathbf{x}_k(2)$ is linear. The updated state estimate and its uncertainty at time step $k-1$ are:

$$\hat{\mathbf{x}}_{k-1|k-1} = \begin{bmatrix} 10 \\ 15 \end{bmatrix}; \qquad \mathbf{P}_{k-1|k-1} = \begin{bmatrix} 36 & 0 \\ 0 & 3600 \end{bmatrix}.$$

Monte Carlo Simulation

The mean value and the covariance of the true pdf $p(x_k|\hat{\mathbf{Z}}_{k-1})$ are calculated with a (computationally expensive) Monte Carlo simulation based on a Gaussian pdf $p(x_{k-1}|\hat{\mathbf{Z}}_{k-1})$ with mean $\hat{\mathbf{x}}_{k-1|k-1}$ and covariance matrix $\mathbf{P}_{k-1|k-1}$. The results of this computation are used to illustrate the (in)consistency and information content of the state estimates of the different KF variants. The mean and covariance matrix of the $p(x_k|\hat{\mathbf{Z}}_{k-1})$ calculated by the Monte Carlo algorithm are:

$$\hat{\mathbf{x}}_{k|k-1} = \begin{bmatrix} 136 \\ 55 \end{bmatrix}; \qquad \mathbf{P}_{k|k-1} = \begin{bmatrix} 16994 & 721 \\ 721 & 32436 \end{bmatrix}.$$

(I)EKF

Figure 4.2 shows the updated and (I)EKF predicted state estimates and their uncertainty ellipses[7]. The dotted line is the uncertainty ellipse of the distribution obtained by Monte Carlo simulation. The IEKF state prediction and its covariance matrix are:

$$\hat{\mathbf{x}}_{k|k-1} = \begin{bmatrix} 100 \\ 55 \end{bmatrix}; \qquad \mathbf{P}_{k|k-1} = \begin{bmatrix} 14400 & 720 \\ 720 & 32436 \end{bmatrix}.$$

Due to the neglected linearization errors, the state estimate is *inconsistent*: the covariance $\mathbf{P}_{k|k-1}$ is smaller than the covariance calculated by the Monte Carlo simulation for the first state component $\mathbf{x}(1)$ which had a nonlinear process update. For consistent results this covariance should even be larger because the IEKF estimate $\hat{\mathbf{x}}_{k|k-1}$ differs from the mean of the pdf $p(x_k|\hat{\mathbf{Z}}_{k-1})$, calculated by the Monte Carlo simulation.

LRKF

Figure 4.3 shows the $\mathcal{X}^j_{k-1|k-1}$ points (top figure) and the $\mathcal{X}^j_{k|k-1}$ points (bottom figure), the updated state estimate (top) and the predicted state estimate (bottom) and their uncertainty ellipses for the LRKF. The $\mathcal{X}^j_{k-1|k-1}$

[7] The uncertainty ellipsoid

$$(\mathbf{x}_k - \hat{\mathbf{x}}_{k|i})^T \mathbf{P}_{k|i}^{-1} (\mathbf{x}_k - \hat{\mathbf{x}}_{k|i}) = 1$$

is a graphical representation of the uncertainty on the state estimate $\hat{\mathbf{x}}_{k|i}$. Starting from the point $\hat{\mathbf{x}}_{k|i}$, the distance to the ellipse in a direction is a measure for the uncertainty on $\hat{\mathbf{x}}_{k|i}$ in that direction.

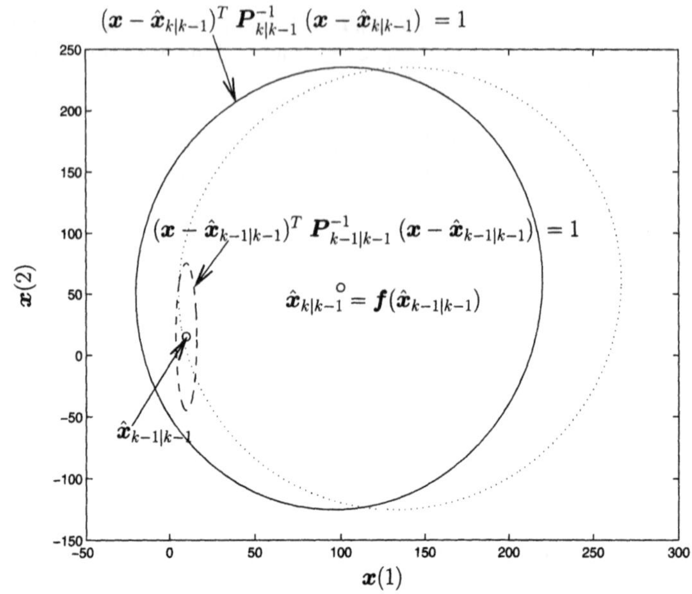

Fig. 4.2. Nonlinear process model. Uncertainty ellipse for the updated state estimate at $k-1$ (*dashed line*), uncertainty ellipse for the (I)EKF predicted state estimate (*full line*) and the Monte Carlo uncertainty ellipse (*dotted line*). The predicted state estimate is inconsistent due to the neglected linearization errors: the uncertainty ellipse of the IEKF predicted estimate is shifted with respect to the Monte Carlo uncertainty ellipse and is somewhat smaller

points are chosen with the UKF algorithm of Julier and Uhlmann [133] where $\kappa = 3 - n = 1$. This corresponds to choosing 6 regression points, including 2 times the point $\hat{\mathbf{x}}_{k-1|k-1}$. The uncertainty ellipse obtained by Monte Carlo simulation coincides with the final uncertainty ellipse of the LRKF predicted state estimate (bottom figure). This indicates *consistent* and *informative* results. The LRKF predicted state estimate and its covariance matrix are

$$\hat{\mathbf{x}}_{k|k-1} = \begin{bmatrix} 136 \\ 55 \end{bmatrix}; \qquad \mathbf{P}_{k|k-1} = \begin{bmatrix} 16992 & 720 \\ 720 & 32436 \end{bmatrix}.$$

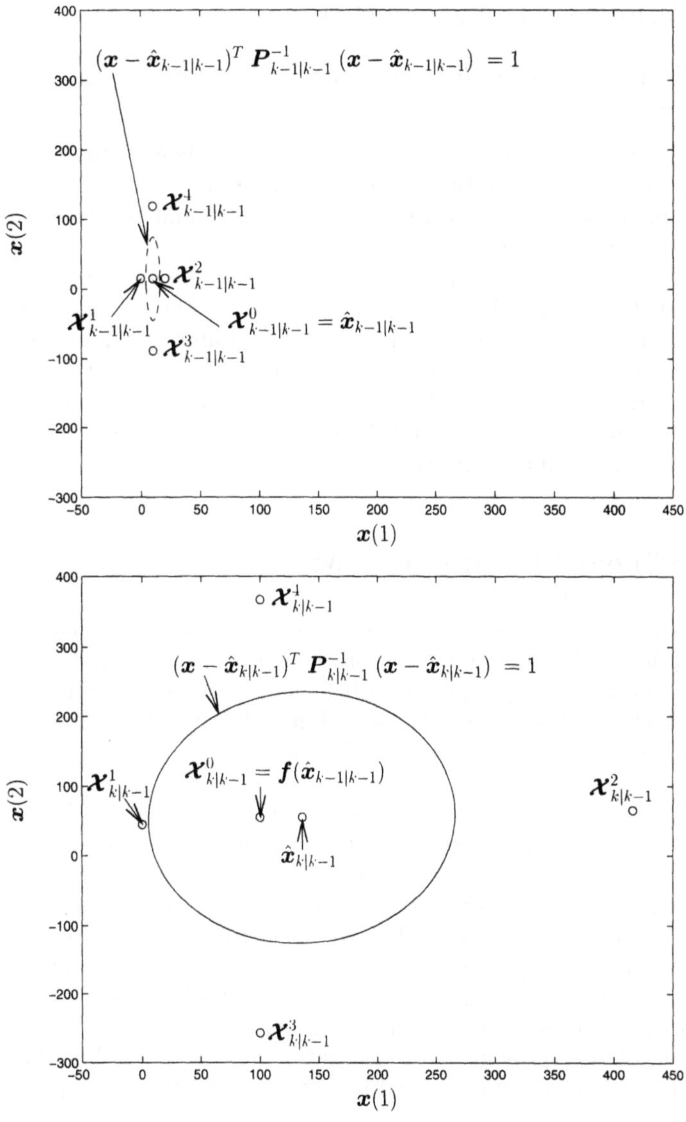

Fig. 4.3. Nonlinear process model. (**top**) Uncertainty ellipse for the updated state estimate at $k-1$ (*dashed line*). (**bottom**) Uncertainty ellipse for the LRKF predicted state estimate (*full line*) and Monte Carlo uncertainty ellipse (*dotted line* which coincides with the full line). The LRKF predicted state estimate is consistent and informative: its uncertainty ellipse coincides with the Monte Carlo uncertainty ellipse

4.4.5 Conclusion: The Process Update

The LRKF performs better than the (I)EKF when dealing with nonlinear process functions:

1. the LRKF linearizes the function based on its behaviour in the "uncertainty region" *around* the updated state estimate. The (I)EKF on the other hand only uses the function evaluation and its Jacobian *in* this state estimate.
2. the LRKF deals with linearization errors in a theoretically founded way (provided that enough regression points are chosen). The (I)EKF on the other hand needs trial and error for each particular example to obtain good values for the covariance matrix representing the linearization errors.
3. unlike the (I)EKF, the LRKF does not need the function Jacobian. This is an advantage where this Jacobian is difficult to obtain or non-existing (e.g., for discontinuous process functions).

4.5 Nonlinear Measurement Models

The previous section contains a comparison between the (I)EKF and LRKF *process* updates; this section focuses on their *measurement* updates for a nonlinear measurement model (4.11) with linearization (4.13). The EKF, IEKF and LRKF choose $\mathbf{H}_k$, $\mathbf{d}_k$ and $\mathbf{R}_k^*$ in a different way. After linearization they use the KF update equations[3,p. 53] (4.5)–(4.9).

The linearization of the measurement model by the IEKF (Sect. 4.5.2) takes the measurement into account; the EKF (Sect. 4.5.1) and LRKF (Sect. 4.5.3) linearize the measurement model based only on the predicted state estimate and its uncertainty. For the latter filters, the linearization errors ($\mathbf{R}_k^*$) are larger, especially when the measurement function is quite nonlinear in the uncertainty region around the predicted state estimate. The large uncertainty on the measurement model $\mathbf{R}_k^* + \mathbf{E}_k\mathbf{R}_k\mathbf{E}_k^T$, due to a large uncertainty on the *state* estimate, results in throwing away the greater part of the information of the (possibly very accurate) *measurement*. The different linearization formulas are summarised in Table 4.2. Section 4.5.5 presents some examples.

4.5.1 The Extended Kalman Filter

The EKF linearizes the measurement model *around the predicted state estimate* $\hat{\mathbf{x}}_{k|k-1}$:

$$\mathbf{H}_k = \left.\frac{\partial \mathbf{h}_k}{\partial \mathbf{x}}\right|_{\hat{\mathbf{x}}_{k|k-1}};$$

$$\mathbf{d}_k = \mathbf{h}_k(\hat{\mathbf{x}}_{k|k-1}) - \mathbf{H}_k\hat{\mathbf{x}}_{k|k-1}.$$

Table 4.2. Summary of the linearization of the measurement model by the Extended Kalman Filter (EKF), the Iterated Extended Kalman Filter (IEKF) and the Linear Regression Kalman Filter (LRKF)

	$\mathbf{H}_k$	$\mathbf{d}_k$	$\mathbf{R}_k^*$
EKF	$\left.\frac{\partial \mathbf{h}_k}{\partial \mathbf{x}}\right\vert_{\hat{\mathbf{x}}_{k\vert k-1}}$	$\mathbf{h}_k(\hat{\mathbf{x}}_{k\vert k-1}) - \mathbf{H}_k\hat{\mathbf{x}}_{k\vert k-1}$	$\mathbf{0}_{m\times m}$
IEKF	$\left.\frac{\partial \mathbf{h}_k}{\partial \mathbf{x}}\right\vert_{\hat{\mathbf{x}}_{k\vert k}}$	$\mathbf{h}_k(\hat{\mathbf{x}}_{k\vert k}) - \mathbf{H}_k\hat{\mathbf{x}}_{k\vert k}$	$\mathbf{0}_{m\times m}$
LRKF	$\min_{(\mathbf{H},\mathbf{d})}\sum_{j=1}^{r}\mathbf{e}_j^T\mathbf{e}_j$	$\min_{(\mathbf{H},\mathbf{d})}\sum_{j=1}^{r}\mathbf{e}_j^T\mathbf{e}_j$	$\frac{1}{r}\sum_{j=1}^{r}\mathbf{e}_j\mathbf{e}_j^T$

The basic EKF algorithm *does not take the linearization errors into account*:

$$\mathbf{R}_k^* \equiv \mathbf{0}_{m\times m};$$

where m is the dimension of the measurement vector $\mathbf{z}_k$. If the measurement model is nonlinear in the "uncertainty region" around the predicted state estimate, the linearization errors are not negligible. This means that the linearized measurement model does not reflect the relation between the true state value and the measurement. I.e., the *true* state value is "far from"[8] the linearized measurement model. After processing the measured value, given the linear measurement model and the measurement uncertainty, the state is believed to be in a region which does not include the true state estimate, i.e., the updated state estimate is *inconsistent*.

4.5.2 The Iterated Extended Kalman Filter

The EKF of the previous section, linearizes the measurement model around the *predicted* state estimate. The IEKF tries to do better by linearizing the measurement model around the *updated* state estimate:

$$\mathbf{H}_k = \left.\frac{\partial \mathbf{h}_k}{\partial \mathbf{x}}\right|_{\hat{\mathbf{x}}_{k|k}};$$

$$\mathbf{d}_k = \mathbf{h}_k(\hat{\mathbf{x}}_{k|k}) - \mathbf{H}_k\hat{\mathbf{x}}_{k|k}.$$

This is achieved by iteration: the filter first linearizes the model around a value $\hat{\mathbf{x}}_{k|k}^0$ (often taken equal to the predicted state estimate $\hat{\mathbf{x}}_{k|k-1}$) and calculates the updated state estimate. Then, the filter linearizes the model around the newly obtained estimate $\hat{\mathbf{x}}_{k|k}^1$ and calculates a new updated state estimate (based on $\hat{\mathbf{x}}_{k|k-1}$, $\mathbf{P}_{k|k-1}$ and the new linearized model). This process is iterated until a state estimate $\hat{\mathbf{x}}_{k|k}^i$ is found for which $\hat{\mathbf{x}}_{k|k}^i$ is close to $\hat{\mathbf{x}}_{k|k}^{i-1}$. The state estimate $\hat{\mathbf{x}}_{k|k}$ and uncertainty $\mathbf{P}_{k|k}$ are calculated starting from the

[8] "Far from" (and "close to") must be understood as: the deviation of the true state with respect to the linearized measurement model is not justified (is justified) by the measurement model uncertainty $\mathbf{R}_k^* + \mathbf{E}_k\mathbf{R}_k\mathbf{E}_k^T$.

state estimate $\hat{\mathbf{x}}_{k|k-1}$ with its uncertainty $\mathbf{P}_{k|k-1}$ and the measurement model linearized around $\hat{\mathbf{x}}^i_{k|k}$.

Like the EKF algorithm, the basic IEKF algorithm *does not take the linearization errors into account*:

$$\mathbf{R}^*_k \equiv \mathbf{0}_{m\times m}.$$

If the measurement model is nonlinear in the "uncertainty region" around the updated state estimate $\hat{\mathbf{x}}_{k|k}$, state estimates will be *inconsistent.* In case of a measurement model that instantaneously *fully observes* the state (or at least the part of the state that causes the nonlinearities in the measurement model), the linearization errors will be small[9] in the "uncertainty region" around $\hat{\mathbf{x}}_{k|k}$. The true state estimate is then "close to" the linearized measurement function and the updated state estimate is *consistent.* The result is also *informative* because no uncertainty due to linearization errors needs to be added.

4.5.3 The Linear Regression Kalman Filter

The LRKF algorithm uses a linearized measurement function (4.2) where $\mathbf{H}_k$, $\mathbf{d}_k$ and $\mathbf{R}^*_k$ are obtained by statistical linear regression through a number of regression points $\left(\boldsymbol{\mathcal{X}}^j_{k|k-1}, \boldsymbol{\mathcal{Z}}^j_k\right)$, $j = 1, \dots, r$ (see Sect. 4.2.2). The literature on the CDF, DD1 and UKF filters proposes different choices of the $\boldsymbol{\mathcal{X}}^j_{k|k-1}$ points for which —in case of a *continuous* measurement function— the calculated $\bar{\boldsymbol{\mathcal{Z}}}_k$ with covariance $\mathbf{P}_{\boldsymbol{\mathcal{Z}}_k\boldsymbol{\mathcal{Z}}_k}$ (the mean and covariance matrix of the points $\boldsymbol{\mathcal{Z}}^j_k$) is equally or less inconsistent than the measurement prediction calculated by the EKF (estimate $\mathbf{H}_k\hat{\mathbf{x}}_{k|k-1} + \mathbf{d}_k$ with covariance matrix $\mathbf{H}_k\mathbf{P}_{k|k-1}\mathbf{H}^T_k$).

Intuitively we feel that *when enough*[6, p. 62] *regression points* $(\boldsymbol{\mathcal{X}}^j_{k|k-1}, \boldsymbol{\mathcal{Z}}^j_k)$ *are taken* the state estimates of the LRKF measurement update are *consistent* because $\mathbf{R}^*_k$ gives a well founded approximation of the linearization errors, see (4.17). However, if the measurement model is highly nonlinear in the "uncertainty region" around $\hat{\mathbf{x}}_{k|k-1}$, the $(\boldsymbol{\mathcal{X}}^j_{k|k-1}, \boldsymbol{\mathcal{Z}}^j_k)$ points deviate substantially from a hyperplane. This results in a large $\mathbf{R}^*_k$ and *non-informative* updated state estimates (see the example in Sect. 4.5.5).

4.5.4 Extra Measurement Model Uncertainty

In order to make the state estimates consistent, the user can tune an inconsistent filter by *adding extra measurement model uncertainty* $\mathbf{R}^*_k$.

Only off-line tuning or on-line parameter learning can lead to a good value for $\mathbf{R}^*_k$ for a particular problem. In many practical cases consistency is assured by choosing the added uncertainty too large, e.g., by taking a constant $\mathbf{R}^*$ over time which compensates for decreasing linearization errors. This reduces the information content of the results.

[9] This assumes that the iterations lead to an accurate $\hat{\mathbf{x}}^i_{k|k}$. The linearizations are started around a freely chosen $\hat{\mathbf{x}}^0_{k|k}$. In order to assure quick and correct iteration, (part of) this value can be chosen based on the measurement information if this information is more accurate than the predicted state estimate.

4.5.5 Examples

First Example

The comparison between the different measurement updates is illustrated with the measurement function $z_k = h_1(\mathbf{x}_k) + \rho_{m,k}$;

$$h_1(\mathbf{x}_k) = (\mathbf{x}_k(1))^2 + (\mathbf{x}_k(2))^2 . \tag{4.19}$$

$\mathbf{x}_k = \begin{bmatrix} 15 \\ 20 \end{bmatrix}$ is the true value and $\hat{\mathbf{x}}_{k|k-1} = \begin{bmatrix} 10 \\ 15 \end{bmatrix}$ is the predicted state estimate with covariance matrix $\mathbf{P}_{k|k-1} = \begin{bmatrix} 36 & 0 \\ 0 & 3600 \end{bmatrix}$. The processed measurement is $\hat{z}_k = 630$ and the measurement covariance is $R_k = 400$.

Second Example

To illustrate the consistency of the state estimate of an IEKF when the measurement observes the state completely, a second example is used. The measurement function is

$$\mathbf{z}_k = \mathbf{h}(\mathbf{x}_k) + \boldsymbol{\rho}_{m,k} = \begin{bmatrix} h_1(\mathbf{x}_k) + \boldsymbol{\rho}_{m,k}(1) \\ h_2(\mathbf{x}_k) + \boldsymbol{\rho}_{m,k}(2) \end{bmatrix} ; \tag{4.20}$$

with

$$\begin{cases} h_1(\mathbf{x}_k) = (\mathbf{x}_k(1))^2 + (\mathbf{x}_k(2))^2 ; \\ h_2(\mathbf{x}_k) = 3\,(\mathbf{x}_k(2))^2 / \mathbf{x}_k(1); \end{cases} \tag{4.21}$$

$\mathbf{x}_k = \begin{bmatrix} 15 \\ 20 \end{bmatrix}$ is the true value and $\hat{\mathbf{x}}_{k|k-1} = \begin{bmatrix} 10 \\ 15 \end{bmatrix}$ is the predicted state estimate with covariance matrix $\mathbf{P}_{k|k-1} = \begin{bmatrix} 36 & 0 \\ 0 & 3600 \end{bmatrix}$. The processed measurement and the measurement covariance matrix are:

$$\hat{\mathbf{z}}_k = \begin{bmatrix} 630 \\ 85 \end{bmatrix} ; \qquad \mathbf{R}_k = \begin{bmatrix} 400 & 0 \\ 0 & 400 \end{bmatrix} .$$

In all figures, the true state value $\mathbf{x}_k$ is plotted; if this value is "far" outside the uncertainty ellipse of a state estimate, the corresponding estimate is inconsistent. Because the measurement is accurate and the initial estimate is not, the uncertainty on the state estimate should drop considerably when the measurement is processed. The updated state estimate is not informative if this is not the case.

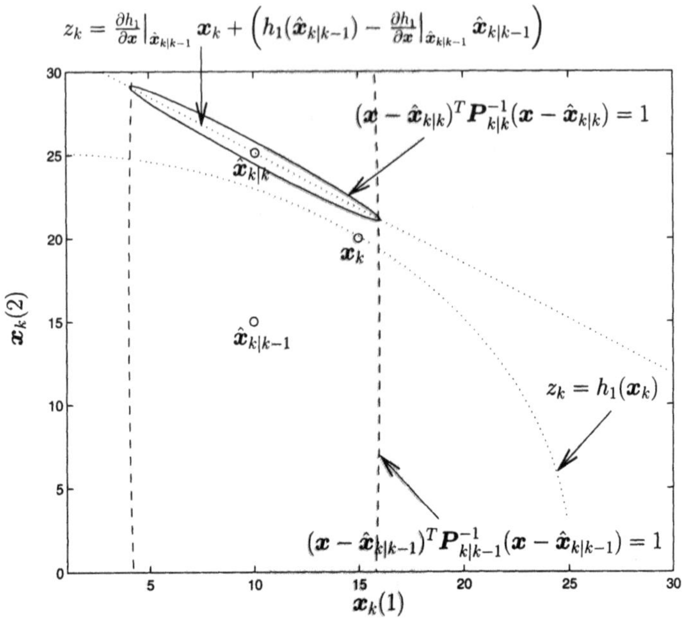

Fig. 4.4. Nonlinear measurement model $z = h_1(\mathbf{x}_k)$ and EKF linearization around $\hat{\mathbf{x}}_{k|k-1}$ (*dotted lines*). The true state $\mathbf{x}_k$ is "far from" this linearization and the obtained state estimate $\hat{\mathbf{x}}_{k|k}$ (uncertainty ellipse in *full line*) is inconsistent

EKF

Figure 4.4 shows the state estimates, uncertainty ellipses and measurement functions for the EKF applied to the first example (4.19). The true measurement function is nonlinear. $\mathbf{x}_k$ is the true value of the state, and is "close to" this function. The linearization around the uncertain predicted state estimate is not a good approximation of the function around the true state value: the true state value is "far from" the linearized measurement function. The resulting updated state estimate

$$\hat{\mathbf{x}}_{k|k} = \begin{bmatrix} 10 \\ 25 \end{bmatrix}; \qquad \mathbf{P}_{k|k} = \begin{bmatrix} 36 & -24 \\ -24 & 16 \end{bmatrix};$$

is *inconsistent.*

IEKF

Figure 4.5 shows the measurement function, the linearized measurement function around the point $\hat{\mathbf{x}}^i_{k|k}$, the true state value $\mathbf{x}_k$ and the state estimates for the IEKF applied to the first example (4.19). The measurement model does not fully observe the state. This results in an uncertain updated state estimate $\hat{\mathbf{x}}^i_{k|k}$ around which the filter linearizes the measurement function. As

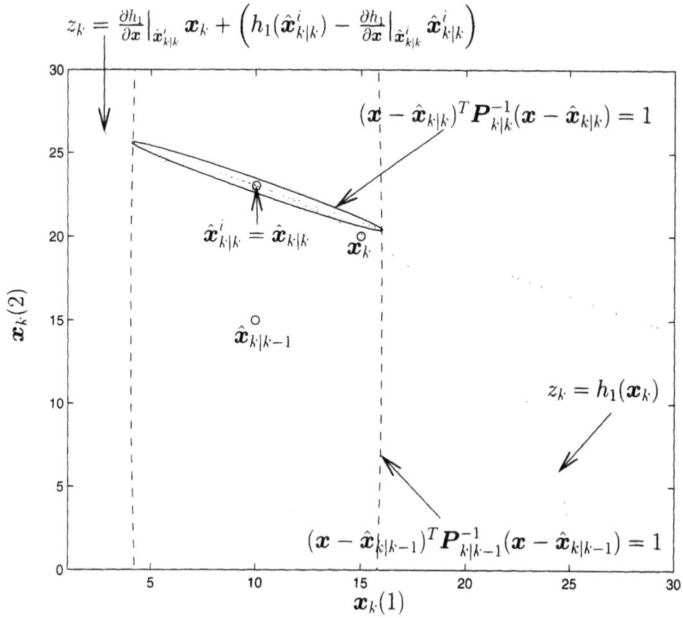

Fig. 4.5. Nonlinear measurement model $z = h_1(\mathbf{x}_k)$ that does *not* observe the full state, and its IEKF linearization around $\hat{\mathbf{x}}_{k|k}$ (*dotted lines*). The true state $\mathbf{x}_k$ is "far from" this linearization, leading to an inconsistent state estimate $\hat{\mathbf{x}}_{k|k}$ (uncertainty ellipse in *full line*)

was the case for the EKF, the linearization errors are not negligible and the true value is "far from" the linearized measurement function. The updated state estimate

$$\hat{\mathbf{x}}_{k|k} = \begin{bmatrix} 10 \\ 23 \end{bmatrix}; \qquad \mathbf{P}_{k|k} = \begin{bmatrix} 36 & -16 \\ -16 & 7.0 \end{bmatrix};$$

is *inconsistent.*

If however the measurement model fully observes the state, the IEKF updated state estimate is accurately known; hence, the linearization errors are small and the true state value is "close to" the linearized measurement function. In this case, the updated state estimate is *consistent.* Figure 4.6 shows the measurement function, the linearized measurement function, the true state value $\mathbf{x}_k$, the state estimates and the uncertainty ellipses for the IEKF applied to the second example (4.20)–(4.21). The updated state estimate and covariance matrix

$$\hat{\mathbf{x}}_{k|k} = \begin{bmatrix} 14 \\ 21 \end{bmatrix}; \qquad \mathbf{P}_{k|k} = \begin{bmatrix} 2.6 & -1.7 \\ -1.7 & 1.3 \end{bmatrix};$$

are *consistent and informative* due to the small, ignored, linearization errors.

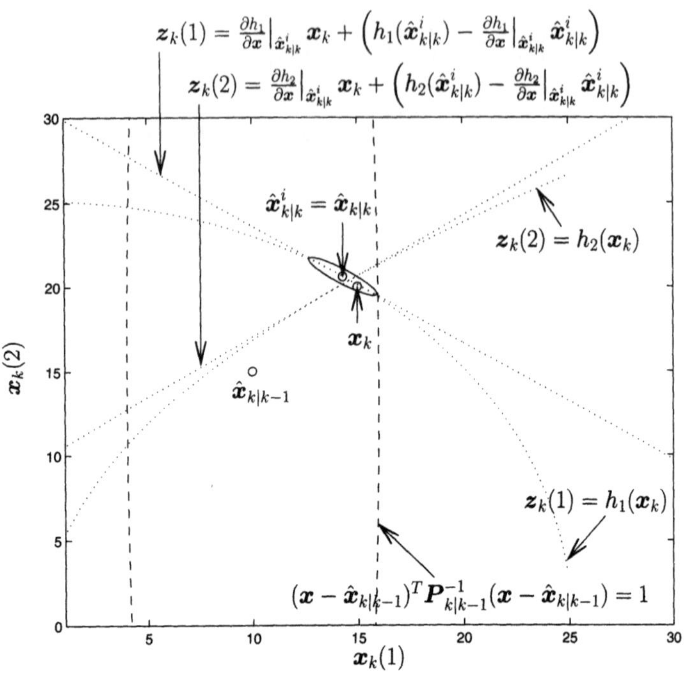

Fig. 4.6. Nonlinear measurement model $\mathbf{z} = \mathbf{h}(\mathbf{x}_k)$ that fully observes the state, and its IEKF linearization around $\hat{\mathbf{x}}_{k|k}$ (*dotted lines*). The true state $\mathbf{x}_k$ is "close to" this linearization, leading to a consistent state estimate (uncertainty ellipse in *full line*)

LRKF

A LRKF is run on the first example (4.19). The $\mathcal{X}^j_{k|k-1}$ points are chosen with the UKF algorithm with $\kappa = 3 - n = 1$ [133]. This corresponds to choosing 6 regression points, including 2 times the point $\hat{\mathbf{x}}_{k|k-1}$. Figures 4.7 and 4.8 show the nonlinear measurement function, the $\mathcal{X}^j_{k|k-1}$-points and the linearization. The predicted state estimate is uncertain, hence the $\mathcal{X}^j_{k|k-1}$-points are widespread. R^*_k is large ($R^*_k = 2.6 \times 10^7$) due to the large deviations between the ($\mathcal{X}^j_{k|k-1}$, $\mathcal{Z}^j_k$) points and the linearized measurement function (see Fig. 4.8). The updated state estimate and its covariance matrix are

$$\hat{\mathbf{x}}_{k|k} = \begin{bmatrix} 10 \\ 2.6 \end{bmatrix}; \qquad \mathbf{P}_{k|k} = \begin{bmatrix} 36 & 0 \\ 0 & 3600 \end{bmatrix}.$$

Figure 4.9 shows the $\mathcal{X}^j_{k|k-1}$ points, the measurement function, the LRKF linearized measurement function, the true state value $\mathbf{x}_k$, the state estimates and the uncertainty ellipses. The updated state estimate is consistent, however, it can hardly be called an improvement over the previous state estimate

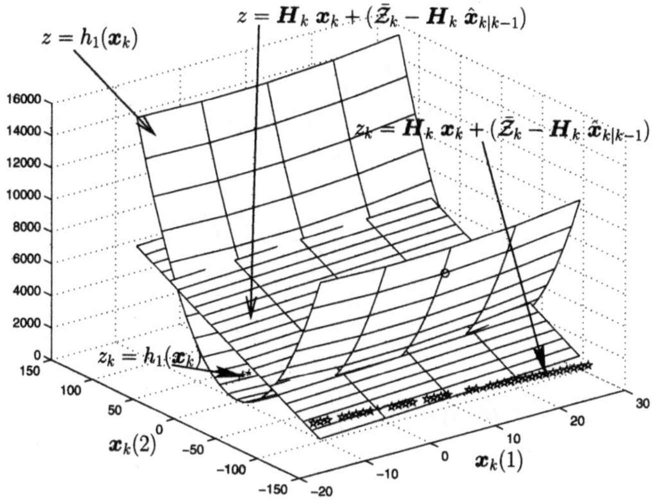

Fig. 4.7. Nonlinear measurement model $z = h_1(\mathbf{x})$ and LRKF linearization. The linearization errors are large

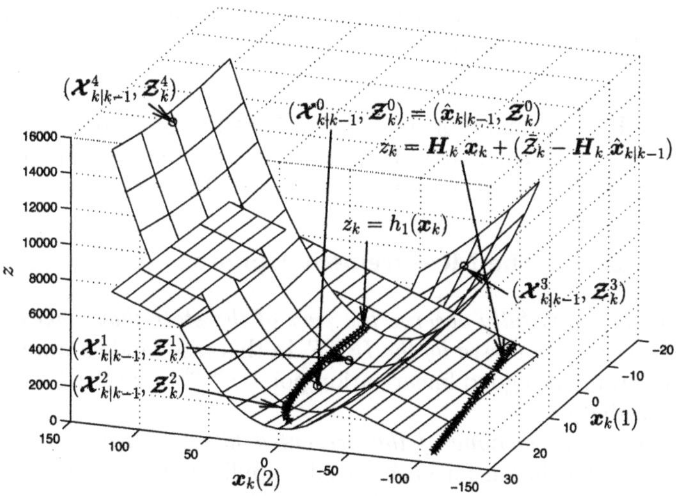

Fig. 4.8. Figure 4.7 seen from another angle

$(\mathbf{P}_{k|k} \approx \mathbf{P}_{k|k-1})$. *The information in the measurement is neglected* due to the high "measurement model" uncertainty $\mathbf{R}_k^* + \mathbf{E}_k \mathbf{R}_k \mathbf{E}_k^T$ on the linearized function.

Note that some kind of *Iterated* LRKF (similar to the Iterated EKF) would not solve this problem: the updated state estimate $\hat{\mathbf{x}}_{k|k}$ and its covari-

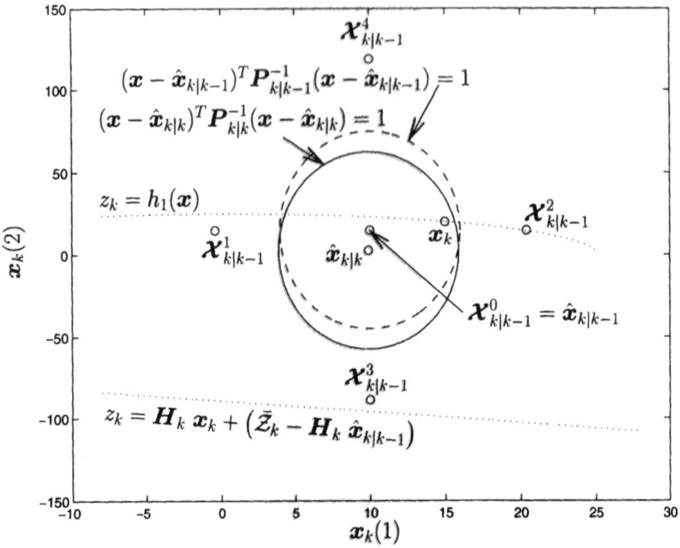

Fig. 4.9. Nonlinear measurement model $z = h_1(\mathbf{x})$ and LRKF linearization (*dotted lines*). The large linearization errors result in a large measurement model uncertainty $\mathbf{R}^*_k + \mathbf{E}_k \mathbf{R}_k \mathbf{E}^T_k$. The updated state estimate (uncertainty ellipse in *full line*) is consistent but non-informative

ance matrix $\mathbf{P}_{k|k}$ are more or less the same as the predicted state estimate $\hat{\mathbf{x}}_{k|k-1}$ and its covariance matrix $\mathbf{P}_{k|k-1}$. Hence, the regression points and the linearization would approximately be the same after iteration.

4.5.6 Conclusion: The Measurement Update

Measurements which fully observe the part of the state that makes the model nonlinear, are best processed by the IEKF. In this case (and assuming that the algorithm iterates to a good linearization point[9, p. 68]), the IEKF linearization errors are negligible.

In the other cases, none of the presented filters outperforms the others. A filter should be chosen for each specific application: the LRKF makes an estimate of its linearization errors ($\mathbf{R}^*_k$), the EKF and IEKF on the other hand require off-line tuning or on-line parameter learning of $\mathbf{R}^*_k$ to yield consistent state estimates. Because the IEKF additionally takes the measurement into account when linearizing the measurement model, its linearization errors are smaller than those of the EKF and LRKF. This means that once a well-tuned IEKF is available, the state estimates it returns can be far more informative than those of the LRKF or a well-tuned EKF.

Finally, note that the LRKF does not use the Jacobian of the measurement function, which makes it possible to process discontinuous measurement functions.

4.6 Conclusions

This chapter describes how the Extended Kalman Filter (EKF), the Iterated Extended Kalman Filter (IEKF) and the Linear Regression Kalman Filter (LRKF) differ in the linearization of the process and measurement models and how they deal with linearization errors. This insight is important to understand the advantages and drawbacks of each of these filters.

This chapter presents the Central Difference Filter (CDF), the first-order Divided Difference Filter (DD1) and the Unscented Kalman Filter (UKF) in a unified way as the LRKF. The LRKF (i) linearizes the process and measurement function by statistical linear regression of the function through a number of regression points in the "uncertainty region" around the state estimate, and (ii) approximates the linearization errors by the sample covariance matrix of the deviations between the function values of the nonlinear and the linearized function in these points. An advantage of the LRKF is that it does not need the Jacobian of the nonlinear function.

The quality of the state estimates is expressed by two criteria: the consistency and the information content of the estimates. This work relates the consistency and information content of the estimates to (i) how the linearization is performed and (ii) how the linearization errors are taken into account. The understanding of the linearization processes allows us to make a well-founded choice of filter for a specific application.

The performance of the different filters is compared for the process and measurement updates separately, because a good performance for one of these updates does not necessarily mean a good performance for the other update. This makes it interesting in some cases to use different filters for both updates.

The difference in performance for the process and measurement updates is due to the fact that for the linearization of the process model, the state estimate and its uncertainty are the only available information; while for the linearization of the measurement model, also the measurement and its uncertainty can be used.

For *process* updates the LRKF performs better than the other mentioned KF variants because (i) the LRKF linearizes the process model based on its behaviour in the "uncertainty region" *around* the updated state estimate. The (I)EKF on the other hand only uses the function evaluation and its Jacobian *in* this state estimate; and (ii) the LRKF deals with linearization errors in a theoretically founded way, provided that enough regression points are chosen. The (I)EKF on the other hand needs trial and error for each particular application to obtain a good covariance matrix representing the linearization errors.

The IEKF is the best way to handle nonlinear *measurement* models that fully observe the part of the state that makes the measurement model nonlinear. In the other cases, *none of the presented filters* outperforms the others:

the LRKF makes an estimation of the linearization errors, the EKF and IEKF on the other hand require extensive off-line tuning or on-line parameter learning in order to yield consistent state estimates. However, unlike the EKF and LRKF, the IEKF additionally uses the measurement value in order to linearize the measurement model. Hence, its linearization errors are smaller and once a well-tuned IEKF is available, the state estimates it returns can be far more informative than those of the LRKF or a well-tuned EKF.

Kalman Filtering in this Work

This book presents the estimation of inaccurately known positions, orientations and dimensions of contacting objects during compliant motion. These positions, orientations and dimensions are *static* state variables, hence, the filters do not need a process update step. For small uncertainties the measurement update is performed with the *IEKF* because this filter is informative and consistent for small linearization errors. For larger uncertainties the IEKF yields inconsistent results because the measurements do not fully observe the state. Chapter 5 presents a new *exact Bayesian* filter for static systems with additive Gaussian measurement model uncertainty. The filter can be interpreted as a KF working in a higher-dimensional state space in which the measurement equations are linear. The filter is computationally more expensive than the IEKF, but yields good results for the large uncertainties that the IEKF cannot handle.

5

The Non-Minimal State Kalman Filter

On voit que la théorie des probabilités n'est,
au fond, que le bon sens réduit au calcul.
(Probability theory is nothing but
common sense reduced to calculation.)

Pierre-Simon Laplace, *Essai*
Philosophique sur les Probabilités, 1819.

5.1 Introduction

Exact finite-dimensional Bayesian filters exist only for a small class of systems. The previous chapter discussed the best known example, i.e., the Kalman Filter (KF) for linear systems subject to additive Gaussian uncertainties. Other examples are the filters of Beneš [25], which requires the measurement model to be linear, and Daum [61], applicable to a more general class of systems with nonlinear process and measurement models for which the posterior pdf is any exponential distribution.

This chapter presents a *new exact Bayesian filter* able to deal with static systems (parameter estimation) with any kind of nonlinear measurement model subject to additive Gaussian measurement uncertainty (Sect. 5.2). Section 5.3 extends the results to a limited class of dynamic systems. The filter is experimentally validated on a contour matching experiment.

The new filter is situated between the Kalman Filter and the Daum Filter: the Kalman Filter solves a subclass of problems solved by the new filter and the new filter solves a subclass of problems solved by the Daum Filter. Also in terms of computational complexity the new filter is situated in between these two.

An advantage of the new filter is its theoretical simplicity for people familiar with the KF algorithm. The filter can be interpreted as a KF linearizing the process and measurement models in a higher-dimensional state space. Therefore, we call the new filter the *"Non-Minimal State Kalman Filter" (NMSKF)*. The "mean vector" and "covariance matrix" of the Gaussian in the higher-dimensional state space are sufficient statistics of the posterior pdf over the original state. However, they do ***not*** represent the mean and covariance of this posterior.

The previous chapter presented approximations to the KF algorithm for nonlinear systems by linearizing the process and measurement models around the (original) state variables. Because these filters are only approximations, their performance (consistency and information content) is not optimal. In contrast with these KF variants, the new filter presented in this chapter is an *exact Bayesian filter* calculating the full, non-Gaussian, posterior pdf over the state.

5.2 The NMSKF for Static Systems

Section 5.2.1 presents the Non-Minimal State Kalman Filter (NMSKF) measurement update equations for static systems (i.e., parameter estimation). These equations describe the sufficient statistics of the posterior pdf over the state. The calculation of an estimated *value* for the state based on this pdf is discussed in Sect. 5.2.2. If the prior information is Gaussian, the expression of the posterior pdf can be simplified (Sect. 5.2.3). Section 5.2.4 contains a simple example of parameter estimation.

5.2.1 Analytical Posterior Pdf

The NMSKF is applicable to static systems ($\mathbf{x}_k = \mathbf{x}$) with any kind of nonlinear measurement model

$$\mathbf{z}_k = \mathbf{h}_k(\mathbf{x}) + \mathbf{E}_k \boldsymbol{\rho}_{m,k};$$

subject to additive, zero mean, Gaussian uncertainty $\boldsymbol{\rho}_{m,k}$ which is uncorrelated over time. The subscript k refers to the order in which the measurements are received by the filter. Theorem 5.1 states that the sufficient statistics of the posterior pdf over the original state $\mathbf{x}$ are obtained through linear Kalman filtering for *a higher-dimensional state vector* $\mathbf{x}' = \mathbf{g}(\mathbf{x})$ in which the measurement model is linear:

$$\mathbf{z}_k = \mathbf{H}_k\, \mathbf{x}' + \mathbf{d}_k + \mathbf{E}_k \boldsymbol{\rho}_{m,k}.$$

The posterior pdf belongs to a subclass of the exponential distributions. Before presenting the theorem, we first discuss the choice of the higher-dimensional state vector $\mathbf{x}' = \mathbf{g}(\mathbf{x})$.

When dealing with a *limited* number N of measurement models (sensors), we can always find an $\mathbf{x}' = \mathbf{g}(\mathbf{x})$ which linearizes them; for instance by choosing

$$\mathbf{x}' = \begin{bmatrix} \mathbf{h}_1(\mathbf{x}) \\ \dots \\ \mathbf{h}_N(\mathbf{x}) \end{bmatrix}.$$

The choice of the linearizing vector $\mathbf{x}'$ is not uniquely defined. For example for the following system description

$$\mathbf{x} = \begin{bmatrix} u \\ v \\ \alpha \end{bmatrix}; \qquad \mathbf{z} = \mathbf{h}(\mathbf{x}) = \begin{bmatrix} e^v + uv^2 + 4 \\ uv^2 \cos\alpha + 3e^v + 2u \\ 5uv^2 + 2 \\ e^v - u + 0.9 \\ e^v \end{bmatrix};$$

possible choices of $\mathbf{x}'$ (and the corresponding $\mathbf{H}_k$ and $\mathbf{d}_k$) are:

$$\mathbf{x}' = \begin{bmatrix} e^v + uv^2 + 4 \\ uv^2 \cos\alpha + 3e^v + 2u \\ 5uv^2 + 2 \\ e^v - u + 0.9 \\ e^v \end{bmatrix}; \quad \mathbf{H}_k = \mathbf{I}_{5\times5}; \quad \mathbf{d}_k = \mathbf{0}_{5\times1};$$

$$\mathbf{x}' = \begin{bmatrix} uv^2 \\ uv^2 \cos\alpha \\ 5uv^2 \\ e^v - u \\ e^v \end{bmatrix}; \quad \mathbf{H}_k = \begin{bmatrix} 1 & 0 & 0 & 0 & 1 \\ 0 & 1 & 0 & -2 & 5 \\ 0 & 0 & 1 & 0 & 0 \\ 0 & 0 & 0 & 1 & 0 \\ 0 & 0 & 0 & 0 & 1 \end{bmatrix}; \quad \mathbf{d}_k = \begin{bmatrix} 4 \\ 0 \\ 2 \\ 0.9 \\ 0 \end{bmatrix};$$

$$\mathbf{x}' = \begin{bmatrix} e^v \\ uv^2 \\ uv^2 \cos\alpha \\ u \end{bmatrix}; \quad \mathbf{H}_k = \begin{bmatrix} 1 & 1 & 0 & 0 \\ 3 & 0 & 1 & 2 \\ 0 & 5 & 0 & 0 \\ 1 & 0 & 0 & -1 \\ 1 & 0 & 0 & 0 \end{bmatrix}; \quad \mathbf{d}_k = \begin{bmatrix} 4 \\ 0 \\ 2 \\ 0.9 \\ 0 \end{bmatrix}. \tag{5.1}$$

In two cases, the dimension of the new state $\mathbf{x}'$ is *not larger* than that of the state $\mathbf{x}$:

1. If a number of (combinations) of the original states $\mathbf{x}$ stay unobserved, the dimension of $\mathbf{x}'$ can be lower than that of $\mathbf{x}$. In this case, the original estimation problem can equally be described by less parameters $\mathbf{x}$. For this new problem description, $\mathbf{x}'$ is of equal (see case 2) or larger dimension than $\mathbf{x}$.
2. $\mathbf{x}'$ has the same dimension as $\mathbf{x}$ if the original measurement model can be linearized by applying a one-to-one transformation to the state $\mathbf{x}$. In this case, by expressing the measurement model as a function of the transformed state $\mathbf{x}'$, the system is linear and no nonlinear filter is needed.

Theorem 5.1 (Analytical posterior pdf for static systems). *The posterior pdf over the state* $\mathbf{x}$*, given* k *measurements* $\hat{\mathbf{Z}}_k = \{\hat{\mathbf{z}}_1, \hat{\mathbf{z}}_2, \ldots, \hat{\mathbf{z}}_k\}$ *and the measurement model*

$$\mathbf{z}_k = \mathbf{h}_k(\mathbf{x}) + \mathbf{E}_k \boldsymbol{\rho}_{m,k} = \mathbf{H}_k\, \mathbf{g}(\mathbf{x}) + \mathbf{d}_k + \mathbf{E}_k \boldsymbol{\rho}_{m,k};$$

with zero mean Gaussian uncertainty $\boldsymbol{\rho}_{m,k}$ *which is uncorrelated over time and has covariance matrix* $\mathbf{R}_k$*, can be expressed in analytical form as*

$$p(\mathbf{x}|\hat{\mathbf{Z}}_k) \sim p(\mathbf{x}) \exp\Big(-\frac{1}{2}\big(\mathbf{g}(\mathbf{x}) - \hat{\mathbf{x}}'_k\big)^T \mathbf{P}_k^{'-1} \big(\mathbf{g}(\mathbf{x}) - \hat{\mathbf{x}}'_k\big)\Big). \qquad (5.2)$$

$p(\mathbf{x})$ *is the prior pdf over* $\mathbf{x}$*. The sufficient statistics* $\hat{\mathbf{x}}'_k$ *and* $\mathbf{P}'_k$ *of pdf (5.2) are defined by the Kalman Filter equations for the state estimation of the* n'*-dimensional state* $\mathbf{x}' = \mathbf{g}(\mathbf{x})$ *(the iteration starts with* $\mathbf{P}_0^{'-1} = \mathbf{0}_{n' \times n'}$*, i.e., this filter has no prior information):*

$$\hat{\mathbf{x}}'_k = \hat{\mathbf{x}}'_{k-1} + \mathbf{K}_k \boldsymbol{\eta}_k; \qquad (5.3)$$

$$\mathbf{P}'_k = (\mathbf{I}_{n' \times n'} - \mathbf{K}_k \mathbf{H}_k)\, \mathbf{P}'_{k-1}; \qquad (5.4)$$

where

$$\boldsymbol{\eta}_k = \hat{\mathbf{z}}_k - \mathbf{H}_k \hat{\mathbf{x}}'_{k-1} - \mathbf{d}_k; \qquad (5.5)$$

$$\mathbf{S}_k = \mathbf{E}_k \mathbf{R}_k \mathbf{E}_k^T + \mathbf{H}_k \mathbf{P}'_{k-1} \mathbf{H}_k^T; \qquad (5.6)$$

$$\mathbf{K}_k = \mathbf{P}'_{k-1} \mathbf{H}_k^T \mathbf{S}_k^{-1}. \qquad (5.7)$$

The proof of this theorem is given in Appendix B.

5.2.2 Retrieving an Estimate for the Original State x

The NMSKF calculates the posterior *pdf* $p(\mathbf{x}|\hat{\mathbf{Z}}_k)$, (5.2). However, an application usually needs *an estimated value* for $\mathbf{x}$. This value can be the mean value of the pdf, the maximum likelihood value of the pdf, the positions of several peaks of a multi-model pdf, etc. Sometimes also a description of the uncertainty on the estimated value (e.g., a covariance matrix) is needed.

For Gaussian distributions, the mean value (which coincides with the maximum likelihood value) and the covariance matrix are directly obtained from the analytical pdf description. For other distributions, obtaining a single estimated value for $\mathbf{x}$ and its uncertainty is more difficult.

Here, we propose two approximate methods in case no analytical solution is found:

1. The first method is to *draw samples* from the pdf (5.2). The sample mean and covariance are an approximation of the mean of the pdf and its covariance matrix.

2. The second method is based on the observation that the same posterior pdf (5.2) is obtained for an estimation problem of $\mathbf{x}$ with prior pdf $p(\mathbf{x})$, measurement $\hat{\mathbf{x}}'_k$ and measurement model

$$\mathbf{x}'_k = \mathbf{g}(\mathbf{x}) + \boldsymbol{\rho}'_{m,k};$$

subject to additive, zero-mean, Gaussian measurement model uncertainty $\boldsymbol{\rho}'_{m,k}$ with covariance matrix $\mathbf{P}'_k$.
This nonlinear filtering problem can for instance be solved by an Iterated Extended Kalman Filter (IEKF, Chap. 4). An IEKF returns a consistent and informative approximation of the mean value and covariance matrix of the posterior pdf *if* the measurements $\hat{\mathbf{Z}}_k$, whose information about $\mathbf{x}$ is stored in $\hat{\mathbf{x}}'_k$ and $\mathbf{P}'_k$, fully observe the system. The IEKF algorithm for this problem iterates:

$$\begin{aligned}
\mathbf{H}^i &= \left.\frac{\partial \mathbf{g}(\mathbf{x})}{\partial \mathbf{x}}\right|_{\hat{\mathbf{x}}^{i-1}};\\
\mathbf{d}^i &= \mathbf{g}(\hat{\mathbf{x}}^{i-1}) - \mathbf{H}^i\hat{\mathbf{x}}^{i-1};\\
\boldsymbol{\eta}^i &= \hat{\mathbf{x}}'_k - \mathbf{H}^i\hat{\mathbf{x}}_0 - \mathbf{d}^i;\\
\mathbf{S}^i &= \mathbf{P}'_k + \mathbf{H}^i\mathbf{P}_0\mathbf{H}^{iT};\\
\mathbf{K}^i &= \mathbf{P}_0\mathbf{H}^{iT}(\mathbf{S}^i)^{-1};\\
\hat{\mathbf{x}}^i &= \hat{\mathbf{x}}_0 + \mathbf{K}^i\boldsymbol{\eta}^i;
\end{aligned}$$

where $\hat{\mathbf{x}}_0$ and $\mathbf{P}_0$ are the mean and covariance of the initial estimate. The iteration is started from a value $\hat{\mathbf{x}}^0$ (e.g., chosen equal to $\hat{\mathbf{x}}_0$) until a value $\hat{\mathbf{x}}^i$ is obtained close to $\hat{\mathbf{x}}^{i-1}$. The estimated value for $\mathbf{x}$ and its covariance are:

$$\begin{aligned}
\hat{\mathbf{x}}_k &= \hat{\mathbf{x}}^i;\\
\mathbf{P}_k &= (\mathbf{I}_{n\times n} - \mathbf{K}^i\mathbf{H}^i)\mathbf{P}_0.
\end{aligned}$$

We call the combination of a NMSKF for the estimation of the pdf (5.2) and an IEKF for the calculation of a state value $\hat{\mathbf{x}}_k$ and its uncertainty $\mathbf{P}_k$ the *"Cascaded Kalman filter" (CKF)* .

Figure 5.1 shows the information flow in the Non-Minimal State Kalman Filter. The calculation of a state value for $\mathbf{x}$ is drawn in dashed lines. This calculation can be done at any time, but is optional. The NMSKF does not use the estimates of $\mathbf{x}$ further on, neither to make new estimates for $\mathbf{x}$, nor to estimate $\mathbf{x}'$. Hence, the filtering from $\mathbf{x}'$ to $\mathbf{x}$ does *not* need to be performed, and when performed, it does need to be at the sampling rate.

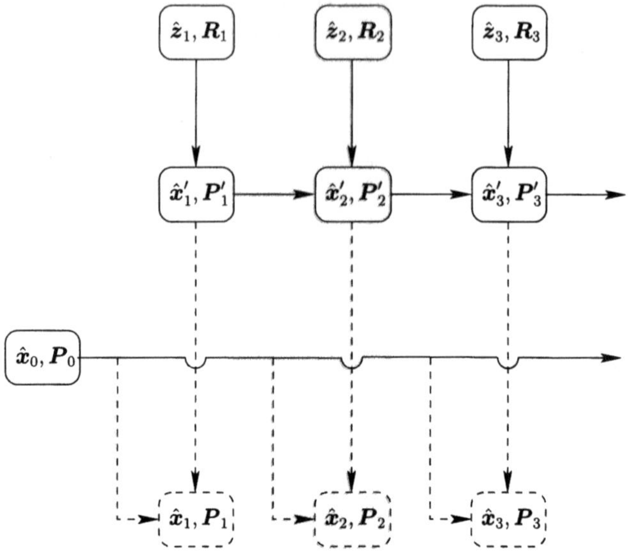

Fig. 5.1. Information flow in the Non-Minimal State Kalman Filter for static systems. The calculations represented in dashed lines are optional

5.2.3 Prior Information

A Gaussian prior $p(\mathbf{x})$ with mean $\hat{\mathbf{x}}_0$ and covariance matrix $\mathbf{P}_0$ can be introduced in (5.2) in an alternative way by augmenting $\mathbf{x}'$ with the state vector: $\mathbf{x}' \leftarrow \begin{bmatrix} \mathbf{x}' \\ \mathbf{x} \end{bmatrix}$ and replacing (5.2) by

$$p(\mathbf{x}|\hat{\mathbf{Z}}_k) \sim \exp\Big(-\frac{1}{2}(\mathbf{g}(\mathbf{x}) - \hat{\mathbf{x}}'_k)^T \mathbf{P}'^{-1}_k (\mathbf{g}(\mathbf{x}) - \hat{\mathbf{x}}'_k)\Big);$$

$$\hat{\mathbf{x}}'_0 \leftarrow \begin{bmatrix} \mathbf{0}_{(n'-n)\times 1} \\ \hat{\mathbf{x}}_0 \end{bmatrix}; \qquad \mathbf{P}'^{-1}_0 \leftarrow \begin{bmatrix} \mathbf{0}_{(n'-n)\times(n'-n)} & \mathbf{0}_{(n'-n)\times n} \\ \mathbf{0}_{n\times(n'-n)} & \mathbf{P}_0^{-1} \end{bmatrix}.$$

E.g., for Example (5.1) this means that a possible choice for $\mathbf{x}'$ is:

$$\mathbf{x}' = \mathbf{g}(\mathbf{x}) = \begin{bmatrix} e^v \\ uv^2 \\ uv^2\cos\alpha \\ u \\ v \\ \alpha \end{bmatrix}; \qquad \hat{\mathbf{x}}'_0 = \begin{bmatrix} 0 \\ 0 \\ 0 \\ u_0 \\ v_0 \\ \alpha_0 \end{bmatrix}; \qquad \mathbf{P}'^{-1}_0 = \begin{bmatrix} \mathbf{0}_{3\times 3} & \mathbf{0}_{3\times 3} \\ \mathbf{0}_{3\times 3} & \mathbf{P}_0^{-1} \end{bmatrix}.$$

Figure 5.2 shows the information flow in this NMSKF. Because the prior information is introduced in the estimation of $\mathbf{x}'$, it should not be taken into account a second time when estimating a value for $\mathbf{x}$.

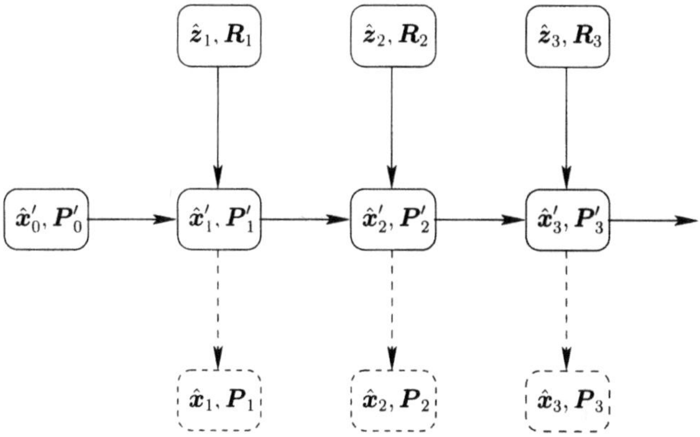

Fig. 5.2. Information flow in the Non-minimal State Kalman Filter for static systems where the Gaussian prior information on the state $\mathbf{x}$ is introduced as prior information on an augmented state $\mathbf{x}'$. The calculations represented in dashed lines are optional

5.2.4 Example

This section presents a small example of the use of the NMSKF and compares the results to the results of the common KF variants described in the previous chapter.

Problem Formulation

The example is the estimation of an angle x based on measurements z which are linear combinations of the cosine and sine of this angle:

$$z_k = a_k \cos(x) + b_k \sin(x) + \rho_{m,k}.$$

This measurement model is linearized by taking a higher-dimensional state

$$\mathbf{x}' = \mathbf{g}(x) = \begin{bmatrix} \cos(x) \\ \sin(x) \end{bmatrix}; \qquad z_k = \begin{bmatrix} a_k & b_k \end{bmatrix} \mathbf{x}' + \rho_{m,k}.$$

The true state value is $x^* = 90°$. The prior pdf over x is a Gaussian with mean $\hat{x}_0 = 180°$ and covariance $P_0 = (60°)^2$. In this section we do not augment $\mathbf{x}'$ with the state x in order to include this prior (Sect. 5.2.3), because the presentation in figures is easier for a 2D than for a 3D state $\mathbf{x}'$.

Results of the Non-Minimal State Kalman Filter

Figure 5.3 (left) shows the likelihood over *the non-minimal state* $\mathbf{x}'$ after one measurement $\hat{z}_1$. The likelihood function is

$$p(\hat{z}_1|\mathbf{x}') \sim \exp\Big(-\frac{1}{2}(\mathbf{x}' - \hat{\mathbf{x}}'_1)^T \mathbf{P}'^{-1}_1 (\mathbf{x}' - \hat{\mathbf{x}}'_1)\Big);$$

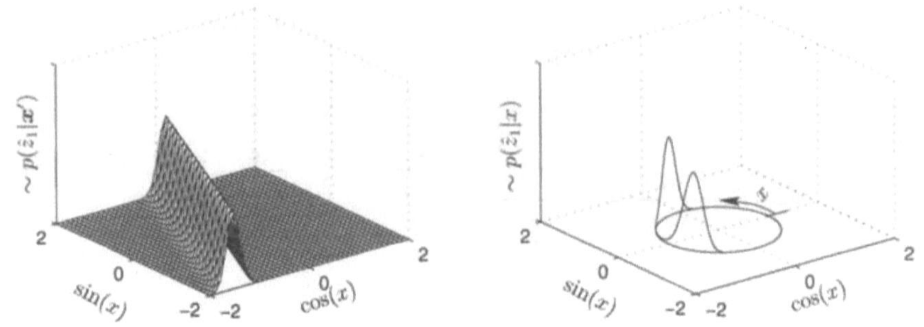

Fig. 5.3. The Non-minimal State Kalman Filter. **(left)** Likelihood $p(\hat{z}_1|\mathbf{x}')$ over the non-minimal state space $\mathbf{x}'$. **(right)** Likelihood $p(\hat{z}_1|x)$ over the original state. This likelihood is obtained by looking at the subspace of values $\mathbf{x}'$ for which an x exists such that $\mathbf{x}' = \mathbf{g}(x)$

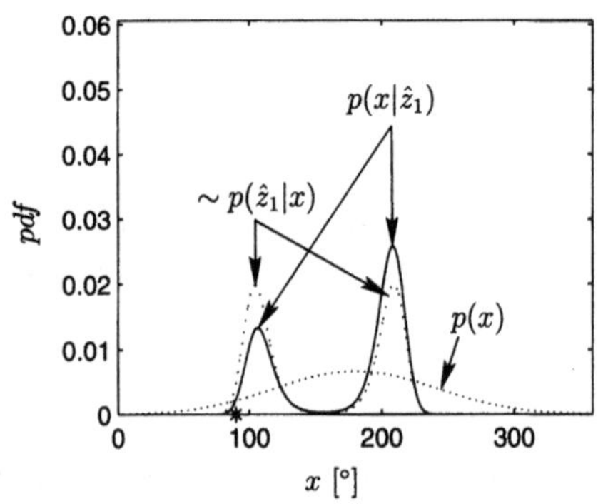

Fig. 5.4. Prior pdf $p(x)$ and likelihood $p(\hat{z}_1|x)$ (*dotted lines*). The posterior pdf $p(x|\hat{z}_1)$ (*full line*) is proportional to the product of the prior pdf and likelihood

where $\hat{\mathbf{x}}_1'$ and $\mathbf{P}_1'$ are obtained by the KF algorithm. This likelihood is *not* constrained to the subspace of values $\mathbf{x}'$ for which an x exists such that $\mathbf{x}' = \mathbf{g}(x)$, i.e., the likelihood plotted on Fig. 5.3 (left) is valid for all values of $\mathbf{x}'$ in $\Re^2$, including values which do not correspond to the cosine and sine of an angle. The nonlinear constraint between the variables $\mathbf{x}'$ is only applied when estimating the original state x.

The likelihood of the measurement over *the original state* x is

$$p(\hat{z}_1|x) \sim \exp\Big(-\frac{1}{2}\big(\mathbf{g}(x) - \hat{\mathbf{x}}_1'\big)^T \mathbf{P}_1'^{-1}\big(\mathbf{g}(x) - \hat{\mathbf{x}}_1'\big)\Big).$$

This likelihood is obtained by looking at the subspace of values $\mathbf{x}'$ for which an x exists such that $\mathbf{x}' = \mathbf{g}(x)$, Fig. 5.3 (right).

Figure 5.4 shows the likelihood $p(\hat{z}_1|x)$, the prior pdf $p(x)$ and the posterior pdf $p(x|\hat{z}_1) \sim p(\hat{z}_1|x)p(x)$.

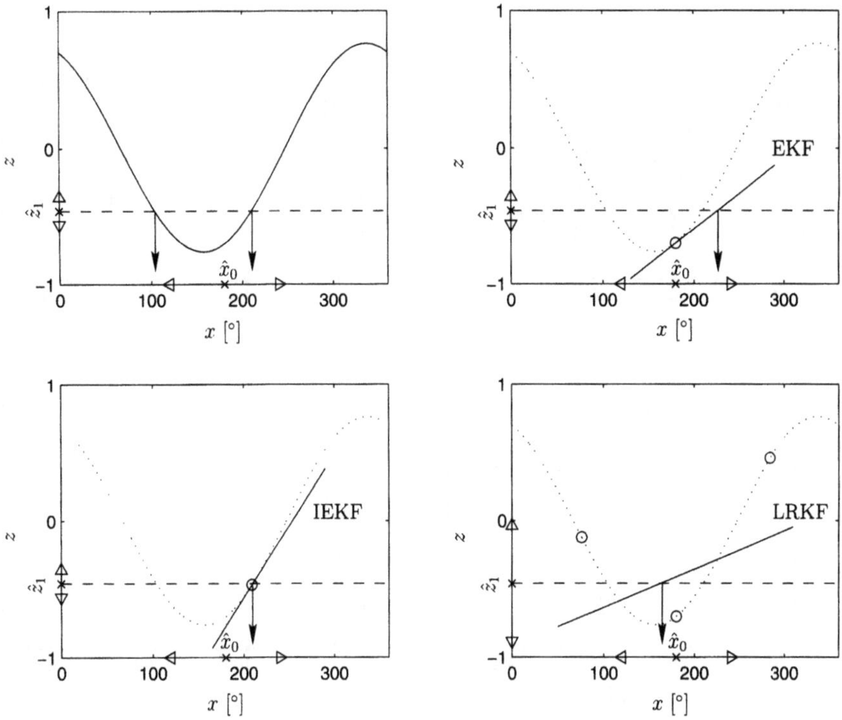

Fig. 5.5. (top-left) Nonlinear measurement model. **(top-right)** Linearized model by the Extended Kalman Filter (EKF). **(bottom-left)** Linearized model by the Iterated Extended Kalman Filter (IEKF). **(bottom-right)** Linearized model by the Linear Regression Kalman Filter (LRKF). The prior estimate $\hat{x}_0$ and an uncertainty interval of once its standard deviation are plotted on the horizontal axes. The measurement $\hat{z}_1$ and an uncertainty interval of once the measurement model standard deviation are plotted on the vertical axes

Results of the Common Kalman Filter Variants

Figures 5.5 and 5.6 compare these results to the estimates of the Extended Kalman Filter (EKF), the Iterated Extended Kalman Filter (IEKF) and the Linear Regression Kalman Filter (LRKF). The regression points of the LRKF are chosen with the algorithm of the Unscented Kalman Filter for κ=3-n=2, [133].

Figure 5.5 (top-left) shows the nonlinear measurement model $z(x)$. On the horizontal axis, the initial estimate $\hat{x}_0$ is drawn with an uncertainty interval of once its standard deviation. On the vertical axis the measurement $\hat{z}_1$ is indicated with an uncertainty interval of once the measurement model standard deviation. The measurement indicates that the true value of x lies around 105° or 210° (marked on the figure with arrows). Figures 5.5 (top-right, bottom-left, bottom-right) show the original measurement model (dotted lines) and

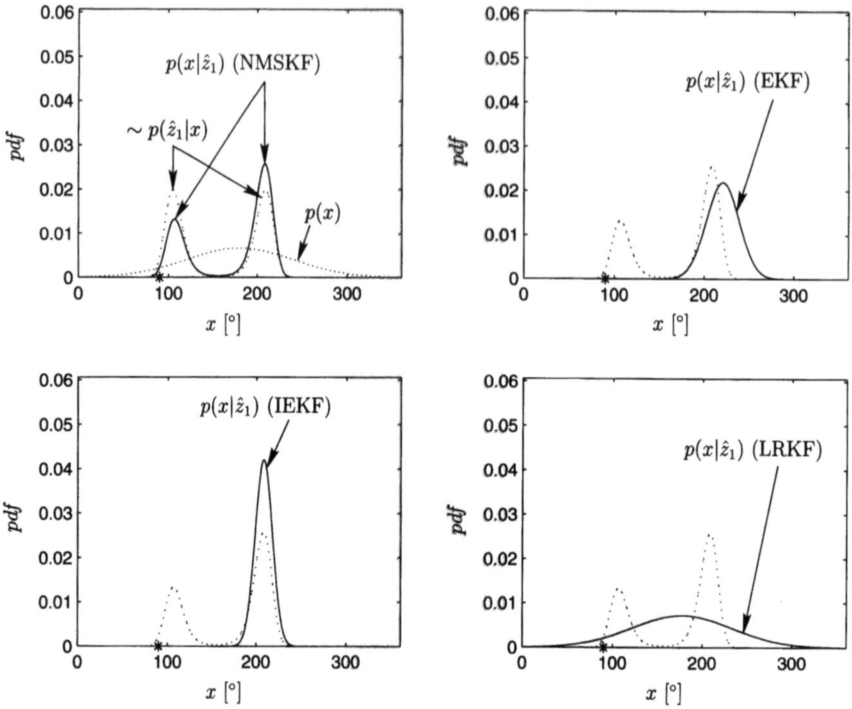

Fig. 5.6. (top-left) Posterior pdf of the Non-minimal State Kalman Filter (MNSKF, this is equal to Fig. 5.4). **(top-right)** Posterior pdf of the Extended Kalman Filter (EKF). **(bottom-left)** Posterior pdf of the Iterated Extended Kalman Filter (IEKF). **(bottom-right)** Posterior pdf of the Linear Regression Kalman Filter (LRKF). The true value of x is indicated with a star on the horizontal axis. The true posterior pdf, i.e., the one calculated by the NMSKF, is drawn in dotted lines in the three other figures

the linearized measurement model by the EKF, IEKF and LRKF respectively (full lines). The LRKF takes the linearization errors into account by an increased measurement model uncertainty.

In Fig. 5.6 (top-left) the posterior pdf calculated by the NMSKF is shown, this is the true posterior pdf. This pdf is drawn in dotted lines in the top-right, bottom-left and bottom-right figures. These figures show the posterior pdfs calculated by the EKF, IEKF and LRKF respectively (full lines). By neglecting the linearization errors, the EKF and the IEKF return inconsistent results: originally probable values are considered to be improbable (the left peak is missing). The LRKF returns consistent results because it takes the linearization errors into account. However, because these linearization errors are large, the LRKF loses almost all information about the measurement. This results in a posterior pdf which is almost equal to the prior pdf.

The NMSKF calculates the true posterior pdf over $\mathbf{x}$, (5.2). However, if an IEKF is applied to obtain an estimated value for x at this point (the CKF algorithm, Sect. 5.2.2), the result will look like Fig. 5.6 (bottom-left) because the measurement does not fully observe the state yet (bimodal posterior pdf $p(x|\hat{z}_1)$). Contrary to the standard IEKF, this inconsistent estimate is *not used in the next measurement update.* The estimation of $\hat{\mathbf{x}}'_k$ and $\mathbf{P}'_k$ at the next time instant k starts from the estimate of $\mathbf{x}'$, as given in Fig. 5.3 (left).

Results of the Non-minimal State Kalman Filter After a Second Measurement

Figure 5.7 shows the likelihood calculated by the NMSKF after processing a second measurement. This new measurement is inferred with the pdf over $\mathbf{x}'$ as given in Fig. 5.3 (left). After processing the second measurement, one of both peaks is already largely reduced.

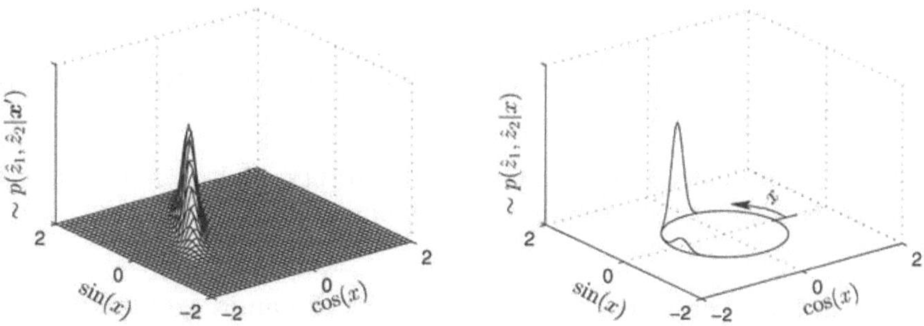

Fig. 5.7. The Non-minimal State Kalman Filter after a measurement $\hat{z}_2$. **(left)** Likelihood $p(\hat{z}_1, \hat{z}_2|\mathbf{x}')$ over the non-minimal state space $\mathbf{x}'$. **(right)** Likelihood $p(\hat{z}_1, \hat{z}_2|x)$ as a function of x after applying the constraint

5.3 The NMSKF for Dynamic Systems

Section 5.3.1 describes the NMSKF update equations for *dynamic* systems. The calculation of an estimated *value* for the original state vector is discussed in Sect. 5.3.2. Section 5.3.3 presents experimental results.

The discussion of the NMSKF for dynamic systems assumes that the prior information is introduced in the prior pdf over $\mathbf{x}'_0$ (see Sect. 5.2.3).

5.3.1 Analytical Posterior Pdf

For a number of dynamic systems, a higher-dimensional state $\mathbf{x}'_k = \mathbf{g}(\mathbf{x}_k)$ can be found in which both the process and measurement models are linear

$$\mathbf{x}'_k = \mathbf{F}_{k-1}\mathbf{x}'_{k-1} + \mathbf{b}_{k-1} + \mathbf{C}_{k-1}\boldsymbol{\rho}_{p,k-1};$$
$$\mathbf{z}_k = \mathbf{H}_k\mathbf{x}'_k + \mathbf{d}_k + \mathbf{E}_k\boldsymbol{\rho}_{m,k};$$

and subject to uncorrelated, zero mean Gaussian uncertainties $\boldsymbol{\rho}_{p,k-1}$ and $\boldsymbol{\rho}_{m,k}$. An example of a dynamic system for which a linearizing $\mathbf{x}'$ can be found is a systems with a linear process model with *no* uncertainty and with a polynomial measurement model subject to Gaussian uncertainty. E.g., for a 2D state $\mathbf{x} = [u,\ v]^T$, the model

$$\mathbf{x}_k = \mathbf{F}_{k-1}\mathbf{x}_{k-1} + \mathbf{b}_{k-1};$$
$$\mathbf{z}_k = \sum_{i=1}^{i=N}\sum_{j=1}^{j=i} \mathbf{h}_{ij,k} u_k^{i-j} v_k^j + \mathbf{d}_k + \mathbf{E}_k\boldsymbol{\rho}_{m,k};$$

is linear in the non-minimal state

$$\mathbf{x}'^T = \mathbf{g}(\mathbf{x}) = \left[u^N,\ u^{N-1}v,\ \ldots,\ uv^{N-1}, v^N,\ \ldots,\ u^2,\ uv,\ v^2,\ u,\ v\right].$$

The experiment of Sect. 5.3.3 presents another (1D) dynamic system for which a linearizing $\mathbf{x}'$ can be found.

Theorem 5.2 derives the analytical posterior $p(\mathbf{x}_k|\hat{\mathbf{Z}}_k)$ over the original state for dynamic systems for which an appropriate $\mathbf{x}'$ is found.

Theorem 5.2 (Analytical posterior pdf for dynamic systems). *Consider an estimation problem with measurements $\hat{\mathbf{Z}}_i = \{\hat{\mathbf{z}}_1, \hat{\mathbf{z}}_2, \ldots, \hat{\mathbf{z}}_i\}$ and process and measurement models*

$$\mathbf{x}'_k = \mathbf{F}_{k-1}\mathbf{x}'_{k-1} + \mathbf{b}_{k-1} + \mathbf{C}_{k-1}\boldsymbol{\rho}_{p,k-1};$$
$$\mathbf{z}_k = \mathbf{H}_k\ \mathbf{x}'_k + \mathbf{d}_k + \mathbf{E}_k\boldsymbol{\rho}_{m,k};$$

where $\mathbf{x}'_k = \mathbf{g}(\mathbf{x}_k) = \begin{bmatrix} * \\ \mathbf{x}_k \end{bmatrix}$ and for which the zero mean Gaussian process and measurement uncertainties $\boldsymbol{\rho}_{p,k-1}$ and $\boldsymbol{\rho}_{m,k}$ are mutually uncorrelated and uncorrelated over time. For this system, the posterior pdf over the state $\mathbf{x}_k$ at time step k, can be expressed in analytical form as

$$p(\mathbf{x}_k|\hat{\mathbf{Z}}_{k-1}) \sim \exp\Big(-\frac{1}{2}\big(\mathbf{g}(\mathbf{x}_k) - \hat{\mathbf{x}}'_{k|k-1}\big)^T \mathbf{P}'^{-1}_{k|k-1}\big(\mathbf{g}(\mathbf{x}_k) - \hat{\mathbf{x}}'_{k|k-1}\big)\Big); \quad (5.8)$$
$$p(\mathbf{x}_k|\hat{\mathbf{Z}}_k) \sim \exp\Big(-\frac{1}{2}\big(\mathbf{g}(\mathbf{x}_k) - \hat{\mathbf{x}}'_{k|k}\big)^T \mathbf{P}'^{-1}_{k|k}\big(\mathbf{g}(\mathbf{x}_k) - \hat{\mathbf{x}}'_{k|k}\big)\Big). \quad (5.9)$$

The sufficient statistics $\hat{\mathbf{x}}'_{k|k-1}$, $\mathbf{P}'_{k|k-1}$, $\hat{\mathbf{x}}'_{k|k}$ and $\mathbf{P}'_{k|k}$ are defined by the Kalman Filter equations for the state estimation of the n'-dimensional state $\mathbf{x}'_k$:

$$\hat{\mathbf{x}}'_{k|k-1} = \mathbf{F}_{k-1}\hat{\mathbf{x}}'_{k-1|k-1} + \mathbf{b}_{k-1}; \quad (5.10)$$
$$\mathbf{P}'_{k|k-1} = \mathbf{F}_{k-1}\mathbf{P}'_{k-1|k-1}\mathbf{F}^T_{k-1} + \mathbf{C}_{k-1}\mathbf{Q}_{k-1}\mathbf{C}^T_{k-1}; \quad (5.11)$$
$$\hat{\mathbf{x}}'_{k|k} = \hat{\mathbf{x}}'_{k|k-1} + \mathbf{K}_k\boldsymbol{\eta}_k; \quad (5.12)$$
$$\mathbf{P}'_{k|k} = (\mathbf{I}_{n'\times n'} - \mathbf{K}_k\mathbf{H}_k)\,\mathbf{P}'_{k|k-1}; \quad (5.13)$$

where

$$\boldsymbol{\eta}_k = \hat{\mathbf{z}}_k - \mathbf{H}_k\hat{\mathbf{x}}'_{k|k-1} - \mathbf{d}_k; \quad (5.14)$$
$$\mathbf{S}_k = \mathbf{E}_k\mathbf{R}_k\mathbf{E}^T_k + \mathbf{H}_k\mathbf{P}'_{k|k-1}\mathbf{H}^T_k; \quad (5.15)$$
$$\mathbf{K}_k = \mathbf{P}'_{k|k-1}\mathbf{H}^T_k\mathbf{S}^{-1}_k. \quad (5.16)$$

$\mathbf{Q}_{k-1}$ *and* $\mathbf{R}_k$ *are the covariance matrices of the process and measurement model uncertainties. If no measurement* $\hat{\mathbf{z}}_k$ *is available at a certain time step* k*, then (5.12)-(5.16) reduce to* $\hat{\mathbf{x}}'_{k|k} = \hat{\mathbf{x}}'_{k|k-1}$, $\mathbf{P}'_{k|k} = \mathbf{P}'_{k|k-1}$. *The iteration starts with the prior uncertainty (see Sect. 5.2.3)*

$$\hat{\mathbf{x}}'_{0|0} = \begin{bmatrix} \mathbf{0}_{(n'-n)\times 1} \\ \hat{\mathbf{x}}_{0|0} \end{bmatrix}; \qquad \mathbf{P}'^{-1}_{0|0} = \begin{bmatrix} \mathbf{0}_{(n'-n)\times(n'-n)} & \mathbf{0}_{(n'-n)\times n} \\ \mathbf{0}_{n\times(n'-n)} & \mathbf{P}^{-1}_{0|0} \end{bmatrix}.$$

The proof of this theorem is given in Appendix B.

5.3.2 Retrieving an Estimate for the Original State x

Figure 5.8 shows the information flow in the Non-minimal State Kalman Filter. The optional calculation of a single state estimate value for **x** is drawn

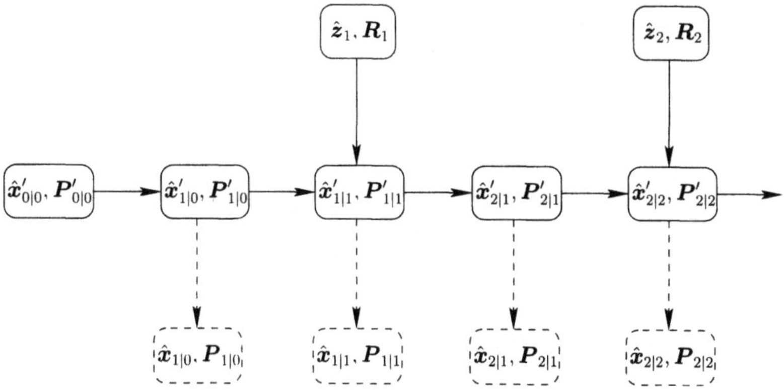

Fig. 5.8. Information flow in the Non-minimal State Kalman Filter for dynamic systems. The calculations represented in dashed lines are optional

in dashed lines. This calculation is similar to the static case (see Sect. 5.2.2). Because the prior information is introduced in the estimation of $\mathbf{x}'$, it should not be taken into account a second time when estimating a value for $\mathbf{x}$.

5.3.3 Experiment

This section contains a 2D contour tracking experiment which illustrates the use of the NMSKF for a dynamic system.

Problem Formulation

In the experiment, a probe is tracking a *known 2D contour*. The goal is to *estimate the contact point position of the probe* on the contour, i.e., to estimate the arc length s with respect to some reference $s = 0$ mm, Fig. 5.9 (left).

At different time steps k, the absolute position $\mathbf{p}_k = \left[p_{x,k}\ p_{y,k}\right]^T$ of the probe with respect to some world frame $\{\mathrm{w}\}$, is measured with an accurate Krypton K600 measuring device, Fig. 5.9 (right). The position and orientation of the contour-object with respect to this world frame is unknown.

The localisation problem is made independent of the contour's position and orientation by identifying the contact point based on

1. the Cartesian distance between the successive probe positions Δs_{k-1},

$$\widehat{\Delta s}_{k-1} = \sqrt{(p_{x,k} - p_{x,k-1})^2 + (p_{y,k} - p_{y,k-1})^2}.$$

   These distances approximate the arc length differences between the probe positions.
2. the curvature κ of the contour at the successive probe positions. These curvature measurements are calculated based on three successive probe position measurements.

$$\hat{\kappa}_k = \frac{v_{x,k}a_{y,k} - a_{x,k}v_{y,k}}{\left(\sqrt{v_{x,k}^2 + v_{y,k}^2}\right)^3};$$

   where

$$v_{x,k} = \frac{p_{x,k} - p_{x,k-1}}{\Delta s_{k-1}}; \qquad v_{y,k} = \frac{p_{y,k} - p_{y,k-1}}{\Delta s_{k-1}};$$

$$a_{x,k} = \frac{2\left(v_{x,k} - v_{x,k-1}\right)}{\Delta s_k + \Delta s_{k-1}}; \qquad a_{y,k} = \frac{2\left(v_{y,k} - v_{y,k-1}\right)}{\Delta s_k + \Delta s_{k-1}}.$$

Process and Measurement Models

The example approximates a system for which a linearizing $\mathbf{x}'$ can be found. The models are a function of a 1D state $x = s$ for which the process model is linear with no uncertainty and the measurement model is an Nth-order sine/cosine series subject to Gaussian uncertainty:

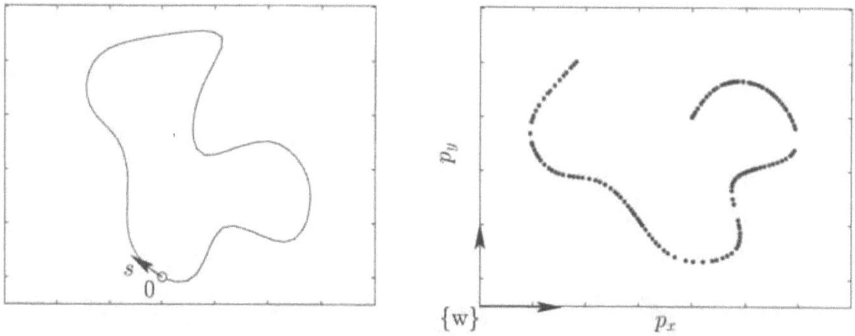

Fig. 5.9. (left) Contour model. **(right)** Measured contact probe positions

1. The *process model* expresses the evolution of the contact point over time:

$$s_k = s_{k-1} + \widehat{\Delta s}_{k-1}.$$

   No process uncertainty is considered. This corresponds to the process model of the system for which a linearizing $\mathbf{x}'$ can be found. The experimental results show that the approximation is valid.
2. Figure 5.10 (full line) shows the contour *measurement model*, i.e., the curvature κ as a function of the arc length s. The contour has a total arc length of $L = 1579$ mm. This model was manually derived beforehand. It is an approximation of the true contour curvature by a $N = 17$th-order sine/cosine series:

$$z_k = \kappa_k = a_0 + \sum_{i=1}^{i=N} \left(a_i \cos(\frac{2i\pi s_k}{L}) + b_i \sin(\frac{2i\pi s_k}{L}) \right) + v_k.$$

   The stars on Fig. 5.10 indicate the curvature *measurements* $\hat{z}_k = \hat{\kappa}_k$, which were taken starting from an initial contact point at arc length 988 mm. The uncertainty v_k is due to the inaccuracy of both the model and the measurements. It is modelled as a zero mean Gaussian with covariance $(0.02 \text{ mm}^{-1})^2$.

Linearized Models for the NMSKF

The process and measurement model are linear when expressed in the $2N$-dimensional state vector $\mathbf{x}'$

$$\mathbf{x}' = \mathbf{g}(s) = \left[\cos(\tfrac{2\pi s}{L}), \ldots, \cos(\tfrac{2N\pi s}{L}), \sin(\tfrac{2\pi s}{L}), \ldots, \sin(\tfrac{2N\pi s}{L})\right]^T. \quad (5.17)$$

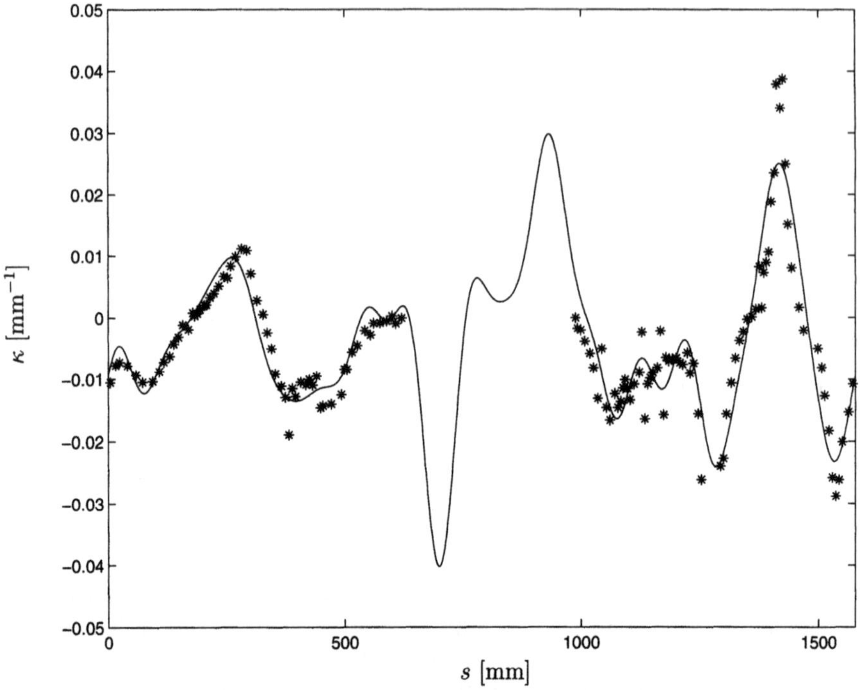

Fig. 5.10. (*full line*) 17th-order sine/cosine curvature model of the contour. (*stars*) Curvature measurements. The initial contact point occurred at a contour length of 988 mm. This initial contact point is unknown to the estimator

The linear models are:

$$\mathbf{x}'_k = \mathbf{F}_{k-1}\mathbf{x}'_{k-1};$$
$$z_k = \mathbf{H}\mathbf{x}'_k + a_0 + v_k;$$

where $(i = 1, \ldots, N)$

$$\mathbf{F}_{k-1}(i,i) = \cos(\frac{2i\pi\widehat{\Delta s}_{k-1}}{L});$$
$$\mathbf{F}_{k-1}(i,i+N) = -\sin(\frac{2i\pi\widehat{\Delta s}_{k-1}}{L});$$
$$\mathbf{F}_{k-1}(i+N,i+N) = \cos(\frac{2i\pi\widehat{\Delta s}_{k-1}}{L});$$
$$\mathbf{F}_{k-1}(i+N,i) = \sin(\frac{2i\pi\widehat{\Delta s}_{k-1}}{L});$$
$$\mathbf{H}(i) = a_i;$$
$$\mathbf{H}(i+N) = b_i.$$

Estimator

Due to the highly nonlinear measurement model (see Fig. 5.10), the "classical" KFs (EKF, IEKF and LRKF, Chap. 4) are found not to be consistent and/or informative unless started with a good initial estimate.

A NMSKF, on the other hand, solves the problem even for an uniformly distributed prior pdf $p(s_0)$. The 34-dimensional state $\mathbf{x}'$, given in (5.17), linearizes the problem.

Results

Figure 5.11 shows the pdf over the contact point position s_k

$$p(s_k|\hat{\mathbf{Z}}_k) \sim \exp\Big(-\frac{1}{2}\big(\mathbf{g}(s_k) - \hat{\mathbf{x}}'_{k|k}\big)^T \mathbf{P}'^{-1}_{k|k}\big(\mathbf{g}(s_k) - \hat{\mathbf{x}}'_{k|k}\big)\Big);$$

after $k = 1, 50, 100$ and 134 measurements. The star on the horizontal axis indicates the arc length at the true contour contact point. The top figure shows that the first measurement possibly originates from about any contact point. After 50 and 100 measurements, the probability is mainly concentrated in two regions. Note that after 100 measurements the true contour contact point is not situated in the region with the highest probability. After 134 measurements, only one contact region is still probable. The standard deviation of this "peak" is 6 mm.

5.4 Conclusions

This chapter presents a new finite-dimensional Bayesian filter for static systems with any kind of nonlinear measurement model subject to additive Gaussian uncertainty and for a limited class of dynamic systems. The core idea of the filter is to linearize the process and measurement models by choosing a suitable non-minimal state description. The sufficient statistics of the posterior pdf over the original (minimal) state are obtained through Kalman filtering in this non-minimal state space. The filter is experimentally validated on a dynamic system estimating the position of a contact point on a known contour based on curvature measurements.

Chapter 7 describes the application of the new filter to the estimation of the geometrical parameters during autonomous compliant motion. Good estimation results are obtained for large initial uncertainties, a situation which previously none of the Kalman Filter variants could handle.

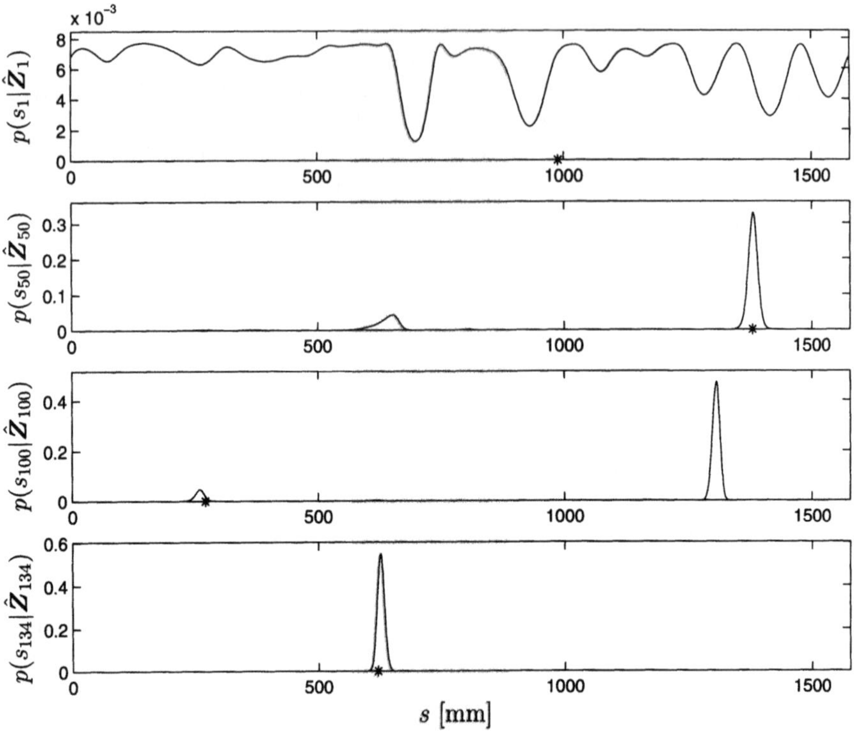

Fig. 5.11. Pdf $p(s_k|\hat{\mathbf{Z}}_k)$ after 1, 50, 100 and 134 measurements. The star on the horizontal axis indicates the true contour contact point

6

Contact Modelling

As far as the laws of mathematics refer to reality,
they are not certain; and as far as they are certain,
they do not refer to reality.

Albert Einstein, (1879–1955).

6.1 Introduction

This chapter describes the contact modelling for autonomous compliant motion assuming quasi-static, rigid, frictionless contacts. The contact models are needed in the force controller, the estimator and the planner of the system. The models are different for each contact formation (CF), and are a function of the geometrical parameters (i.e., the positions, orientations and dimensions of the contacting objects).

The contact models described in the literature of autonomous compliant motion (Chap. 2) are mainly *manually* derived sets of contact constraints (*closure equations*). An exception is the description of the constraints for *polyhedral CFs* (i.e., CFs between polyhedral parts of the objects): these equations are generated based on the equations of the elementary vertex-face, face-vertex and edge-edge CFs in which the CF can be decomposed.

In order to *estimate* the uncertainties, contact models *as a function of the geometrical parameters* are needed. Besides pose measurements, also wrench and twist measurements provide information. Using these measurements requires that, besides the closure equations, we have *twist and wrench equations* relating these measurements to the geometrical parameters. Bruyninckx [41] described the contact equations (closure, twist and wrench equations) as a function of the geometrical parameters for CFs between objects with generally curved surfaces, however, the equations were to be derived manually for each CF.

This work presents the latest evolution in contact modelling for geometrical parameter estimation:

1. The description of the contact models does not need the "Virtual Contact Manipulator" [44], describing the *relative* degrees of freedom between the manipulated object and the environment objects. This results in a considerably *reduced modelling effort* and a *minimal set of closure equations.* The Virtual Contact Manipulator is not needed because the contact points and normals at every time instant are expressed as a function of the geometrical parameters and the manipulated object pose. In previous research, the contact points and normals were described as a function of the joint values of the Virtual Contact Manipulator, i.e., the relative degrees of freedom between the contacting objects.
2. The contact models for multiple-contact CFs are generated based on a *library of elementary CFs*. For instance, we describe the generation of the closure, twist and wrench equations of *polyhedral* CFs, based on elementary vertex-face, face-vertex and edge-edge CFs.
   Contacts between generally curved surfaces can be added to the library if *analytical* surface descriptions are available which allow an analytical description of the contact point(s) and normal(s).

Section 6.3 describes the contact equations. It also presents some ideas about the usefulness of the quasi-static, frictionless, rigid contact models in real-world situations. Section 6.4 presents the automatic generation of the equations based on an elementary contact library. Sections 6.5 and 6.6 provide the detailed contact descriptions of the polyhedral and some non-polyhedral elementary CFs. The contact modelling based on these elementary CFs is illustrated by examples in Sect. 6.7. First of all, Sect. 6.2 presents the problem statement and the definitions of the frames, vectors and matrices used in this work.

6.2 Problem Statement and Definitions

Frames

We consider the following reference frames (Fig. 1.1): {w} attached to the world, {g} attached to the gripper, {m} attached to the manipulated object and {e} attached to the environment object.

Geometrical Parameters

The inaccurately known *geometrical parameters* are the position and orientation of {g} with respect to {m} (*grasping uncertainties* $\mathbf{x}^m$) and of {w} with respect to {e} (*environment uncertainties* $\mathbf{x}^e$); these positions and orientations do not change during the task execution. If more than one environment object contacts the manipulated object during the task execution, an environment frame {i} is attached ($i = 1, 2, \ldots$) to each of them and the environment uncertainties $\mathbf{x}^i$ define the inaccurately known position and orientation of {w} with respect to {i}. Additionally, the *physical dimensions* of the manipulated object ($\mathbf{x}^{dm}$, e.g., the side length of the cube in Fig. 1.1) and of the environment objects ($\mathbf{x}^{de}$ or $\mathbf{x}^{di}$, $i = 1, 2, \ldots$) might be inaccurately known. All the uncertain geometrical parameters are collected in the *state vector* $\mathbf{x}$

$$\mathbf{x} = \left[(\mathbf{x}^m)^T \; (\mathbf{x}^{dm})^T \; (\mathbf{x}^1)^T \; (\mathbf{x}^{d1})^T \; (\mathbf{x}^2)^T \; (\mathbf{x}^{d2})^T \; \ldots \right]^T$$

The state is *static*, i.e., its actual value does not change during the task execution.

Twist, Wrench and Pose

Following *six-vectors* are measured:

1. *twists* $\mathbf{t}$ of the manipulated object (translational velocities $\mathbf{v}$ and rotational velocities $\boldsymbol{\omega}$), derived from the robot joint positions and velocities;

$$\mathbf{t} = \begin{bmatrix} \mathbf{v} \\ \boldsymbol{\omega} \end{bmatrix}. \tag{6.1}$$

2. contact *wrenches* $\mathbf{w}$ (forces $\mathbf{f}$ and moments $\mathbf{m}$), measured by a wrist force sensor;

$$\mathbf{w} = \begin{bmatrix} \mathbf{f} \\ \mathbf{m} \end{bmatrix}. \tag{6.2}$$

3. *poses* $\mathbf{t}_d$ of the manipulated object (positions and orientations), derived from the robot joint positions.

Wrench and Twist Spanning Set

The vector space of all possible contact wrenches is called *wrench space*. Every contact wrench can be written as a linear combination of a fixed set of vectors $\mathbf{G}_i$, $i = 1, \ldots, n_G$, called a *spanning set [56] of the wrench space* or a *wrench spanning set*. We denote a spanning set by a matrix $\mathbf{G}$ whose columns are the spanning vectors $\mathbf{G}_i$. Once a wrench spanning set $\mathbf{G}$ is chosen, every wrench $\mathbf{w}$ can be represented by a coordinate vector $\boldsymbol{\phi}$ describing the linear dependency of $\mathbf{w}$ on $\mathbf{G}$: $\mathbf{w} = \mathbf{G}\boldsymbol{\phi}$. The elements of the wrench spanning set have units; $\boldsymbol{\phi}$ is dimensionless. $\boldsymbol{\phi}$ is only uniquely defined if the spanning vectors are linearly independent, i.e., if the spanning set is a *base* for the wrench space.

The treatment of twists is similar: a *twist spanning set* $\mathbf{J}$ ($\mathbf{J}_i$, $i = 1, \ldots, n_J$) is chosen for the vector space of all possible manipulated object twists (*twist space*). Following this, every twist $\mathbf{t}$ can be represented by a coordinate vector

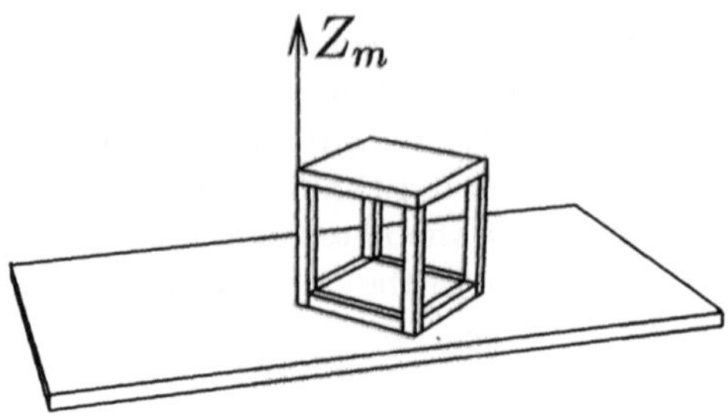

Fig. 6.1. Face-face CF with Z-axis perpendicular to the contacting faces

χ such that $\mathbf{t} = \mathbf{J}\chi$. The elements of the wrench spanning set have units; χ is dimensionless. χ is only uniquely defined if the spanning vectors are linearly independent, i.e., if the spanning set is a *base* for the twist space.

Each CF has different twist and wrench spaces, e.g., possible twist and wrench spanning sets expressed in {m} for the face-face CF of Fig. 6.1 with the Z-axis of {m} perpendicular to the contacting faces, are

$$\mathbf{J}^m = \begin{bmatrix} 1\ \text{m/s} & 0\ \text{m/s} & 0\ \text{m/s} \\ 0\ \text{m/s} & 1\ \text{m/s} & 0\ \text{m/s} \\ 0\ \text{m/s} & 0\ \text{m/s} & 0\ \text{m/s} \\ 0\ \text{rad/s} & 0\ \text{rad/s} & 0\ \text{rad/s} \\ 0\ \text{rad/s} & 0\ \text{rad/s} & 0\ \text{rad/s} \\ 0\ \text{rad/s} & 0\ \text{rad/s} & 1\ \text{rad/s} \end{bmatrix};$$

$$\mathbf{G}^m = \begin{bmatrix} 0\ \text{N} & 0\ \text{N} & 0\ \text{N} \\ 0\ \text{N} & 0\ \text{N} & 0\ \text{N} \\ 1\ \text{N} & 0\ \text{N} & 0\ \text{N} \\ 0\ \text{Nm} & 1\ \text{Nm} & 0\ \text{Nm} \\ 0\ \text{Nm} & 0\ \text{Nm} & 1\ \text{Nm} \\ 0\ \text{Nm} & 0\ \text{Nm} & 0\ \text{Nm} \end{bmatrix}.$$

The first two columns of **J** represent the translational motion freedom directions parallel to the contacting face. The third column corresponds to the rotational motion freedom direction about the normal to the face. The first column of **G** represents the contact force direction perpendicular to the face, and the last two columns represent the contact moment directions about the axes of {m} which are parallel to the face. The choice of these spanning sets, of course, is not unique.

Transformation Matrices

Transformations of *wrenches and wrench spanning sets* to other frames (e.g., {g}, {m}, {i} and {w}) are possible by pre-multiplying them with a *screw transformation matrix*[1] ${}^a_b\mathbf{S}_w$ [208]:

$$\mathbf{w}^b = {}^a_b\mathbf{S}_w\,\mathbf{w}^a; \qquad \mathbf{G}^b = {}^a_b\mathbf{S}_w\,\mathbf{G}^a; \qquad {}^a_b\mathbf{S}_w = \begin{bmatrix} {}^a_b\mathbf{R} & \mathbf{0}_{3\times3} \\ {}^a_b\mathbf{R}\,[{}_a\mathbf{p}^{b,a}\times] & {}^a_b\mathbf{R} \end{bmatrix};$$

${}^a_b\mathbf{R}$ is the rotation matrix to rotate a vector from {a} to {b}, i.e., its columns are the components of a unit vector along the X, Y and Z axis of {a} as seen from {b};

$$[{}_a\mathbf{p}^{b,a}\times] = \begin{bmatrix} 0 & -{}_ap_z^{b,a} & {}_ap_y^{b,a} \\ {}_ap_z^{b,a} & 0 & -{}_ap_x^{b,a} \\ -{}_ap_y^{b,a} & {}_ap_x^{b,a} & 0 \end{bmatrix};$$

${}_c\mathbf{p}^{b,a} = \left[{}_cp_x^{b,a}\ {}_cp_y^{b,a}\ {}_cp_z^{b,a}\right]^T$ is the vector from the origin of {b} to the origin of {a} expressed in the frame {c} (see Fig. 6.2).

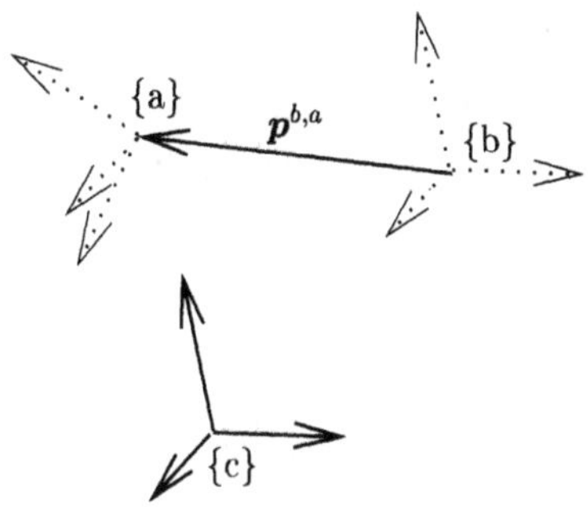

Fig. 6.2. $\mathbf{p}^{b,a}$ is the vector from the origin of {b} to the origin of {a}

On the other hand, transformations of *twists and twist spanning sets* to other frames are possible by pre-multiplying them with a *screw transformation matrix*[1] ${}^a_b\mathbf{S}_t$:

$$\mathbf{t}^b = {}^a_b\mathbf{S}_t\,\mathbf{t}^a; \qquad \mathbf{J}^b = {}^a_b\mathbf{S}_t\,\mathbf{J}^a; \qquad {}^a_b\mathbf{S}_t = \begin{bmatrix} {}^a_b\mathbf{R} & {}^a_b\mathbf{R}\,[{}_a\mathbf{p}^{b,a}\times] \\ \mathbf{0}_{3\times3} & {}^a_b\mathbf{R} \end{bmatrix}.$$

[1] The form of the screw transformation matrices ${}^a_b\mathbf{S}_w$ and ${}^a_b\mathbf{S}_t$ depends on the definitions of $\mathbf{w}$ and $\mathbf{t}$. The given transformation matrices are based on definitions (6.2) and (6.1).

The coordinate transformation of the coordinates of a point d from frame $\{a\}$ to frame $\{b\}$ is expressed using the *homogeneous transformation matrix* ${}^a_b\mathbf{T}$:

$$\begin{bmatrix} {}_b\mathbf{p}^{b,d} \\ 1 \end{bmatrix} = {}^a_b\mathbf{T} \begin{bmatrix} {}_a\mathbf{p}^{a,d} \\ 1 \end{bmatrix};$$

where

$${}^a_b\mathbf{T} = \begin{bmatrix} {}^a_b\mathbf{R} & -{}^a_b\mathbf{R}\,{}_a\mathbf{p}^{a,b} \\ \mathbf{0}_{1\times 3} & 1 \end{bmatrix} = \begin{bmatrix} {}^a_b\mathbf{R} & {}_b\mathbf{p}^{b,a} \\ \mathbf{0}_{1\times 3} & 1 \end{bmatrix}.$$

6.3 Contact Modelling

This section describes the closure, twist and wrench equations for frictionless, rigid, quasi-static contacts (Sects. 6.3.1–6.3.2). Section 6.3.3 discusses the validity of these equations in real-world applications.

6.3.1 Closure Equations

The closure equations relate the manipulated object pose $\mathbf{t}_d$ to the geometrical parameters $\mathbf{x}$, given the known geometry of the contacting objects:

$$\mathbf{h}_d(\mathbf{x}, \mathbf{t}_d) = \mathbf{0}_{n_c\times 1}; \tag{6.3}$$

where n_c is the number of constraints. The closure equations express that the contact is really established, e.g., for a vertex-face CF, the closure equation expresses that the vertex lies in the face (the distance between the vertex and the face is zero).

In previous work [68], the closure equations include equations that express the dependency of the joint variables of the Virtual Contact Manipulator on the geometrical parameters $\mathbf{x}$, i.e., the closure equations are written for all six degrees of freedom (three position and three orientation constraint equations).

Here we do not use the joint variables of the Virtual Contact Manipulator. This reduces the number of closure equations to the number of constraints on the relative motion degrees of freedom between the objects in contact, i.e., the rank of the wrench space. E.g., a face-face CF constrains the translation perpendicular to the face and the rotations about axes in the face. Hence, the contact involves a 3D wrench space spanned by the force direction perpendicular to the face and the two moment directions about axes in the face. The closure equations express that the face of the manipulated object *coincides with* the face of the environment object (3 equations). A possible choice of closure equations is to define $\mathbf{h}_d(\mathbf{x}, \mathbf{t}_d)$ as the Cartesian *distance* between the contacting parts of the objects (see Sects. 6.5–6.6). Other functions such as the square of the distance are also possible.

6.3.2 Twist and Wrench Equations

Bruyninckx [41] describes two equivalent sets of contact equations for twists and wrenches: the *reciprocity* and the *consistency* equations.

The *reciprocity* based twist and wrench equations express that no power is dissipated in the compliant motion. In other words, the equations express that the manipulated object twists $\mathbf{t}$ are reciprocal to the n_G wrench spanning vectors $\mathbf{G}$, and that the contact wrenches $\mathbf{w}$ are reciprocal to the n_J twist spanning vectors $\mathbf{J}$. The reciprocity based twist and wrench equations are:

$$(\mathbf{G}(\mathbf{x},\mathbf{t}_d))^T \ \mathbf{t} = \mathbf{0}_{n_G\times 1}; \tag{6.4}$$
$$(\mathbf{J}(\mathbf{x},\mathbf{t}_d))^T \ \mathbf{w} = \mathbf{0}_{n_J\times 1}; \tag{6.5}$$

where $\mathbf{G}$ are $\mathbf{J}$ are expressed with respect to the same reference frame as $\mathbf{t}$, respectively $\mathbf{w}$. Note that these equations are only valid if the units of all components yield consistent (scaled) power terms. E.g., this is the case if all force components of $\mathbf{w}$ and $\mathbf{G}$ are expressed in N, all moment components of $\mathbf{w}$ and $\mathbf{G}$ in Nm, all translational velocity components of $\mathbf{t}$ and $\mathbf{J}$ in m/s and all rotational velocity components of $\mathbf{t}$ and $\mathbf{J}$ in rad/s.

The *consistency* based twist and wrench equations express that the twist and wrench can be decomposed in the twist and wrench spanning sets:

$$\mathbf{J}(\mathbf{x},\mathbf{t}_d)\ \boldsymbol{\chi} - \mathbf{t} = \mathbf{0}_{6\times 1}. \tag{6.6}$$
$$\mathbf{G}(\mathbf{x},\mathbf{t}_d)\ \boldsymbol{\phi} - \mathbf{w} = \mathbf{0}_{6\times 1}; \tag{6.7}$$

where $\mathbf{J}$ and $\mathbf{G}$ are expressed with respect to the same reference frame as $\mathbf{t}$, respectively $\mathbf{w}$.

6.3.3 Effect of Friction

The model described in the above section is a model for quasi-static, rigid, frictionless contacts. In many compliant motion tasks, such as assembly, the contact is rigid and the motions are (still) slow enough to be considered quasi-static. However, *friction* introduces large deviations between the real world and the modelled wrench equations. This section contains a short analysis of the effect of unmodelled friction on the controller, the estimator and the planner.

In *hybrid velocity/force control*, friction does not really matter because the desired motions are specified in velocity controlled directions. Force measurements in these directions are ignored. Friction is only a problem if the desired velocity control command is too small to overcome sticking.

The *task planner* generating the hybrid control commands is not much influenced by friction either. It only needs to prevent sticking by avoiding small desired motions in possible friction directions.

The *estimation component* is most affected by friction. The wrench measurements do not correspond to (6.5) nor (6.7). This work takes the unmodelled friction into account by putting *extra uncertainty* on the wrench measurement equations (see Chap. 7).

6.4 Modelling Based on a Library of Elementary CFs

The previous section described the wrench, twist and closure equations, which are expressed in the wrench spanning set $\mathbf{G}(\mathbf{x}, \mathbf{t}_d)$, the twist spanning set $\mathbf{J}(\mathbf{x}, \mathbf{t}_d)$ and the distance function $\mathbf{h}_d(\mathbf{x}, \mathbf{t}_d)$. This section contains the derivation of the spanning sets and the distance functions for different CFs.

Previous contact models for geometrical parameter estimation [41] are built on the concept of a Virtual Contact Manipulator. This Virtual Contact Manipulator has as joint values the *relative* position and orientation degrees of freedom between the manipulated object and the environment objects. Besides the geometrical parameters of the contacting objects, the twist, wrench and closure equations also depend on these joint values, i.e., the state vector is augmented with the (variable) joint values of the Virtual Contact Manipulator.

This section presents the calculation of the distance functions, wrench and twist spanning sets as functions of the geometrical parameters of the contacting objects only, without explicitly modelling the relative degrees of freedom between them. The presented work does not need the Virtual Contact Manipulator because the contact points and normals in every contact are expressed as a function of the geometrical parameters and the manipulated object pose (see Sects. 6.5 and 6.6). This *reduces the modelling effort* and *allows the automatic generation of the equations based on the models of the elementary CFs.*

Section 6.4.1 describes how contact models for multiple-contact CFs can be constituted from the models of the CFs in which they can be decomposed. Section 6.4.2 presents the decomposition of polyhedral CFs into elementary vertex-face, edge-edge and face-vertex CFs. Section 6.4.3 introduces the problem of non-minimal contact descriptions and its consequences.

6.4.1 Multiple-Contact CFs

Figure 6.3 shows a multiple-contact CF. The closure equation expresses that the contact is established in all of the l individual CFs ("contact areas") in which it can be decomposed:

$$\mathbf{h}_d(\mathbf{x}, \mathbf{t}_d) = \begin{bmatrix} \mathbf{h}_{d1}(\mathbf{x}, \mathbf{t}_d) \\ \dots \\ \mathbf{h}_{dl}(\mathbf{x}, \mathbf{t}_d) \end{bmatrix} = \mathbf{0}_{n_c \times 1}.$$

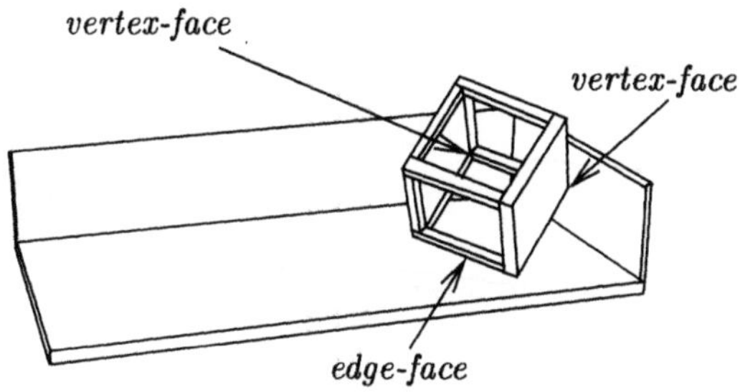

Fig. 6.3. Multiple-contact CF between the manipulated object and the environment object: the cube makes an edge-face CF with the "horizontal face" and two vertex-face CFs with the "vertical faces" of the environment object

The contact wrench is the resultant of the contact wrenches in the individual CFs, hence, the wrench space of the multiple-contact CF is the *union* of the wrench spaces of its l individual CFs. A wrench spanning set of the CF is obtained by merging all the columns of the l individual wrench spanning sets into a single matrix:

$$\mathbf{G} = \left[\mathbf{G}_1 \ldots \mathbf{G}_l\right].$$

The twist space of the CF is the *intersection* of the twist spaces of its l individual CFs. Calculating a spanning set for this intersection (analytically) can be a tedious task. On the other hand, *numerical* calculation of a twist spanning set based on the wrench space of the multiple-contact CF is straightforward. Therefore, in this work, *the description of the contact models of the individual CFs is limited to the description of their closure equations and wrench spanning sets.*

6.4.2 Polyhedral CFs

A polyhedral CF can be decomposed into a number of elementary vertex-face, face-vertex and/or edge-edge CFs. This simplification, of course, is only valid if one is not interested in the *distribution* of the contact pressure in the contact area(s). This section contains some examples, Sect. 6.5 describes a possible distance function and wrench spanning set for each of these elementary CFs.

Figure 6.4 shows the multiple-contact CF of Fig. 6.3 decomposed into four vertex-face CFs (1 to 4). The edge-face CF on the horizontal face is described by two vertex-face CFs (3 and 4). The only restriction on the choice of these vertices is that they are located on the edge, e.g. the two vertex-face CFs 3 and 5 equally represent the edge-face CF.

A face-face CF can be decomposed into three vertex-face and/or face-vertex CFs. Figure 6.5 shows a possible decomposition (contacts 1 to 3). The

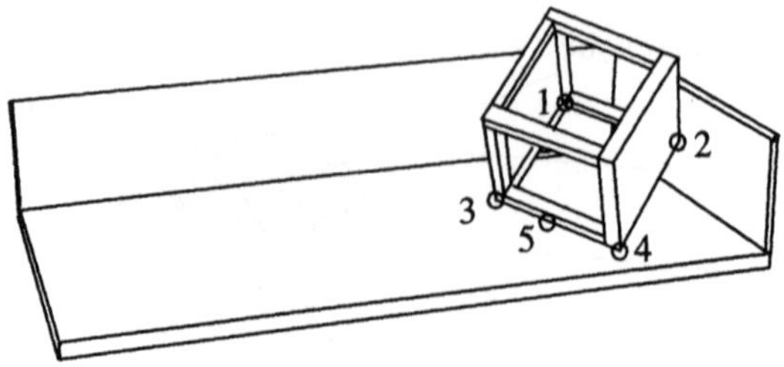

Fig. 6.4. A multiple-contact CF decomposed into four vertex-face CFs (e.g., 1 to 4)

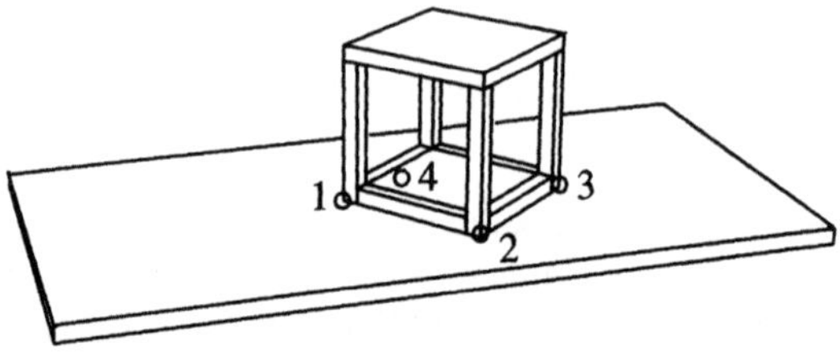

Fig. 6.5. A face-face CF can be modelled by three vertex-face CFs (e.g., 1 to 3)

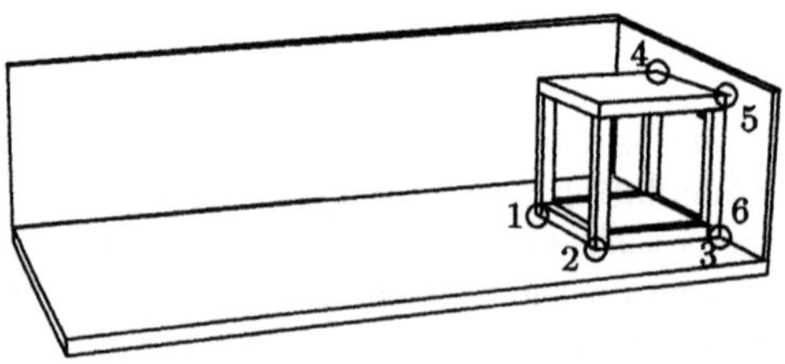

Fig. 6.6. A double face-face CF can be modelled by five vertex-face CFs (e.g., 1 to 5)

restrictions on the three chosen vertices are that (i) they are located on the contacting face and (ii) they are not collinear, e.g., the decomposition into vertex-face CFs 2 to 4 is also valid.

Figure 6.6 presents a possible choice for the location of the five vertex-face CFs (1 to 5) defining a double face-face CF. The "horizontal" face-face contact is represented by three vertex-face CFs (1 to 3). The "vertical" face-face contact is represented by only two vertex-face CFs (4 and 5), because one of the directions in which a moment can be exerted is already modelled by the vertex-face CFs 1 to 3.

6.4.3 Non-Minimal Contact Descriptions

Automatic generation of the closure, twist and wrench equations for general CFs is somewhat *easier if* a non-minimal number of elementary CFs can be used. E.g., it should be possible to represent the double face-face CF of Fig. 6.6 with elementary vertex-face CFs 1 to 6. Using a non-minimal number of elementary CFs results in a non-minimal wrench spanning set and a non-minimal number of closure equations.

Allowing non-minimal contact descriptions makes the *automatic generation of the equations easier*, however, it *requires that the system components using the non-minimal description (the controller, the estimator and the planner) are robust against the non-minimality.* The estimator (Chap. 7) and the planner (Chap. 9) described in this work are robust to non-minimal modelling.

6.5 Polyhedral Elementary CFs

This section presents the contact models of the elementary vertex-face, edge-edge and face-vertex CFs in which a polyhedral CF can be decomposed.

Each elementary CF is described by (i) a frame {c} on the contact, (ii) a wrench spanning set expressed in this frame {c}, and (iii) the (signed) distance function between the two contacting features. Appendix C describes the detailed analytical expressions for the used transformation matrices. Table 6.1 lists the dependencies of the rotation matrices $\mathbf{R}$ and homogeneous transformation matrices $\mathbf{T}$ on the geometrical parameters and the robot pose $\mathbf{t}_d$. For each of the following elementary CF descriptions a table is added which summarises the dependencies of the used variables on the geometrical parameters.

Table 6.1. Dependencies of rotation and point transformation matrices on the geometrical parameters and the robot pose

	$\mathbf{x}^m$	$\mathbf{x}^i$	$\mathbf{x}^{dm}$	$\mathbf{x}^{di}$	$\mathbf{t}_d$
${}^g_m\mathbf{R}$, ${}^g_m\mathbf{T}$	x	-	-	-	-
${}^w_g\mathbf{R}$, ${}^w_g\mathbf{T}$	-	-	-	-	x
${}^i_w\mathbf{R}$, ${}^i_w\mathbf{T}$	-	x	-	-	-

6.5.1 Vertex-Face CF

This section describes a wrench spanning set and the (signed) distance function for a vertex-face contact, Fig. 6.7. The wrench spanning set is expressed in a frame {c} in the contact. The dependencies of the used variables on the geometrical parameters and the robot pose are summarised in Tables 6.1 and 6.2.

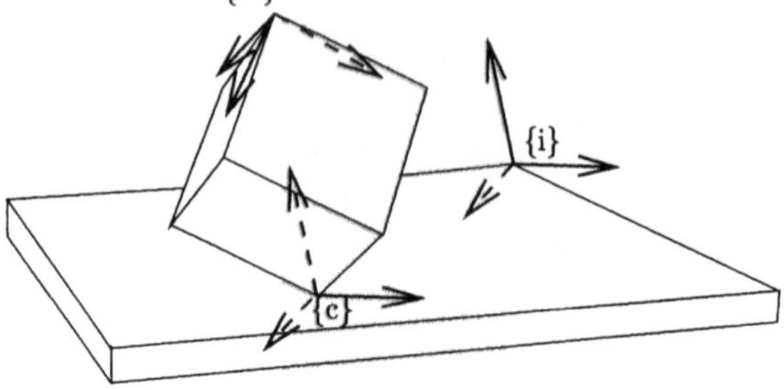

Fig. 6.7. Vertex-face CF

Geometric Description of the Contacting Object Features

The position of the contacting *vertex* in {m}, ${}_m\mathbf{p}^{m,c}$, is a function of the inaccurately known dimensions $\mathbf{x}^{dm}$ of the manipulated object. The *face* is defined in {i} by a normal vector $\mathbf{n}$ and a (signed) distance to the origin d. Both are functions of the inaccurately known dimensions $\mathbf{x}^{di}$ of the ith environment object. A point y corresponds to the contacting face if $\mathbf{n}^T\ {}_i\mathbf{p}^{i,y} = d$.

Definition of a Frame {c}

Consider a frame {c} in the contact point with axes parallel to the axes of {i}, see Fig. 6.7. The transformation of the {c} frame with respect to the {m} frame is defined by:

1. the position of the origin of {c} in {m}, ${}_m\mathbf{p}^{m,c}$, which is the position of the contacting vertex.
2. the rotation matrix from {c} to {m} ${}^c_m\mathbf{R}$ which equals ${}^g_m\mathbf{R}\ {}^w_g\mathbf{R}\ {}^i_w\mathbf{R}\ {}^c_i\mathbf{R}$ with ${}^c_i\mathbf{R} \equiv \mathbf{I}_{3\times3}$.

Description of a Wrench Spanning Set in {c}

The wrench spanning set $\mathbf{G}^c$ can be written as:

$$\mathbf{G}^c = \begin{bmatrix} \mathbf{n} \\ \mathbf{0}_{3\times1} \end{bmatrix}.$$

Description of the (Signed) Distance Function

The distance from the vertex to the face is:

$$h_d = \frac{\mathbf{n}^T\ {}_i\mathbf{p}^{i,c} - d}{\sqrt{\mathbf{n}^T\mathbf{n}}};$$

where ${}_i\mathbf{p}^{i,c}$ is the vertex expressed in {i}:

$$\begin{bmatrix} {}_i\mathbf{p}^{i,c} \\ 1 \end{bmatrix} = {}^w_i\mathbf{T}\,{}^g_w\mathbf{T}\,{}^m_g\mathbf{T} \begin{bmatrix} {}_m\mathbf{p}^{m,c} \\ 1 \end{bmatrix}.$$

Note that also the less complicated equation

$$\mathbf{n}^T\ {}_i\mathbf{p}^{i,c} - d = 0;$$

is a valid closure equation.

Table 6.2. Vertex-face CF: summary of the dependencies of the used variables on the geometrical parameters and the robot pose

	$\mathbf{x}^m$	$\mathbf{x}^i$	$\mathbf{x}^{dm}$	$\mathbf{x}^{di}$	$\mathbf{t}_d$
${}_m\mathbf{p}^{m,c}$	-	-	x	-	-
n	-	-	-	x	-
d	-	-	-	x	-
${}^c_m\mathbf{R}$	x	x	-	-	x
$\mathbf{G}^c$	-	-	-	x	-
${}_i\mathbf{p}^{i,c}$	x	x	x	-	x
h_d	x	x	x	x	x

6.5.2 Edge-Edge CF

This section describes a wrench spanning set and the (signed) distance function for an edge-edge contact, Fig. 6.8. The wrench spanning set is expressed in a frame {c} in the contact. The dependencies of the used variables

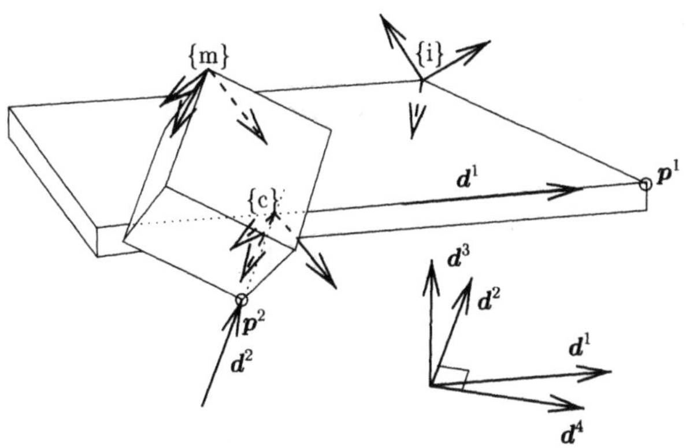

Fig. 6.8. Edge-edge CF

on the geometrical parameters and the robot pose are summarised in Tables 6.1 and 6.3.

Geometric Description of the Contacting Object Features

The *contacting edge of the manipulated object* is described by a point $\mathbf{p}^{2m}$ on the edge and a vector $\mathbf{d}^{2m}$ parallel to the edge. Both are expressed in {m} and are functions of the dimensions $\mathbf{x}^{dm}$ of the manipulated object. The *contacting edge of the ith environment object* is given by a point $\mathbf{p}^{1i}$ on the edge and a vector $\mathbf{d}^{1i}$ parallel to the edge. Both are expressed in {i} and are functions of the dimensions $\mathbf{x}^{di}$ of the ith environment object.

Definition of a Frame {c}

Consider a frame {c} in the contact point with axes parallel to the axes of {m}, see Fig. 6.8. The transformation of the {c} frame with respect to the {m} frame is defined by:

1. the position of the origin of {c} in {m} ${}_m\mathbf{p}^{m,c}$, which is given below in (6.8);
2. the rotation matrix from {c} to {m} ${}^c_m\mathbf{R} = \mathbf{I}_{3\times 3}$.

The point $\mathbf{p}^{1i}$ and the vector $\mathbf{d}^{1i}$ expressed with respect to {m} are:

$$\begin{bmatrix}\mathbf{p}^{1m}\\ 1\end{bmatrix} = {}^g_m\mathbf{T}\,{}^w_g\mathbf{T}\,{}^i_w\mathbf{T}\begin{bmatrix}\mathbf{p}^{1i}\\ 1\end{bmatrix};$$
$$\mathbf{d}^{1m} = {}^g_m\mathbf{R}\,{}^w_g\mathbf{R}\,{}^i_w\mathbf{R}\,\mathbf{d}^{1i}.$$

Both edges together form the "contacting face". The vector perpendicular to both edges (i.e., perpendicular to the contacting face) in {m} is $\mathbf{d}^{3m} = \mathbf{d}^{1m} \times \mathbf{d}^{2m}$. $\mathbf{d}^{4m} = \mathbf{d}^{2m} \times \mathbf{d}^{3m}$ is the vector perpendicular to $\mathbf{d}^{2m}$ *in* the contacting face. The origin of the frame {c} lies on the contacting edge of the environment object:

$${}_m\mathbf{p}^{m,c} = \mathbf{p}^{1m} + \alpha\mathbf{d}^{1m}.$$

α is calculated based on the equivalence of the projection of $\mathbf{p}^2$ and of ${}_m\mathbf{p}^{m,c}$ on $\mathbf{d}^4$:

$$(\mathbf{d}^{4m})^T\mathbf{p}^{2m} = \mathbf{d}^{4mT}(\mathbf{p}^{1m} + \alpha\mathbf{d}^{1m});$$
$$\alpha = \frac{(\mathbf{d}^{4m})^T(\mathbf{p}^{2m} - \mathbf{p}^{1m})}{\mathbf{d}^{4mT}\mathbf{d}^{1m}}.$$

Hence, ${}_m\mathbf{p}^{m,c}$ equals

$${}_m\mathbf{p}^{m,c} = \mathbf{p}^{1m} + \frac{(\mathbf{d}^{4m})^T(\mathbf{p}^{2m} - \mathbf{p}^{1m})}{\mathbf{d}^{4mT}\mathbf{d}^{1m}}\mathbf{d}^{1m}. \tag{6.8}$$

Note that the product $\mathbf{d}^{4mT}\mathbf{d}^{1m}$ only equals zero if the contacting edges are parallel. This is a degenerate contact case.

Description of a Wrench Spanning Set in {c}

The wrench spanning set $\mathbf{G}^c$ can be written as:

$$\mathbf{G}^c = \begin{bmatrix} \mathbf{d}^{3m} \\ \mathbf{0}_{3\times 1} \end{bmatrix}.$$

Description of the (Signed) Distance Function

The distance between the two edges is

$$h_d = \frac{\mathbf{d}^{3mT}\mathbf{p}^{2m} - \mathbf{d}^{3mT}\mathbf{p}^{1m}}{\sqrt{\mathbf{d}^{3mT}\mathbf{d}^{3m}}};$$

i.e., the difference in the (signed) distances from both edges to the same face which is parallel to the contacting face. Note that also the less complicated equation

$$\mathbf{d}^{3mT}\mathbf{p}^{2m} - \mathbf{d}^{3mT}\mathbf{p}^{1m} = 0;$$

is a valid closure equation.

Table 6.3. Edge-edge CF: summary of the dependencies of the used variables on the geometrical parameters and the robot pose

	$\mathbf{x}^m$	$\mathbf{x}^i$	$\mathbf{x}^{dm}$	$\mathbf{x}^{di}$	$\mathbf{t}_d$
$\mathbf{p}^{2m}$	-	-	x	-	-
$\mathbf{d}^{2m}$	-	-	x	-	-
$\mathbf{p}^{1i}$	-	-	-	x	-
$\mathbf{d}^{1i}$	-	-	-	x	-
${}_m\mathbf{p}^{m,c}$	x	x	x	x	x
${}^c_m\mathbf{R}$	-	-	-	-	-
$\mathbf{p}^{1m}$	x	x	-	x	x
$\mathbf{d}^{1m}$	x	x	-	x	x
$\mathbf{d}^{3m}$	x	x	x	x	x
$\mathbf{d}^{4m}$	x	x	x	x	x
$\mathbf{G}^c$	x	x	x	x	x
h_d	x	x	x	x	x

6.5.3 Face-Vertex CF

This section describes a wrench spanning set and the (signed) distance function for a face-vertex contact, Fig. 6.9. The wrench spanning set is expressed in a frame {c} in the contact. The dependencies of the used variables on the geometrical parameters and the robot pose are summarised in Tables 6.1 and 6.4.

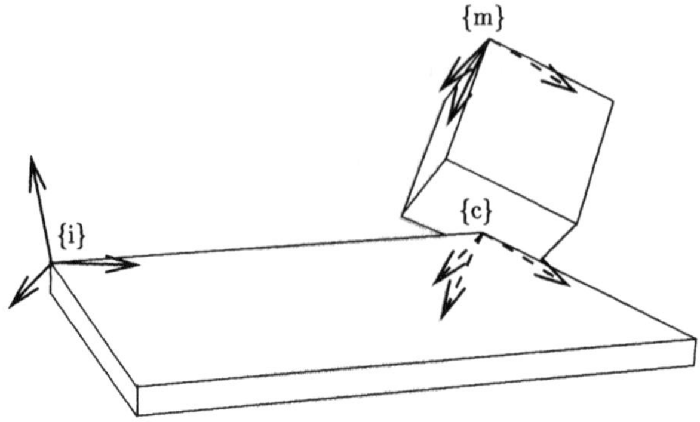

Fig. 6.9. Face-vertex CF

Geometric Description of the Contacting Object Features

The *face* is defined in {m} by a normal vector $\mathbf{n}$ and a (signed) distance to the origin d. Both are functions of the inaccurately known dimensions $\mathbf{x}^{dm}$ of the manipulated object. A point y corresponds to the contacting face if $\mathbf{n}^T\, {}_m\mathbf{p}^{m,y} = d$. The position of the contacting *vertex* ${}_i\mathbf{p}^{i,c}$ in {i} is a function of the inaccurately known dimensions $\mathbf{x}^{di}$ of the ith environment object.

Definition of a Frame {c}

Consider a frame {c} in the contact point with axes parallel to the axes of {m}, see Fig. 6.7. The transformation of the {c} frame with respect to the {m} frame is defined by:

1. the position of the origin of {c} in {m}, ${}_m\mathbf{p}^{m,c}$,

$$ {}_m\mathbf{p}^{m,c} = {}^{g}_{m}\mathbf{T}\, {}^{w}_{g}\mathbf{T}\, {}^{i}_{w}\mathbf{T}\, {}_i\mathbf{p}^{i,c}; \tag{6.9} $$

2. the rotation matrix from {c} to {m} ${}^{c}_{m}\mathbf{R} = \mathbf{I}$.

Description of a Wrench Spanning Set in {c}

The wrench spanning set $\mathbf{G}^c$ can be written as:

$$ \mathbf{G}^c = \begin{bmatrix} \mathbf{n} \\ \mathbf{0}_{3\times 1} \end{bmatrix}. $$

Description of the (Signed) Distance Function

The distance from the vertex to the face is:

$$h_d = \frac{\mathbf{n}^T \; {}_m\mathbf{p}^{m,c} - d}{\sqrt{\mathbf{n}^T\mathbf{n}}}.$$

Note that also the less complicated equation

$$\mathbf{n}^T \; {}_m\mathbf{p}^{m,c} - d = 0;$$

is a valid closure equation.

Table 6.4. Face-vertex CF: summary of the dependencies of the used variables on the geometrical parameters and the robot pose

	$\mathbf{x}^m$	$\mathbf{x}^i$	$\mathbf{x}^{dm}$	$\mathbf{x}^{di}$	$\mathbf{t}_d$
n	-	-	x	-	-
d	-	-	x	-	-
${}_i\mathbf{p}^{i,c}$	-	-	-	x	-
${}_m\mathbf{p}^{m,c}$	x	x	-	x	x
${}^c_m\mathbf{R}$	-	-	-	-	-
$\mathbf{G}^c$	-	-	x	-	-
h_d	x	x	x	x	x

6.6 Non-Polyhedral Elementary CFs

The previous section describes the polyhedral elementary CFs. Of course, the elementary contact library can be augmented with *non-polyhedral* CFs. Contacts between generally curved surfaces can be added if *analytical* surface descriptions are available which allow an analytical description of the contact point(s) and normal(s). This section presents the models of the non-polyhedral elementary CFs used in the examples, i.e., the vertex-cylinder CF, the ball-face CF and the ball-cylinder CF.

The structure of the CF descriptions is similar to the one used in the previous section: each elementary CF is described by (i) a frame {c} on the contact, (ii) a wrench spanning set expressed in this frame {c}, and (iii) the (signed) distance function between the two contacting features.

6.6.1 Vertex-Cylinder CF

This section describes a wrench spanning set and the (signed) distance function for a vertex-cylinder contact, Fig. 6.10. The wrench spanning set is expressed in a frame {c} in the contact. The dependencies of the used variables on the geometrical parameters and the robot pose are summarised in Tables 6.1 and 6.5.

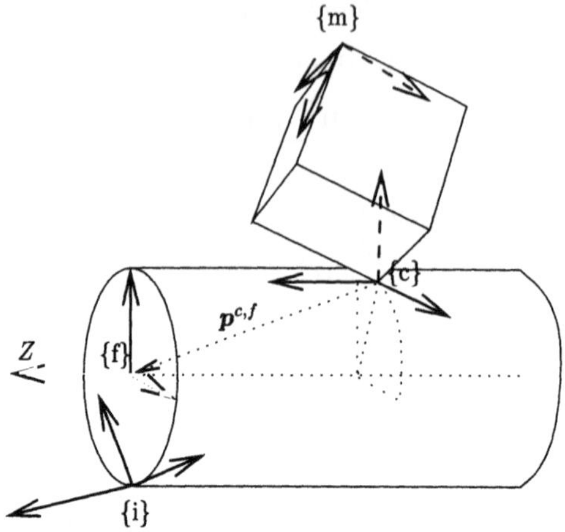

Fig. 6.10. Vertex-cylinder CF

Geometric Description of the Contacting Object Features

The position of the contacting *vertex* ${}_m\mathbf{p}^{m,c}$ in {m} is a function of the inaccurately known dimensions $\mathbf{x}^{dm}$ of the manipulated object. The *cylinder* is defined by the rotation of its axis[2] ${}^f_i\mathbf{R}$, the coordinate of the centre bottom point ${}_i\mathbf{p}^{i,f}$, its radius r and height h. All of these can be a function of the inaccurately known dimensions $\mathbf{x}^{di}$ of the ith environment object.

Definition of a Frame {c}

Consider a frame {c} in the contact point with axes parallel to the axes of the cylinder frame {f}, see Fig. 6.10. The transformation of the {c} frame with respect to the {m} frame is defined by:

1. the position of the origin of {c} in {m} ${}_m\mathbf{p}^{m,c}$, i.e., the position of the contacting vertex;
2. the rotation matrix from {c} to {m}: ${}^c_m\mathbf{R} = {}^g_m\mathbf{R}\ {}^w_g\mathbf{R}\ {}^i_w\mathbf{R}\ {}^f_i\mathbf{R}\ {}^c_f\mathbf{R}$; with ${}^c_f\mathbf{R} = \mathbf{I}$.

Description of a Wrench Spanning Set in {c}

The wrench spanning set $\mathbf{G}^c$ indicates that the contact force is normal to the cylinder surface. This normal can be described by the first two components of ${}_c\mathbf{p}^{c,f}$:

[2] The formulas here assume that the frame {f} is defined such that the Z-axis lies along the cylinder axis, see Fig. 6.10. The orientation of the axis with respect to {i} can be described by two parameters (the orientation around the cylinder axis is not relevant). Independent of the choice of the representation of this orientation, the rotation matrix ${}^f_i\mathbf{R}$ can always be calculated.

$$\mathbf{G}^c = \begin{bmatrix} {}_c p_x^{c,f} \\ {}_c p_y^{c,f} \\ \mathbf{0}_{4\times 1} \end{bmatrix};$$

where

$$ {}_c\mathbf{p}^{c,f} = {}^f_c\mathbf{R}\,{}^i_f\mathbf{R}\left({}_i\mathbf{p}^{i,f} - {}_i\mathbf{p}^{i,c}\right); \tag{6.10}$$

$$\begin{bmatrix} {}_i\mathbf{p}^{i,c} \\ 1 \end{bmatrix} = {}^w_i\mathbf{T}\,{}^g_w\mathbf{T}\,{}^m_g\mathbf{T}\begin{bmatrix} {}_m\mathbf{p}^{m,c} \\ 1 \end{bmatrix}. \tag{6.11}$$

Description of the (Signed) Distance Function

The distance between the vertex and the cylinder is

$$h_d = \text{sign}(d)\sqrt{|d|}; \tag{6.12}$$

where

$$d = \left({}_c p_x^{c,f}\right)^2 + \left({}_c p_y^{c,f}\right)^2 - r^2. \tag{6.13}$$

Note that also the less complicated equation

$$\left({}_c p_x^{c,f}\right)^2 + \left({}_c p_y^{c,f}\right)^2 - r^2 = 0$$

is a valid closure equation.

Table 6.5. Vertex-cylinder CF: summary of the dependencies of the used variables on the geometrical parameters and the robot pose

	$\mathbf{x}^m$	$\mathbf{x}^i$	$\mathbf{x}^{dm}$	$\mathbf{x}^{di}$	$\mathbf{t}_d$
${}_m\mathbf{p}^{m,c}$	-	-	x	-	-
${}^f_i\mathbf{R}$	-	-	-	x	-
${}_i\mathbf{p}^{i,f}$	-	-	-	x	-
r	-	-	-	x	-
h	-	-	-	x	-
${}^c_m\mathbf{R}$	x	x	-	x	x
$\mathbf{G}^c$	x	x	x	x	x
${}_i\mathbf{p}^{i,c}$	x	x	x	-	x
${}_c\mathbf{p}^{c,f}$	x	x	x	x	x
h_d	x	x	x	x	x

6.6.2 Ball-Face CF

The contact between a ball surface and a face (Fig. 6.11) can be simplified to a vertex-face CF with the vertex located at the ball's centre point and the face translated to the ball's centre point. The distance d of the face changes with the ball's radius, hence, it is a function of $\mathbf{x}^{dm}$. The vertex-face CF is described in Sect. 6.5.1. For a ball-face CF, Table 6.2, which describes the dependencies of the used variables on the geometrical parameters and the robot pose, is replaced by Table 6.6.

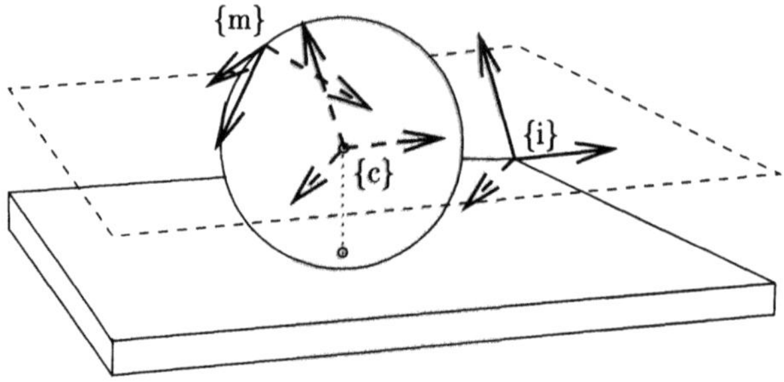

Fig. 6.11. Ball-face CF

Table 6.6. Ball-face CF: summary of the dependencies of the used variables on the geometrical parameters and the robot pose

	$\mathbf{x}^m$	$\mathbf{x}^i$	$\mathbf{x}^{dm}$	$\mathbf{x}^{di}$	$\mathbf{t}_d$
${}_m\mathbf{p}^{m,c}$	-	-	x	-	-
n	-	-	-	x	-
d	-	-	x	x	-
${}^c_m\mathbf{R}$	x	x	-	-	x
$\mathbf{G}^c$	-	-	-	x	-
${}_i\mathbf{p}^{i,c}$	x	x	x	-	x
h_d	x	x	x	x	x

6.6.3 Ball-Cylinder CF

The contact between a ball and a cylinder surface (Fig. 6.12) can be simplified to a vertex-cylinder CF. The vertex is located at the ball's centre point. The cylinder's radius is made larger/smaller with the ball's radius, hence, the cylinder radius r is a function of $\mathbf{x}^{dm}$. The vertex-cylinder CF is described in Sect. 6.6.1. For a ball-cylinder CF, Table 6.5, which describes the dependencies of the used variables on the geometrical parameters and the robot pose, is replaced by Table 6.7.

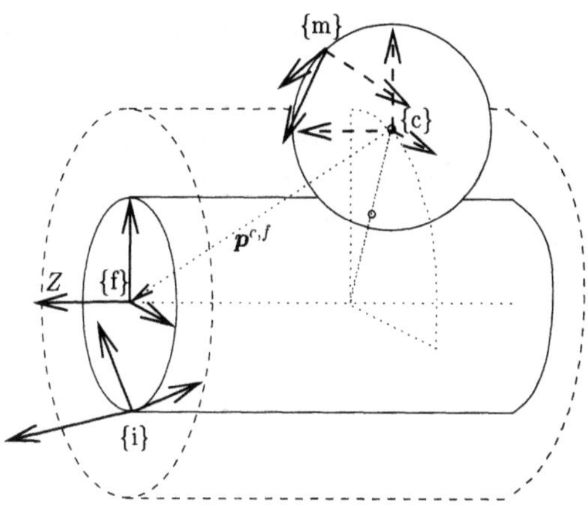

Fig. 6.12. Ball-cylinder CF

Table 6.7. Ball-cylinder CF: summary of the dependencies of the used variables on the geometrical parameters and the robot pose

	$\mathbf{x}^m$	$\mathbf{x}^i$	$\mathbf{x}^{dm}$	$\mathbf{x}^{di}$	$\mathbf{t}_d$
${}_m\mathbf{p}^{m,c}$	-	-	x	-	-
${}^f_i\mathbf{R}$	-	-	-	x	-
${}_i\mathbf{p}^{i,f}$	-	-	-	x	-
r	-	-	x	x	-
h	-	-	-	x	-
${}^c_m\mathbf{R}$	x	x	-	x	x
$\mathbf{G}^c$	x	x	x	x	x
${}_i\mathbf{p}^{i,c}$	x	x	x	-	x
${}_c\mathbf{p}^{c,f}$	x	x	x	x	x
h_d	x	x	x	x	x

6.7 Examples

Figure 6.13 shows the *polyhedral* CF of a ladder leaning against a wall. The CF is modelled by two vertex-face CFs on the ground and two edge-edge CFs on top of the wall. Figures 6.14 and 6.15 illustrate the two degrees of freedom of the ladder that do not brake the contact. The figures are obtained by projecting a desired motion onto the twist space of the CF.

Figure 6.16 shows the motion constraint for a *non-polyhedral* CF: a ball slides along a cylinder surface. The five remaining degrees of freedom of the ball are two translational motion directions along the cylinder surface and the three rotational motion degrees of freedom. Figure 6.17 shows a motion allowed by the contact.

Finally, Figs. 6.18 and 6.19 show the modelling of a CF with *both polyhedral and non-polyhedral contacting elements.* The stick makes a vertex-face CF with one end and a vertex-cylinder CF with the other. Figure 6.19 shows a motion in the twist space of this CF.

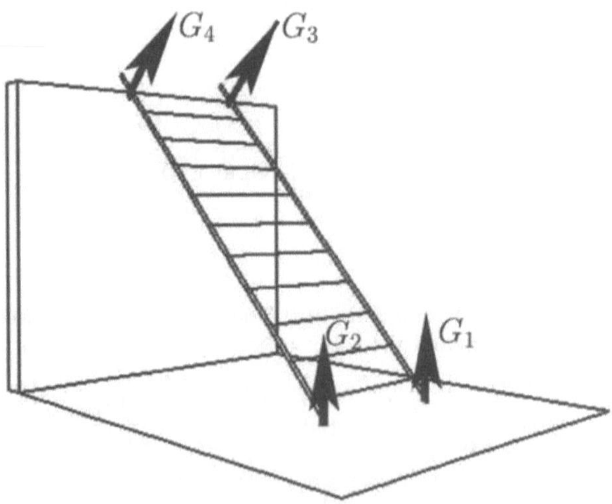

Fig. 6.13. Ladder leaning against a wall. The CF consists of two vertex-face CFs on the ground and two edge-edge CFs on top of the wall

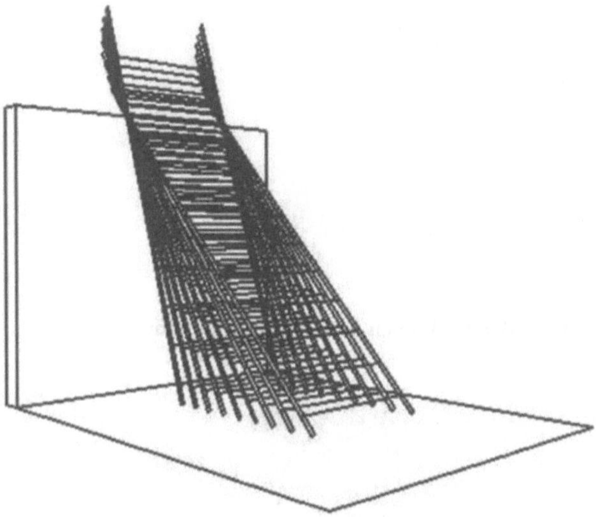

Fig. 6.14. Allowed motion of the ladder

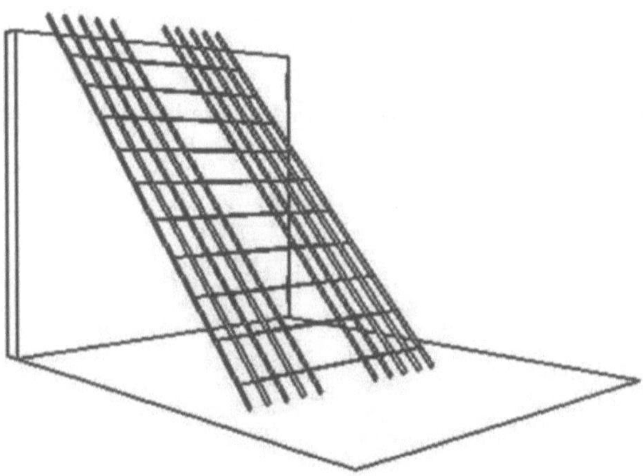

Fig. 6.15. Allowed motion of the ladder

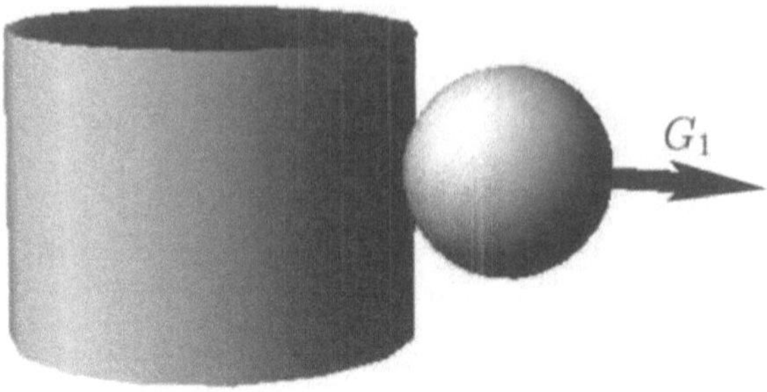

Fig. 6.16. Wrench base for a ball-cylinder contact

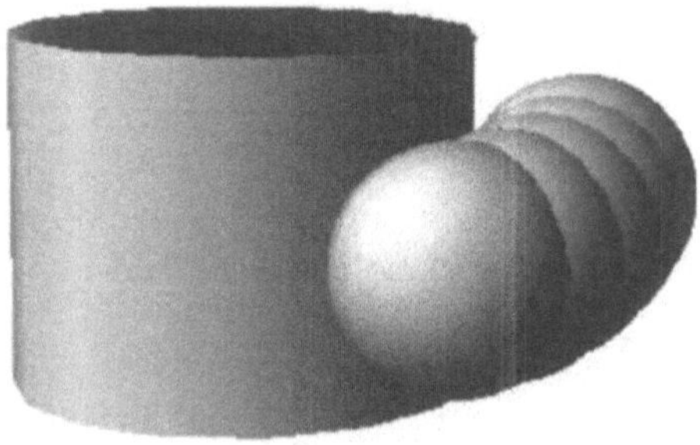

Fig. 6.17. Motion of the ball allowed by the ball-cylinder contact

Fig. 6.18. Combination of a polyhedral and a non-polyhedral contact. The stick has a motion constraint on each of its ends

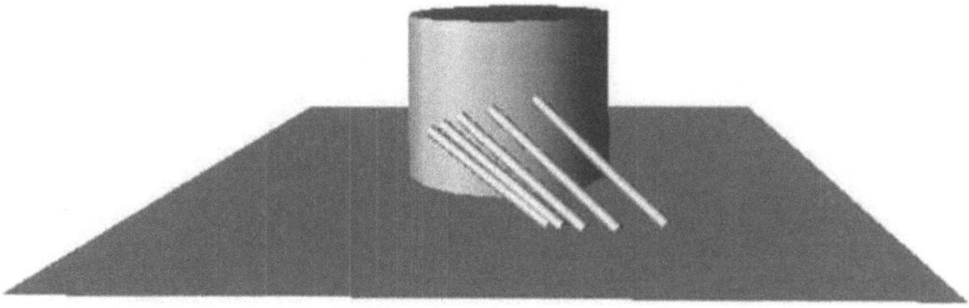

Fig. 6.19. Motion of the stick allowed by both contacts

6.8 Conclusions

This chapter presents the latest evolution in modelling the closure, twist and wrench contact equations as a function of the geometrical parameters:

1. The "Virtual Contact Manipulator", describing the *relative* degrees of freedom between the manipulated object and the environment objects, is not needed. This results in a considerably *reduced modelling effort* and a *minimal set of closure equations.*
   The presented work does not need the Virtual Contact Manipulator because the contact points and normals at every time instant are expressed as a function of the geometrical parameters and the manipulated object pose. Previous research modelled the contact points and normals as a function of the joint values of the Virtual Contact Manipulator.
2. The closure, twist and wrench equations for multiple-contact CFs are automatically generated from a *library of elementary CFs.* E.g., polyhedral CFs can be decomposed into elementary vertex-face, face-vertex and edge-edge CFs.
   Contacts between generally curved surfaces can be added if *analytical* surface descriptions are available which allow an analytical description of the contact point(s) and normal(s).

7

Geometrical Parameter Estimation and CF Recognition

You can observe a lot just by watching.

Yogi Berra, (1925–).

7.1 Introduction

This chapter presents the estimator component of the autonomous compliant motion system (see Fig. 1.2). This component

1. estimates the geometrical parameters (positions, orientations and dimensions of the contacting objects) during autonomous compliant motion.
2. monitors CF transitions and recognises the CF after a transition.

The estimation is based on measurements of the manipulated object *pose* and *twist* and of the contact *wrench*.

The filter equations are *automatically generated* based on the elementary contact library. To this aim, the filters use a *reciprocity based* twist measurement equation and a *consistency based* wrench measurement equation. This avoids the use of the twist spanning set for which an analytical description needs to be derived manually for each CF. The application of the consistency based wrench measurement equation additionally results in a *wrench decomposition* by the estimator. The decomposition can be used as feedback to the force controller.

The nonlinear estimation problem under *large uncertainty* is solved by the *Non-minimal State Kalman Filter*. The possibility of estimation under large uncertainty expands the application area of the explicit contact models to the modelling of the contacting objects "from scratch". The objects are then represented by an unknown number of features (such as vertices, edges and faces for polyhedral objects) with inaccurately known positions, orientations and dimensions.

The *propagation of the estimates* from one CF to the next is straightforward because we always consider all parameters, even the instantaneously unobservable ones, in the estimator.

The Kalman Filter (KF) algorithm is shown to be *robust against non-minimal CF modelling.* We additionally show that considering all parameters does not increase the *computational load* of the KF estimator.

Finally, this chapter presents also *solutions to the practical problems* of (i) the instability of the contact after a transition, and (ii) the numerical instability of the KF algorithm when a part of the parameter values are observed very accurately while others are still inaccurately known.

Section 7.2 describes the measurement models used by the estimator component. The estimation of the geometrical parameters is described in Sect. 7.3. Inconsistency of the measurements with the models is used to monitor CF transitions and to recognise the CF, Sect. 7.4. Section 7.5 concludes. Experimental results are presented in the next chapter.

7.2 Measurement Models

The previous chapter presented the contact models for poses, twists and wrenches. The *closure equation* is given by (6.3),

$$\mathbf{h}_{d,k}(\mathbf{x}, \mathbf{t}_{d,k}) = \mathbf{0}_{n_c \times 1}.$$

For the *twist and wrench measurement equations*, two equivalent models are available: one based on consistency and one based on reciprocity. Section 7.2.1 presents our choice of using the *reciprocity based twist equation* and the *consistency based wrench equation.*

Because the pose and twist measurements are both derived from the robot joint measurements, they are interdependent. This is explained in Sect. 7.2.2. Section 7.2.3 discusses the uncertainty on the measurement models.

7.2.1 Twist and Wrench Measurement Models

The generation of the contact models from an elementary contact library results in an *analytical* description of a *wrench spanning set* and a set of contact *constraints* as functions of the geometrical parameters and the robot pose. For *given* values of the parameters and the pose, the *twist spanning set* can be derived *numerically*, based on the numerical value of the wrench spanning set (Sect. 6.4.1).

The KF algorithm requires the derivative of the measurement models with respect to the geometrical parameters. The choice of a *reciprocity based twist equation* (6.4)

$$(\mathbf{G}_k(\mathbf{x}, \mathbf{t}_{d,k}))^T \, \mathbf{t}_k = \mathbf{0}_{n_G \times 1};$$

and a *consistency based wrench equation* (6.7)

$$\mathbf{G}_k(\mathbf{x}, \mathbf{t}_{d,k})\boldsymbol{\phi}_k - \mathbf{w}_k = \mathbf{0}_{6 \times 1};$$

avoids the use of the twist spanning set $\mathbf{J}_k(\mathbf{x}, \mathbf{t}_{d,k})$, for which no analytical description is available.

Working with the consistency based wrench equation results in the additional estimation of the vector $\boldsymbol{\phi}_k$ describing the linear dependence of the wrench $\mathbf{w}_k$ on the wrench spanning set $\mathbf{G}_k(\mathbf{x}, \mathbf{t}_{d,k})$. This estimate is sent to the force set-point controller, which, for a minimal spanning set $\mathbf{G}_k(\mathbf{x}, \mathbf{t}_{d,k})$, independently controls the contact wrenches in each of the individual modelled contacts. If the spanning set $\mathbf{G}_k(\mathbf{x}, \mathbf{t}_{d,k})$ is non-minimal, only resultant wrenches can be estimated and controlled.

7.2.2 Closure Versus Twist Measurement Equations

In our setup, the information in the pose and twist measurements is *not independent* of each other because *both are derived from the robot joint positions.* The calculation of the twists is possible because the successive increments between the robot joint position measurements are small.

The dependence between the pose and the twist is also visible in their measurement models: the "zeroth-order" closure equation expresses that the *contact is established*, while the "first-order" twist equation provides information on how the manipulated object instantaneously moves while *keeping* the contact with the environment object(s).

Figure 7.1 illustrates the interdependence with position vectors and translational motions of a point object. The position vectors $\mathbf{p}_i$ indicate the *location* of the point, the translational motions $\mathbf{v}_i$ indicate how the point *moves* in between two positions. All position vectors hold the same information as one position vector and all velocity vectors.

This work introduces the *closure equation once at the beginning of a CF* and afterwards applies the *twist equation during the CF-compliant motion.* This is equivalent to applying the closure equation at every time step (and never applying the twist equation).

The choice is limited to the use of the closure equation only when no instantaneous twist measurement is available. This is for instance the case if the increments between the available robot joint position measurements are too large to calculate the instantaneous twist. Another application where no instantaneous twist measurement is available, is the estimation of the geometrical parameters based on spurious contacts obtained by "hopping" over the

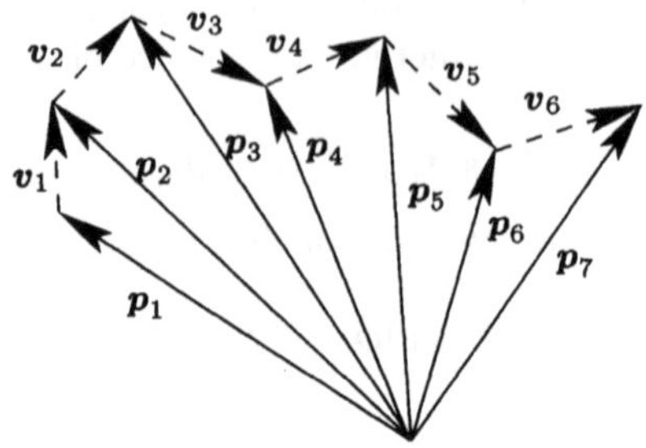

Fig. 7.1. The figure illustrates the interdependence between poses and twists for a point object. The position vectors $\mathbf{p}_i$ indicate a *location* of the point, the translational motions $\mathbf{v}_i$ indicate how the point *moves* in between two positions. All position vectors hold the same information as one position vector and all velocity vectors

environment surfaces, e.g., like a Coordinate Measuring Machine measures different points on a surface. In this case, only the closure equation is applied (at the times of contact) because no motion compliant to the surface is performed.

In other setups, the pose and twist measurements can be (statistically) independent because their measurements originate from different sensors. In such cases, both the closure and twist equations can be applied at all time.

7.2.3 Uncertainty on the Models

For a well-calibrated robot, there is almost no uncertainty on the *pose and twist* measurements. We choose a small uncertainty representing equally small modelling errors on the closure and twist equations (such as small flexibilities). An advantage of considering a small uncertainty compared to *no* uncertainty at all, is that the propagation of this uncertainty to the estimates directly takes the conditioning of the estimation problem into account.

The *wrench measurements* are subject to sensor noise. The covariance of this noise is experimentally determined. Furthermore, the *wrench model* is inaccurate because the assumption of a *friction*less contact is violated in almost all real-world task executions. This work takes the unmodelled friction into account by increasing the experimentally determined wrench covariance matrix.

A more accurate wrench model is obtained by adding only uncertainty $\rho_{m,k}$ in the space of *possible* contact motions[1]

$$\left(\mathbf{I}_{6\times 6} - \mathbf{G}_k(\mathbf{x}, \mathbf{t}_{d,k})(\mathbf{G}_k(\mathbf{x}, \mathbf{t}_{d,k}))^{\#}\right) \rho_{m,k}; \tag{7.1}$$

or even only in the space of *applied* contact motions. However, the *uncertainty* in these models depends on the inaccurately known *geometrical parameters*,

[1] $\mathbf{G}^{\#}$ denotes a generalised inverse of $\mathbf{G}$ [24].

which is not allowed in the applied filters. A possible *approximation* is the use of the *estimated* space of possible contact motions

$$\left(\mathbf{I}_{6\times 6} - \mathbf{G}_k(\hat{\mathbf{x}}, \mathbf{t}_{d,k})(\mathbf{G}_k(\hat{\mathbf{x}}, \mathbf{t}_{d,k}))^{\#}\right) \boldsymbol{\rho}_{m,k}, \tag{7.2}$$

where $\hat{\mathbf{x}}$ denotes an estimate of the geometrical parameter values. This approximation, however, is only valid when the estimate $\mathbf{G}_k(\hat{\mathbf{x}}, \mathbf{t}_{d,k})$ is already accurate.

The estimation process of the geometrical parameters is dominated by the accurate closure and twist equations. The wrench equation dominates only the estimation of the parameter combinations which are not estimated by the closure or twist equations. The estimates of these parameter combinations are less accurate.

Even if the wrench measurements do not observe any other parameter combination than the pose and twist measurements, it is still useful to include the wrench measurements in the estimator:

1. the wrench measurements allow the estimator to detect collisions. When a contact is gained, the wrench measurements are inconsistent with the previous CF model.
2. the estimator performs a wrench decomposition $\boldsymbol{\phi}$ in the wrench spanning set $\mathbf{G}_k(\hat{\mathbf{x}}, \mathbf{t}_{d,k})$. This wrench decomposition is used by the force controller.

7.3 Geometrical Parameter Estimation in a CF

This section describes the KF *estimation of the geometrical parameters* based on measurements during CF-compliant motion. *The CF is assumed to be known.*

Section 7.3.1 discusses the implicit form of the measurement models and the consequences for their use in the KF algorithms. Section 7.3.2 presents the details of the considered algorithms. The algorithms are robust against non-minimal measurement models (Sect. 7.3.3).

When some of the parameter values are observed very accurately and others are still inaccurate, the KF algorithm becomes numerically unstable. We solve this by limiting the accuracy of the estimates (Sect. 7.3.4). Finally, Sect. 7.3.5 describes how the computational load of the KF algorithm can be reduced in case of partial observation of the parameters.

7.3.1 Implicit Measurement Equations

A characteristic of the closure equation and the reciprocity based twist and wrench equations is that they are *implicit*, i.e., of the form $\mathbf{h}_k(\mathbf{x}_k, \mathbf{z}_k) = \mathbf{0}_{n_h \times 1}$ where $\mathbf{z}_k$ represents the measurements $\mathbf{t}_{d,k}$, $\mathbf{t}_k$ and $\mathbf{w}_k$ and where n_h is the number of measurement equations.

Bayesian filtering theory can deal with implicit equations because the measurement update of the pdf, i.e., (3.2), is based on the probability $p(\hat{\mathbf{z}}_k|\mathbf{x}_k)$ of the *measured* value as a function of the state. The KF solutions, however, are derived and only valid for *explicit* estimation problems, i.e., problems of the form $\mathbf{z}_k = \mathbf{H}_k\mathbf{x}_k + \mathbf{d}_k$. In the KF algorithm, this restriction is reflected in the use of the *expected value* of the measurement $\mathbf{H}_k\hat{\mathbf{x}}_k + \mathbf{d}_k$, i.e., the *prediction of the measurement* based on the state estimate $\hat{\mathbf{x}}_k$.

The use of the KF algorithm for implicit measurement equations is made possible by *linearizing* the implicit function $\mathbf{h}_k(\mathbf{x}_k, \mathbf{z}_k)$:

$$\mathbf{h}_k(\mathbf{x}_k, \mathbf{z}_k) \approx \mathbf{D}_k\mathbf{z}_k + \mathbf{H}_k\mathbf{x}_k + \mathbf{d}_k + \boldsymbol{\rho}^*_{m,k} + \mathbf{E}_k\boldsymbol{\rho}_{m,k}. \tag{7.3}$$

E.g., Sorta, Frezza and Perona [238] linearize the equation by a first-order Taylor series around the predicted state estimate $\hat{\mathbf{x}}_{k|k-1}$ and the measurement $\hat{\mathbf{z}}_k$.

$\mathbf{D}_k\mathbf{z}_k$ is considered to be the new measurement, hence, the linearized measurement equation becomes

$$\mathbf{D}_k\mathbf{z}_k = -\mathbf{H}_k\mathbf{x}_k - \mathbf{d}_k - \boldsymbol{\rho}^*_{m,k} - \mathbf{E}_k\boldsymbol{\rho}_{m,k}.$$

The KF update equations (see Chap. 4) for this linearized measurement model are (n is the dimension of the state vector $\mathbf{x}$):

$$\hat{\mathbf{x}}_{k|k} = \hat{\mathbf{x}}_{k|k-1} + \mathbf{K}_k\boldsymbol{\eta}_k; \tag{7.4}$$

$$\begin{aligned}\mathbf{P}_{k|k} &= (\mathbf{I}_{n\times n} - \mathbf{K}_k(-\mathbf{H}_k))\,\mathbf{P}_{k|k-1};\\ &= (\mathbf{I}_{n\times n} + \mathbf{K}_k\mathbf{H}_k)\,\mathbf{P}_{k|k-1};\end{aligned} \tag{7.5}$$

where

$$\begin{aligned}\boldsymbol{\eta}_k &= \mathbf{D}_k\hat{\mathbf{z}}_k - \left(-\mathbf{H}_k\hat{\mathbf{x}}_{k|k-1} - \mathbf{d}_k\right);\\ &= \mathbf{D}_k\hat{\mathbf{z}}_k + \mathbf{H}_k\hat{\mathbf{x}}_{k|k-1} + \mathbf{d}_k;\end{aligned} \tag{7.6}$$

$$\begin{aligned}\mathbf{S}_k &= \mathbf{R}^*_k + \mathbf{E}_k\mathbf{R}_k\mathbf{E}_k^T + (-\mathbf{H}_k)\mathbf{P}_{k|k-1}(-\mathbf{H}_k)^T;\\ &= \mathbf{R}^*_k + \mathbf{E}_k\mathbf{R}_k\mathbf{E}_k^T + \mathbf{H}_k\mathbf{P}_{k|k-1}\mathbf{H}_k^T;\end{aligned} \tag{7.7}$$

$$\begin{aligned}\mathbf{K}_k &= \mathbf{P}_{k|k-1}(-\mathbf{H}_k)^T\mathbf{S}_k^{-1};\\ &= -\mathbf{P}_{k|k-1}\mathbf{H}_k^T\mathbf{S}_k^{-1}.\end{aligned} \tag{7.8}$$

If the uncertainty on the measurement model $\boldsymbol{\rho}_{m,k}$ is due to uncertainty on the measured values $\hat{\mathbf{z}}_k$, then $\mathbf{E}_k = \mathbf{D}_k$.

7.3.2 Kalman Filter Estimators

Estimating the geometrical parameters $\mathbf{x}$ online based on the pose, twist and wrench measurements requires a *recursive estimator*. This section presents the Iterated Extended Kalman Filter (IEKF) which performs well for small uncertainties on the geometrical parameters and the Non-minimal State Kalman Filter (NMSKF) which is computationally more intensive but performs equally well under large uncertainty. The *process* model is linear, hence, the process update is similar for both filters.

The state $\mathbf{x}_k$ in the KF algorithm contains the inaccurately known geometrical parameters $\mathbf{x}^m$, $\mathbf{x}^{dm}$, $\mathbf{x}^1$, $\mathbf{x}^{d1}$, $\mathbf{x}^2$, $\mathbf{x}^{d2}$,... and the wrench decomposition variables $\boldsymbol{\phi}_k$:

$$\mathbf{x}_k = \left[(\mathbf{x}^m)^T \; (\mathbf{x}^{dm})^T \; (\mathbf{x}^1)^T \; (\mathbf{x}^{d1})^T \; (\mathbf{x}^2)^T \; (\mathbf{x}^{d2})^T \ldots \boldsymbol{\phi}_k^T\right]^T.$$

This means that the symbol $\mathbf{x}$, which also represents the vector of geometrical parameters *without* the wrench decomposition variables $\boldsymbol{\phi}$, has a double meaning. The text points out the meaning of $\mathbf{x}$ where confusion is possible.

IEKF Measurement Update

The IEKF linearizes the measurement model (7.3) by a first-order Taylor series around the updated state estimate $\hat{\mathbf{x}}_{k|k}$, which is obtained by iteration. No extra uncertainty is added to compensate for the linearization errors, i.e., $\mathbf{R}_k^* = \mathbf{0}_{n_h \times n_h}$. The algorithm iterates:

$$\begin{aligned}
\mathbf{H}^i &= \left.\frac{\partial \mathbf{h}_k(\mathbf{x}_k, \mathbf{z}_k)}{\partial \mathbf{x}_k}\right|_{\mathbf{x}_k = \hat{\mathbf{x}}^{i-1}; \mathbf{z}_k = \hat{\mathbf{z}}_k}; \\
\mathbf{D}^i &= \left.\frac{\partial \mathbf{h}_k(\mathbf{x}_k, \mathbf{z}_k)}{\partial \mathbf{z}_k}\right|_{\mathbf{x}_k = \hat{\mathbf{x}}^{i-1}; \mathbf{z}_k = \hat{\mathbf{z}}_k}; \\
\Big(\quad \mathbf{d}^i &= \mathbf{h}_k(\hat{\mathbf{x}}^{i-1}, \hat{\mathbf{z}}_k) - \mathbf{D}^i \hat{\mathbf{z}}_k - \mathbf{H}^i \hat{\mathbf{x}}^{i-1}; \quad \Big) \\
\boldsymbol{\eta}^i &= \mathbf{D}^i \hat{\mathbf{z}}_k + \mathbf{H}^i \hat{\mathbf{x}}_{k|k-1} + \mathbf{d}^i; \\
&= \mathbf{h}_k(\hat{\mathbf{x}}^{i-1}, \hat{\mathbf{z}}_k) + \mathbf{H}^i \left(\hat{\mathbf{x}}_{k|k-1} - \hat{\mathbf{x}}^{i-1}\right); \\
\mathbf{S}^i &= \mathbf{D}^i \mathbf{R}_k \mathbf{D}^{iT} + \mathbf{H}^i \mathbf{P}_{k|k-1} \mathbf{H}^{iT}; \\
\mathbf{K}^i &= -\mathbf{P}_{k|k-1} \mathbf{H}^{iT} (\mathbf{S}^i)^{-1}; \\
\hat{\mathbf{x}}^i &= \hat{\mathbf{x}}_{k|k-1} + \mathbf{K}^i \boldsymbol{\eta}^i;
\end{aligned}$$

starting from a value $\hat{\mathbf{x}}^0$ (e.g., chosen equal to $\hat{\mathbf{x}}_{k|k-1}$) until a value $\hat{\mathbf{x}}^i$ is obtained which is close to $\hat{\mathbf{x}}^{i-1}$. The updated state estimate and covariance are:

$$\begin{aligned}
\hat{\mathbf{x}}_{k|k} &= \hat{\mathbf{x}}^i; \\
\mathbf{P}_{k|k} &= \left(\mathbf{I}_{n \times n} + \mathbf{K}^i \mathbf{H}^i\right) \mathbf{P}_{k|k-1}.
\end{aligned}$$

NMSKF Measurement Update

The measurement models corresponding to a CF are linearized by linearizing the wrench spanning sets and closure equations of all elementary CFs in which the CF can be decomposed.

Appendix F describes how for every elementary CF of the contact library a state vector $\mathbf{x}'_j$ can be found. Expressed in this $\mathbf{x}'_j$ both the wrench spanning set $\mathbf{G}_{j,k}(\mathbf{x}'_j, \mathbf{t}_{d,k})$ and the closure equation $h_{dj,k}(\mathbf{x}'_j, \mathbf{t}_{d,k}) = 0$ are linear. The measurement equations, and hence also the linearizing $\mathbf{x}'_j$, depend on the frames in which the measurements are expressed. The appendix presents the detailed formulas for a vertex-face CF when the measurements are expressed in the gripper frame {g}. These formulas are used in the experiment of Chap. 8.

Define the vector $\mathbf{x}'_{CF}$ as the vector containing the $\mathbf{x}'_j$ $(1 \leq j \leq l)$ of the l elementary contacts in which the CF can be decomposed, i.e.,

$$\mathbf{x}'_{CF} = \left[(\mathbf{x}'_1)^T \ldots (\mathbf{x}'_l)^T\right]^T.$$

Define the state vector $\mathbf{x}'_k$ as the vector containing $\mathbf{x}'_{CF}$ and the wrench decomposition $\boldsymbol{\phi}_k$:

$$\mathbf{x}'_k = \left[(\mathbf{x}'_{CF})^T \; (\boldsymbol{\phi}_k)^T\right]^T = \left[(\mathbf{x}'_1)^T \ldots (\mathbf{x}'_l)^T \; (\boldsymbol{\phi}_k)^T\right]^T.$$

Expressed in $\mathbf{x}'_k$, the closure, twist and wrench equations

$$\begin{aligned}
\mathbf{h}_{d,k}(\mathbf{x}'_{CF}, \mathbf{t}_{d,k}) &= \mathbf{0}_{n_c \times 1};\\
\mathbf{G}_k(\mathbf{x}'_{CF}, \mathbf{t}_{d,k}))^T \mathbf{t}_k &= \mathbf{0}_{n_G \times 1};\\
\mathbf{G}_k(\mathbf{x}'_{CF}, \mathbf{t}_{d,k})\boldsymbol{\phi}_k - \mathbf{w}_k &= \mathbf{0}_{6 \times 1};
\end{aligned}$$

are *quasi-linear.* The closure and twist equations are linear in $\mathbf{x}'_k$, but still implicit. The wrench equation contains products of the state variables (products of variables of $\mathbf{x}'_{CF}$ and $\boldsymbol{\phi}_k$). No effort is done to search for a state $\mathbf{x}'_k$, expressed in which $\mathbf{G}_k(\mathbf{x}'_{CF}, \mathbf{t}_{d,k})\boldsymbol{\phi}_k$ is linear. The reason is that no process update is available for the state variables $\mathbf{x}'_k$ which are a function of the wrench decomposition variables $\phi_{j,k}$. By describing the $\phi_{j,k}$ variables as a separate part of $\mathbf{x}'_k$, we do not loose any information about the geometrical parameters when, in the process update step, we make the estimates of the wrench decomposition variables $\phi_{j,k}$ uncertain.

Because the measurement equations are still *implicit* and only *quasi-linear*, the KF algorithm to calculate $\hat{\mathbf{x}}'_{k|k}$ and $\mathbf{P}'_{k|k}$ is performed by an IEKF. I.e., the estimation is performed with the above described algorithm where $\mathbf{x}$ and $\mathbf{P}$ are replaced with $\mathbf{x}'$ and $\mathbf{P}'$.

The pdf over the state $\mathbf{x}$ at each time step k is given by (5.9):

$$p(\mathbf{x}_k|\hat{\mathbf{Z}}_k) \sim \exp\Big(-\frac{1}{2}\big(\mathbf{g}(\mathbf{x}_k) - \hat{\mathbf{x}}'_{k|k}\big)^T \mathbf{P}'^{-1}_{k|k} \big(\mathbf{g}(\mathbf{x}_k) - \hat{\mathbf{x}}'_{k|k}\big)\Big).$$

The experiment in Chap. 8 shows the good performance of the quasi-linear filter for large uncertainties on the geometrical parameters.

Process Update

The part of the state corresponding to the *geometrical parameters* is static, i.e., the true values of the geometrical parameters do not change over time.

The part of the state corresponding to the *wrench decomposition* $\boldsymbol{\phi}$ can not be predicted because no model of the contact dynamics is used. The process update for this part of the state makes the estimates $\boldsymbol{\phi}_{j,k}$ uncertain at each time step k. I.e., the estimate for $\boldsymbol{\phi}_{j,k}$ is determined by the measurement $\mathbf{z}_k$ only, and not by the previous estimate $\boldsymbol{\phi}_{j,k-1}$. To this end, the estimates and their covariances are reset at each time step to the initial estimates $\boldsymbol{\phi}_{0|0}$ and covariances $\mathbf{P}_{\phi,0|0}$.

The process update algorithm for the IEKF is

$$\begin{aligned}
\hat{\mathbf{x}}_{k|k-1} &= \mathbf{F}\,\hat{\mathbf{x}}_{k-1|k-1} + \mathbf{b};\\
\mathbf{P}_{k|k-1} &= \mathbf{F}\,\mathbf{P}_{k-1|k-1}\,\mathbf{F}^T + \mathbf{Q};
\end{aligned}$$

where

$$\mathbf{F} = \begin{bmatrix} \mathbf{I}_{r\times r} & \mathbf{0}_{r\times(n-r)} \\ \mathbf{0}_{(n-r)\times r} & \mathbf{0}_{(n-r)\times(n-r)} \end{bmatrix}; \qquad \mathbf{b} = \begin{bmatrix} \mathbf{0}_{r\times r} \\ \boldsymbol{\phi}_{0|0} \end{bmatrix};$$

$$\mathbf{Q} = \begin{bmatrix} \mathbf{0}_{r\times r} & \mathbf{0}_{r\times(n-r)} \\ \mathbf{0}_{(n-r)\times r} & \mathbf{P}_{\phi,0|0} \end{bmatrix}.$$

r represents the number of state variables corresponding to the geometrical parameters and n is the dimension of the state vector, including the wrench decomposition variables.

The process update algorithm for the NMSKF is similar:

$$\begin{aligned}
\hat{\mathbf{x}}'_{k|k-1} &= \mathbf{F}'\,\hat{\mathbf{x}}'_{k-1|k-1} + \mathbf{b}';\\
\mathbf{P}'_{k|k-1} &= \mathbf{F}'\,\mathbf{P}'_{k-1|k-1}\,\mathbf{F}'^T + \mathbf{Q}';
\end{aligned}$$

where

$$\mathbf{F}' = \begin{bmatrix} \mathbf{I}_{r'\times r'} & \mathbf{0}_{r'\times(n'-r')} \\ \mathbf{0}_{(n'-r')\times r'} & \mathbf{0}_{(n'-r')\times(n'-r')} \end{bmatrix}; \qquad \mathbf{b}' = \begin{bmatrix} \mathbf{0}_{r'\times r'} \\ \boldsymbol{\phi}_{0|0} \end{bmatrix};$$

$$\mathbf{Q}' = \begin{bmatrix} \mathbf{0}_{r'\times r'} & \mathbf{0}_{r'\times(n'-r')} \\ \mathbf{0}_{(n'-r')\times r'} & \mathbf{P}_{\phi,0|0} \end{bmatrix}.$$

r' represents the number of higher-dimensional state variables corresponding to the geometrical parameters and n' is the dimension of the higher-dimensional state vector, including the wrench decomposition variables.

7.3.3 Non-Minimal Contact Description

The *automatic* generation of the measurement models is easier if *non-minimal* contact descriptions are allowed (Sect. 6.4.3). A non-minimal description, however, is only allowed if the estimator using it is able to deal with the non-minimality.

A non-minimal number of elementary CFs results in a non-minimal wrench spanning set and a non-minimal set of closure equations. If the estimator interprets the non-minimal measurement equations as extra equations and *processes the measurement multiple times as independent measurements*, the state estimate is *statistically not correct.* The estimator has to recognise that the measurement equations are non-minimal and has to assure that the interference with the same measurement is not made more than once.

The KF algorithm (and hence the IEKF and NMSKF using this algorithm) recognises a non-minimal set of measurement equations because *the innovation covariance matrix* $\mathbf{S}$ *is singular*, which is never the case for minimal sets. The KF algorithm uses the inverse of this innovation covariance matrix. Whatever generalised inverse $\mathbf{S}^{\#}$ the KF algorithm uses, *the results are the same as the ones obtained by the KF algorithm processing a minimal set of measurement equations*, i.e., the KF algorithm is invariant to non-minimal contact descriptions. Appendix D proves this statement.

A non-minimal wrench spanning set does not only result in non-minimal measurement equations, it also creates extra wrench decomposition variables $\phi_{j,k}$ values in the wrench measurement equation (6.7). The linear combinations of the $\phi_{j,k}$ variables corresponding to the null wrench space do not influence the resultant wrench and are unobservable.

7.3.4 Accuracy Limitation

Large differences in accuracy between the estimates of the observed and the unobserved parameters result in ill-conditioned covariance matrices. In order to avoid numerical problems of the KF algorithm processing these matrices, the accuracy of parameter estimates is limited. This limit should, of course, be chosen smaller than the desired estimation accuracy.

We propose a three-step algorithm to limit the accuracy. The result is invariant to changes in units.

1. For every state variable $\mathbf{x}(i)$, choose an accuracy limit, expressed as a standard deviation $\sigma_{des,i}$. Transform the state $\mathbf{x}$ and its covariance matrix $\mathbf{P}_x$ into a dimensionless state $\mathbf{y} = {}^x_y\mathbf{T}\mathbf{x}$; with covariance $\mathbf{P}_y$,

$$\mathbf{P}_y = {}^x_y\mathbf{T}\,\mathbf{P}_x\,({}^x_y\mathbf{T})^T; \qquad {}^x_y\mathbf{T} = \mathrm{diag}\left(\frac{1}{\sigma_{des,1}}, \ldots, \frac{1}{\sigma_{des,n}}\right).$$

2. The desired accuracy limit in all directions of the $\mathbf{y}$ state space is 1, this is represented by the dashed uncertainty ellipse (a unit circle) in Fig. 7.2 (left). The covariance matrix $\mathbf{P}_y$ needs to be enlarged if the uncertainty $\mathbf{P}_y$ in some directions is smaller than 1. E.g., this is the case for the "original" covariance $\mathbf{P}_y$ whose uncertainty ellipse is drawn in the figure. $\mathbf{P}_y$ is enlarged by setting its eigenvalues which are smaller than 1 to 1:

$$\begin{aligned}\mathbf{P}_y &= \mathbf{U}\boldsymbol{\Sigma}\mathbf{U}^T;\\ \boldsymbol{\Sigma} &= \operatorname{diag}(\sigma_1,\ldots,\sigma_n);\\ \boldsymbol{\Sigma}_{new} &= \operatorname{diag}(\max(1,\sigma_1),\ldots,\max(1,\sigma_n));\\ \mathbf{P}_{y,new} &= \mathbf{U}\boldsymbol{\Sigma}_{new}\mathbf{U}^T.\end{aligned}$$

   The figure shows that the new uncertainty ellipse encloses both the desired minimum uncertainty ellipse (dashed lines) and the original uncertainty ellipse.
3. The resulting uncertainty is transformed back to the state space $\mathbf{x}$

$$\mathbf{P}_{x,new} = {}^{y}_{x}\mathbf{T}\,\mathbf{P}_{y,new}\,({}^{y}_{x}\mathbf{T})^T; \qquad {}^{y}_{x}\mathbf{T} = ({}^{x}_{y}\mathbf{T})^{-1} = \operatorname{diag}(\sigma_{des,1},\ldots,\sigma_{des,n}).$$

   Of course, after transformation, the new uncertainty ellipse still encloses both the desired minimum uncertainty ellipse (dashed line) and the original uncertainty ellipse, Fig. 7.2 (right).

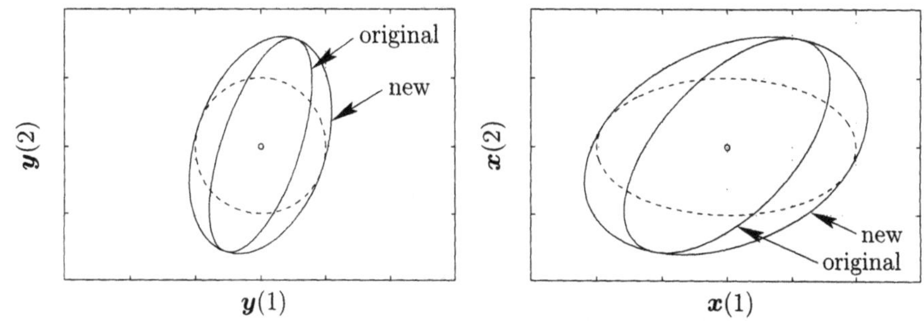

Fig. 7.2. The accuracy of a state estimate $\mathbf{x}$ is limited by enlarging the eigenvalues of the covariance matrix of a transformed state $\mathbf{y}$. The figures illustrate how the uncertainty ellipse changes. The new uncertainty ellipse encloses both the desired minimum uncertainty ellipse (*dashed line*) and the original uncertainty ellipse

The same technique is used to prevent the estimator from becoming too "stiff". I.e., when after a number of measurement updates some estimates are very accurate, they almost do not change any more with future measurements [106]. Generally, this is prevented by adding small process model uncertainty or applying a forgetting factor to the covariance matrix of the estimate, at every time step. These techniques, however, increase the uncertainty of all parameter estimates, even the unobserved (inaccurate) ones.

7.3.5 Partial Observation

The measurement models corresponding to a CF often depend only on part of the state vector. We call the state variables that occur in the measurement equations the *observable state variables* and the others the *unobservable state variables*[2]. Estimators that keep track of the whole state vector are computationally more expensive than estimators that only consider the observable state variables. This section states that in the case of a KF algorithm (and hence the IEKF and NMSKF using this algorithm), the computational load can be reduced to the load of a KF algorithm that only considers the observable state variables. E.g., this is computationally interesting when a lot of environment objects are modelled and only a small number is in contact (is observable).

Lemma 7.1. *The update for the estimate of the observable part of the state vector* $\hat{\mathbf{x}}_o$ *and its covariance matrix* $\mathbf{P}_{oo}$ *are* independent *of the estimate of the unobservable part of the state vector* $\hat{\mathbf{x}}_u$*, its covariance matrix* $\mathbf{P}_{uu}$ *and the correlation* $\mathbf{P}_{uo}$ *between the estimates of the observable and unobservable part.*

This implies that a KF can be run on the reduced state $\mathbf{x}_o$ only. The proof is given in Appendix E.

Lemma 7.2. *Suppose that from a time k until a time k+l only the parameters* $\mathbf{x}_o$ *are observable and a KF calculates* $\hat{\mathbf{x}}_{o,k+l}$ *and* $\mathbf{P}_{oo,k+l}$*. Then the updates for* $\hat{\mathbf{x}}_{u,k+l}$*,* $\mathbf{P}_{uu,k+l}$ *and* $\mathbf{P}_{uo,k+l}$ *are:*

$$\begin{aligned}\hat{\mathbf{x}}_{u,k+l} &= \hat{\mathbf{x}}_{u,k-1} + \mathbf{L}(\hat{\mathbf{x}}_{o,k+l} - \hat{\mathbf{x}}_{o,k-1});\\ \mathbf{P}_{uo,k+l} &= \mathbf{L}\mathbf{P}_{oo,k+l};\\ \mathbf{P}_{uu,k+l} &= \mathbf{P}_{uu,k-1} - \mathbf{L}(\mathbf{P}_{oo,k-1} - \mathbf{P}_{oo,k+l})\mathbf{L}^T;\end{aligned}$$

where $\mathbf{L} = \mathbf{P}_{uo,k-1}\mathbf{P}^{-1}_{oo,k-1}$.

Appendix E contains the proof.

The importance of Lemma 7.2 is that the unobservable part of the state vector and the corresponding covariances can be updated at a CF transition; at that time *all* estimates and covariances are passed on to the estimator corresponding to the next CF (see Sect. 7.4.4).

---

[2] For measurement models which are linear in the state, we can define observable and unobservable *linear combinations* of the state.

For measurement models which are a nonlinear function of the state, the linearization used in the KF algorithm depends on the state *estimate*. Due to changing estimates, $\mathbf{H}_k$ changes and other linear combinations of the observable state variables are observed at each time step.

7.4 Detection of CF Transitions and CF Recognition

The previous section described the estimation of the geometrical parameters during a single CF. This section describes (i) how CF transitions are detected (Sect. 7.4.1), (ii) how the transient control behaviour after these transitions is monitored (Sect. 7.4.2), (iii) how, after a transition detection, the new CF is recognised (Sect. 7.4.3), and (iv) how the propagation of the geometrical parameter estimates from one CF to the next one is performed (Sect. 7.4.4).

7.4.1 Detection of CF Transitions

Each CF corresponds to different measurement models. An estimator recognises a CF *transition* by the statistical inconsistency of the measurements with the current model, i.e., the measurements do not "fit" any more into the CF model and this cannot be explained by the uncertainties on the model or on the parameter estimates. E.g., when a contact is gained, the wrench measurement is inconsistent with the previous measurement model (a wrench is measured in a direction of motion freedom); when a contact is lost, the twist measurement is inconsistent with the previous measurement model (the measured twist deviates from the motion freedom allowed by the contact).

One of the consistency checks of the measurements with respect to the measurement equations in the KF estimator is the *SNIS consistency test*. The Normalised Innovation Squared (NIS) is the square of the innovation $\boldsymbol{\eta}_k$, normalised with the inverse of its covariance $\mathbf{S}_k$:

$$NIS_k = \boldsymbol{\eta}_k^T \mathbf{S}_k^{-1} \boldsymbol{\eta}_k.$$

The SNIS is the Sum over l NIS-values

$$SNIS_k = \sum_{j=k-l+1}^{j=k} \boldsymbol{\eta}_j^T \mathbf{S}_j^{-1} \boldsymbol{\eta}_j.$$

The SNIS value for a linear system subject to Gaussian model uncertainty is χ^2 distributed with $l \times m$ degrees of freedom. l is a trade-off between fast CF transition detection and false alarm. When the SNIS value crosses a chosen confidence boundary, the measurements are said to be inconsistent.

The SNIS-test signals a CF transition one or more time steps *after* it occurred. The state estimate and its covariance which are passed on to the next estimator need to be the ones *before* the CF transition, because introducing inconsistent measurements in the estimator results in bad estimates. A simple solution consists in taking the estimate and covariance corresponding to time step $k - l$ when a CF transition is detected at time step k.

We apply the SNIS consistency test both to the IEKF when dealing with small uncertainties and to the NMSKF when dealing with large uncertainties.

In the latter case, however, the SNIS value checks the consistency of the measurements with respect to the model in $\mathbf{x}'$. This model is more flexible than the contact model because the constraints between the state values $\mathbf{x}'$ are not included. The experiment shows that, by taking a lower confidence boundary, all CF transitions are quickly detected anyway, i.e., the measurements at transitions are inconsistent with this more flexible model too.

7.4.2 Detection of Transient Control Behaviour

Care must be taken not to process any measurements collected during *transient control behaviour*, typically observed after a CF transition. Our quasi-static contact models are not valid during these time instants. The transients show up as oscillations in the measurements and are detected by a check on the measured wrenches. During the transient control behaviour, the state estimate and its uncertainty are kept constant.

If Gaussian distributed wrench uncertainty is assumed, then the value of

$$\sum_{i=k-n_c+1}^{i=k} (\hat{\mathbf{w}}_i - \bar{\mathbf{w}}_i)^T \mathbf{R}_c^{-1} (\hat{\mathbf{w}}_i - \bar{\mathbf{w}}_i) \tag{7.9}$$

is χ^2 distributed with $6n_c$ degrees of freedom. $\mathbf{R}_c$ is the value of the wrench covariance matrix for which the behaviour is said to be transient. n_c is a trade-off between the fast detection of transient behaviour and the false alarm rate. As long as the value of (7.9) is larger than a chosen confidence boundary, we consider the control behaviour as transient and the estimator does not process any measurements. The mean wrench is approximated by the sample mean of the last n_m wrench measurements:

$$\bar{\mathbf{w}}_i = \frac{1}{n_m} \sum_{j=i-n_m+1}^{j=i} \hat{\mathbf{w}}_j.$$

7.4.3 Recognising the CF

For systems with *small* uncertainties, the executed sequence of CFs is assumed to be error-free, i.e., the new CF after a CF transition is the next one in the task plan. This means that after inconsistency detection two CFs are probable: the same CF as before the inconsistency detection (false alarm) and the next CF in the task plan.

For systems with *larger* uncertainties, the true CF sequence can deviate from the planned one. This means that after the inconsistency detection the same CF as before and all of its "neighbouring CFs" are probable. These neighbouring CFs are found in a CF graph [272].

After inconsistency detection and when the contact is stable again, an estimator is constructed for every probable CF and initialised with the current state estimate and uncertainty. The true CF (valid filter) is determined by comparing the probability of the SNIS values of the measurements for each of the filters. Once the probability of the SNIS value of one of the filters exceeds the other probabilities by a user-defined factor, its corresponding CF, its geometrical parameter estimates and their uncertainty are selected for further estimation.

The experiment of the next chapter shows the quick recognition of the CF when the uncertainties on the geometrical parameter estimates are small. For larger uncertainties several CF models stay probable. This result is inherent to the problem and cannot be solved by applying other hypothesis testing techniques. E.g., assume the recognition of a vertex-face CF. The measurements hold the information that we deal with a vertex-face model (one degree of freedom is constrained, five not). With informative measurements, the position of the contacting vertex and the orientation of the contacting face can be identified. However, under large uncertainty of the geometrical parameters, i.e., large uncertainties in the positions, orientations and dimensions of the contacting objects, several of the vertices and faces can correspond to the actual contacting features. Figure 7.3 illustrates this when a vertex-face CF is established with an environment object of which the position and orientation are very uncertain.

The general solution to CF recognition is the calculation of the discrete probability *distribution* over the CFs at all time [105].

7.4.4 Propagation of Geometrical Parameter Estimates

During each CF, different geometrical parameters are observable. The information collected in one CF needs to be propagated to the following CFs. However during these new CFs, some of the previously observable geometrical parameters become unobservable and vice versa. *Propagating* the information through sequences of CFs is done by considering *all geometrical parameters* (even the unobservable ones) at each transition: the full state vector $\mathbf{x}$ and its uncertainty at the end of a CF serve as initial values for the estimator of the next CF. Section 7.3.5 shows that considering unobservable variables does not give any overhead for the KF estimator.

The number of state variables in the KF algorithm is different during different CFs. In the case of the IEKF, only the part of the state corresponding to the wrench decomposition changes. *When a new contact is gained* the state $\mathbf{x}$ is augmented with one variable $\phi_{j,k}$. The covariance matrix $\mathbf{P}$ is augmented accordingly. *When a contact is lost*, the corresponding $\phi_{j,k}$ variable is removed from the state $\mathbf{x}$ and its covariance matrix $\mathbf{P}$. Indeed, the $\phi_{j,k}$ variable does not contain any information which can be useful during future estimation of the state $\mathbf{x}$.

In the case of the NMSKF, we keep track of a higher-dimensional state $\mathbf{x}'$. The part of the state $\boldsymbol{\phi}$ corresponding to the wrench decomposition also

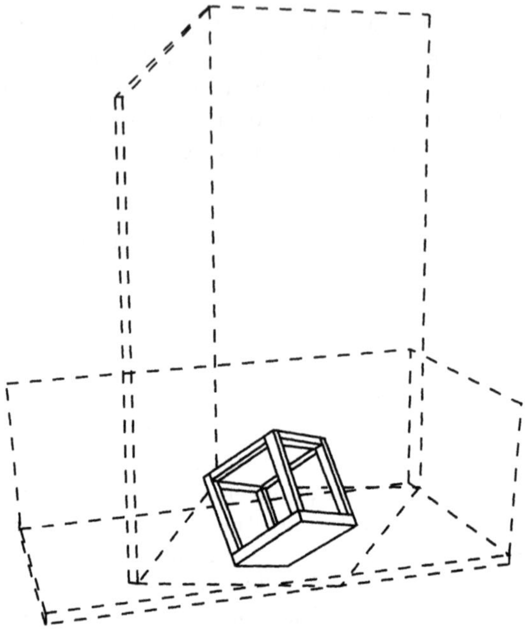

Fig. 7.3. The measurements during a vertex-face CF can identify the position of the contacting vertex and the orientation of the contacting face. The figure shows two possible vertex-face CFs which are consistent with both the measurements and the (large) uncertainties on the position and orientation of the corner object

changes and can be adapted in a similar way as described above. The part of the state $\mathbf{x}'_{CF}$ which corresponds to the geometrical parameters, however, changes too. When a new contact is gained, the state $\mathbf{x}'$ is augmented with the vector $\mathbf{x}'_j$ of the corresponding elementary CF (or at least with the part of $\mathbf{x}'_j$ which is not yet present in the current $\mathbf{x}'$). The covariance matrix $\mathbf{P}'$ is augmented accordingly. When a contact is lost the corresponding $\mathbf{x}'_j$ are *not* removed from the state $\mathbf{x}'$. *The information in these estimates* is still useful because it represents information about the parameters $\mathbf{x}$. Note that having unobserved state variables $\mathbf{x}'$ does not increase the computational load of the filter (Sect. 7.3.5).

7.5 Conclusions and Future Work

This chapter presents the estimator component of an autonomous compliant motion system.

The filter equations are *automatically generated* based on the elementary contact library. The Non-minimal State Kalman Filter is applied to the nonlinear estimation problem under *large uncertainty* of the geometrical parameters.

The possibility of estimation under large uncertainty expands the application area of the explicit contact models to the *identification of the contacting-*

objects models "from scratch". The objects are then represented by an unknown number of features (such as vertices, edges and faces for polyhedral objects) with inaccurately known positions, orientations and dimensions. This is a topic of *future research.*

8

Experiment: A Cube-In-Corner Assembly

For the things we have to learn before we can do them, we learn by doing them.

Aristotle, *Ethica Nichomachea,* ca. 350 B.C.

8.1 Introduction

This chapter validates the theory of the previous chapters with an experiment. The experiment consists of the estimation of the geometrical parameters and the recognition of CFs during a cube-in-corner assembly, Fig. 1.1. This chapter uses the Iterated Extended Kalman Filter (IEKF) described in Chap. 4, the Non-minimal State Kalman Filter (NMSKF) described in Chap. 5 and the contact models of Chap. 6. The details about the application of these filters and models for the estimation of the geometrical parameters and the CF recognition are presented in Chap. 7.

Section 8.2 describes the experiment, Sects. 8.3–8.4 present the estimation results for the IEKF for small and large initial uncertainties on the geometrical parameters, respectively. Section 8.5 shows that, unlike the IEKF, the NMSKF is able to deal with large uncertainties. Section 8.6 concludes.

8.2 Experiment Description

The experiment is an assembly of a cube (manipulated object) into a corner (environment object) executed by a KUKA-IR 361 industrial robot arm with a 6D Schunk force/torque sensor attached to its wrist, see Fig. 1.1. The pose and the twist of the manipulated object and the contact wrench are measured at 10 Hz. The goal of the experiment is to *estimate the geometrical parameters* and to *recognise the CFs* based on these measurements.

The inaccurately known geometrical parameters are:

1. the *six environment uncertainties* $\mathbf{x}^e = [x^e\ y^e\ z^e\ \theta_x^e\ \theta_y^e\ \theta_z^e]^T$. These are defined as the position *of* the {w} frame *in* the {e} frame ${}_e p^{e,w} = [x^e\ y^e\ z^e]^T$ and the orientation $(\theta_x^e, \theta_y^e, \theta_z^e)$ of the {w} frame with respect to the {e} frame. The angles are defined as: rotating the frame {e} about its Z-axis by θ_z^e, followed by rotating about the *new* Y-axis by θ_y^e, followed by rotating about the *new* X-axis by θ_x^e, results in a frame with axes parallel to the {w} frame;
2. the *six grasping uncertainties* $\mathbf{x}^m = [x^m\ y^m\ z^m\ \theta_x^m\ \theta_y^m\ \theta_z^m]^T$. These are defined as the position *of* the {g} frame *in* the {m} frame ${}_m\mathbf{p}^{m,g} = [x^m\ y^m\ z^m]^T$ and the orientation $(\theta_x^m, \theta_y^m, \theta_z^m)$ of the {g} frame with respect to the {m} frame. The definition of these (Euler) angles is: rotating the frame {m} about its Z-axis by θ_z^m, followed by rotating about the *new* Y-axis by θ_y^m, followed by rotating about the *new* X-axis by θ_x^m, results in a frame with axes parallel to the {g} frame.

The executed CF sequence consists of the following CFs: no contact; one vertex-face CF; one edge-face CF; one face-face CF; one face-face plus one edge-face CF; two face-face CF; and finally three face-face CF, see Fig. 1.1. The CF-compliant motions are constant translational and rotational motions of the cube, this is also visible in the measured twists of Fig. 8.1: in free space a translation is executed till a corner of the cube makes contact with the environment, then a rotational motion brings the vertex-face CF to an edge-face CF. A second rotational motion leads to a face-face CF. A pure translation moves the cube to a face-face plus edge-face CF. This motion is followed by a rotation which triggers a two face-face CF. Finally, a translation brings the cube to a three face-face CF with the corner object.

The compliant motion is specified in the Task Frame Formalism [42] and executed by a hybrid force/position controller where force and tracking control loops are closed around an analog velocity controller [69, 70].

The estimation results are obtained by offline processing of the measurements. The estimators are also able to process the measurements online, however, this is not yet integrated in our software architecture.

Section 8.3 presents the IEKF estimation of the parameters when the initial uncertainties are small. The same estimator cannot handle larger uncertainties, Sect. 8.4 shows the results. The NMSKF, on the other hand, is able to solve the estimation problem with the large initial uncertainties, see Sect. 8.5. We first present the data and solution techniques which are common to the three results.

Data Common to the Experiments

- The measurement models are automatically generated based on the elementary contact library (see Chap. 6). The elementary vertex-face CF is sufficient to generate all CFs of the experiment.
- The estimation is based on the processing of the wrench and twist measurements at 10 Hz; one pose measurement is introduced at the beginning of every new CF (see Sect. 7.2.2). Figure 8.1 shows the measured wrenches and twists collected during the experiment.
- In order to cope with the linearization errors, the experimentally determined pose, twist and wrench measurement standard deviations are multiplied by 10. The unmodelled friction is taken into account by multiplying once more the wrench standard deviation by 10.
- After inconsistency detection, a check for the stability of the contact is performed (see Sect. 7.4.2). The check is run with the following parameter values: $\mathbf{R}_c$ is 100 times the experimentally determined wrench measurement covariance matrix, $n_c = 5$ and the confidence boundary is chosen at 60%. The numerical values of these parameters are not critical, because of the very different behaviour of the wrenches during stable and transient control (Fig. 8.1 shows the jumps in the wrench measurements at CF transitions).
- The side length of the cube is 250 mm.
- The true geometrical parameter values are:

$$
\begin{array}{llll}
\theta_x^{e*} = 0.0°; & x^{e*} = 457\ \text{mm}; & \theta_x^{m*} = 0.0°; & x^{m*} = 125\ \text{mm};\\
\theta_y^{e*} = 0.0°; & y^{e*} = -608\ \text{mm}; & \theta_y^{m*} = 0.0°; & y^{m*} = 125\ \text{mm};\\
\theta_z^{e*} = -177.9°; & z^{e*} = -680\ \text{mm}; & \theta_z^{m*} = -45.0°; & z^{m*} = -15\ \text{mm}.
\end{array}
$$

8.3 IEKF Results for Small Uncertainties

Initial Uncertainty

The initial estimator, corresponding to motion in *free space*, is fed with estimates which are rather accurate for the (solidly grasped) manipulated object and less accurate for the (inaccurately positioned) environment object. The initial covariances are:

$$
\begin{aligned}
\sigma^2_{\theta^e_x} = \sigma^2_{\theta^e_y} = \sigma^2_{\theta^e_z} &= (10°)^2;\\
\sigma^2_{x^e} = \sigma^2_{y^e} = \sigma^2_{z^e} &= (100\ \text{mm})^2;\\
\sigma^2_{\theta^m_x} = \sigma^2_{\theta^m_y} = \sigma^2_{\theta^m_z} &= (1°)^2;\\
\sigma^2_{x^m} = \sigma^2_{y^m} = \sigma^2_{z^m} &= (1\ \text{mm})^2.
\end{aligned}
$$

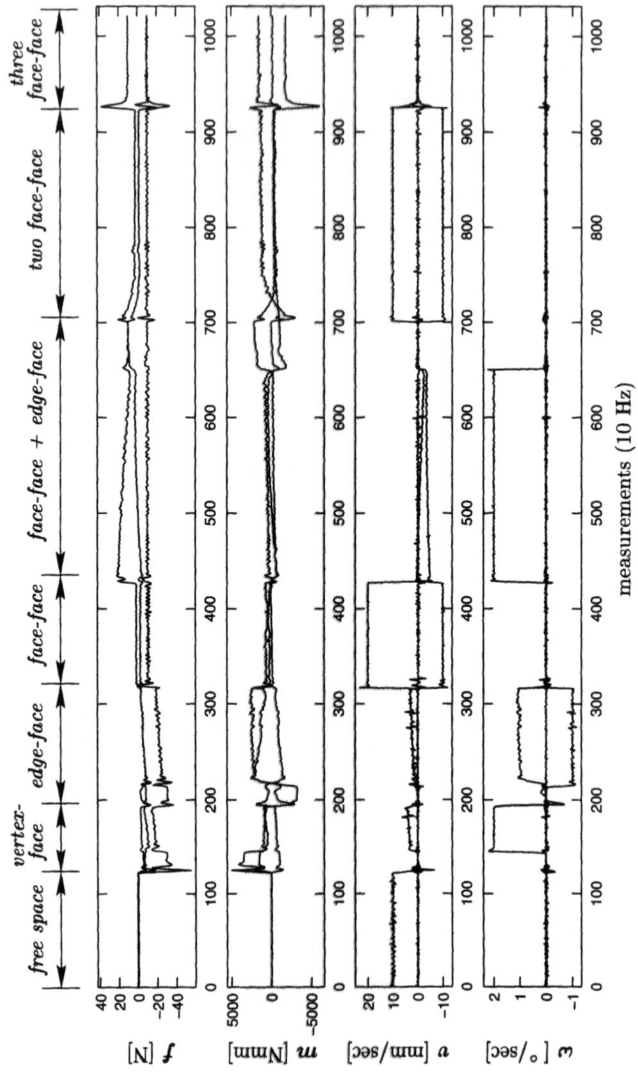

Fig. 8.1. Wrench and twist measurements during experiment expressed with respect to the gripper frame {g}. The measurements are taken at a 10 Hz frequency

The estimation starts with the following initial estimates:

$$\begin{aligned}
\hat{\theta}^e_x &= -4.0°; & \hat{x}^e &= 400 \text{ mm}; & \hat{\theta}^m_x &= 0.6°; & \hat{x}^m &= 122 \text{ mm};\\
\hat{\theta}^e_y &= -3.4°; & \hat{y}^e &= -650 \text{ mm}; & \hat{\theta}^m_y &= 0.6°; & \hat{y}^m &= 127 \text{ mm};\\
\hat{\theta}^e_z &= -183.3°; & \hat{z}^e &= -750 \text{ mm}; & \hat{\theta}^m_z &= -44.7°; & \hat{z}^m &= -16 \text{ mm}.
\end{aligned}$$

CF Transition Monitoring and CF Recognition

Figure 8.3 shows the results of the SNIS consistency test ($l = 3$). The horizontal lines are 60% confidence boundaries. When the SNIS value exceeds this line, an inconsistency is reported. The figure shows that besides all CF transitions, also several false alarms are reported.

After every inconsistency there is a time interval before the CF is recognised. These intervals comprise the stabilisation time and the time necessary to determine if a transition occurred or if the inconsistency is a false alarm. Comparing Fig. 8.3 with Fig. 8.1 learns that the main part of these time intervals are due to stabilisation and that the recognition of the CF is very fast: for this example only one measurement value is needed after every detected inconsistency.

Because the uncertainties on the geometrical parameter estimates are rather small, the executed CF sequence is assumed to be error free. This means that after an inconsistency is detected and the contact has been stabilised, *two* filters are run: one corresponding to the CF before the inconsistency detection and one corresponding to the next CF in the task plan. Whenever the probability of the SNIS value of one filter exceeds the probability of the other one by a factor 2, its corresponding CF, its state estimate and covariance matrix are chosen. The CF is correctly recognised after all inconsistency detections.

Geometrical Parameter Estimation and Wrench Decomposition

Figures 8.4 and 8.5 show the estimates of the geometrical parameters (the environment and the grasping uncertainties). The true values are plotted in dotted lines and the estimates in full lines. The "uncertainty" on the estimates is represented by the dashed lines at a distance of twice the standard deviation, i.e., two times the square root of the corresponding diagonal element of $\mathbf{P}$. For Gaussian distributions this corresponds to a 95% confidence interval.

During the *motion in free space* (first 124 measurements), none of the geometrical parameter estimates change because the motion in free space does not provide any information about the positions and orientations of both objects.

At measurement 124, an inconsistency is detected (Fig. 8.3). Until measurement 141 no measurements are processed because the contact is not stable (peaks in Fig. 8.1). At measurement 141 the measurements indicate that indeed a CF transition occurred and that the vertex-face CF, the next one in the task plan, is valid. From that measurement on, the estimates and uncertainties of the estimator corresponding to this vertex-face CF are taken.

Figures 8.4 and 8.5 show that the uncertainty on several of the geometrical parameter estimates decreases, while for others it stays large. E.g., the orientation of the corner object about the axis perpendicular to the contacting face (θ_z^e) and its translation inside this face (x^e and y^e) are not estimated. The vertex-face CF does not provide any information about these parameters (see Fig. 8.2). θ_z^e and y^e are estimated more accurately from measurement 442 on, i.e., when a contact with one vertical face is made; x^e keeps its initial uncertainty until a three face-face contact is recognised at measurement 941.

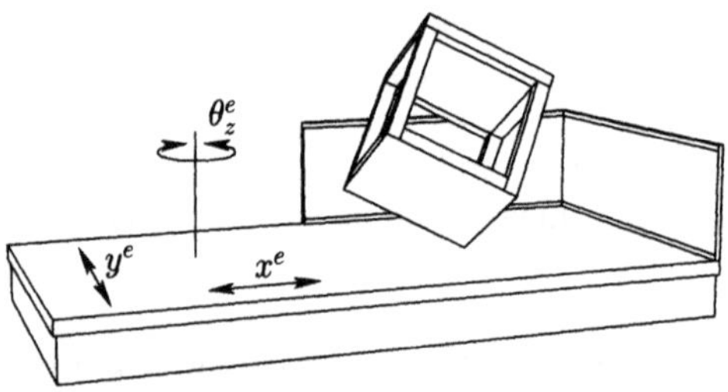

Fig. 8.2. Unobserved parameters during a vertex-face CF. The orientation of the corner object about the axis perpendicular to the contacting face (θ_z^e) and its translation inside this face (x^e and y^e) are unobservable. These parameters stay unobserved until a contact is made with one of the vertical faces of the corner object

During the vertex-face CF, the estimation of the orientation of the table (θ_x^e and θ_y^e) is inaccurate because of friction.

At measurement 196, a new inconsistency is reported (Fig. 8.3) after which the estimates are frozen again until the the edge-face CF is recognised (measurement 208).

A little later, at measurement 211, inconsistency is reported again. This time it is a false alarm and it is recognised as such at measurement 230. The estimation of the geometrical parameters and the detection of the CF sequence continues in the same way until the end of the experiment.

During several CFs, some of the geometrical parameter estimates are not consistent (see Figs. 8.4 and 8.5). E.g., this is clearly the case for the parameters θ_x^e, θ_y^e and z^e during the edge-face CF and the face-face CF, for the parameters θ_x^m and θ_y^m during face-face CF and for the parameter θ_z^e during the face-face plus edge-face CF. These inconsistencies are due to friction and linearization errors. The inconsistencies are small and do not result in inconsistent results during later CFs.

Figure 8.6 shows the estimates of the wrench decomposition variables ϕ. The number of wrench decomposition variables changes with the CFs and corresponds to the number of elementary contacts in which the CFs are decomposed.

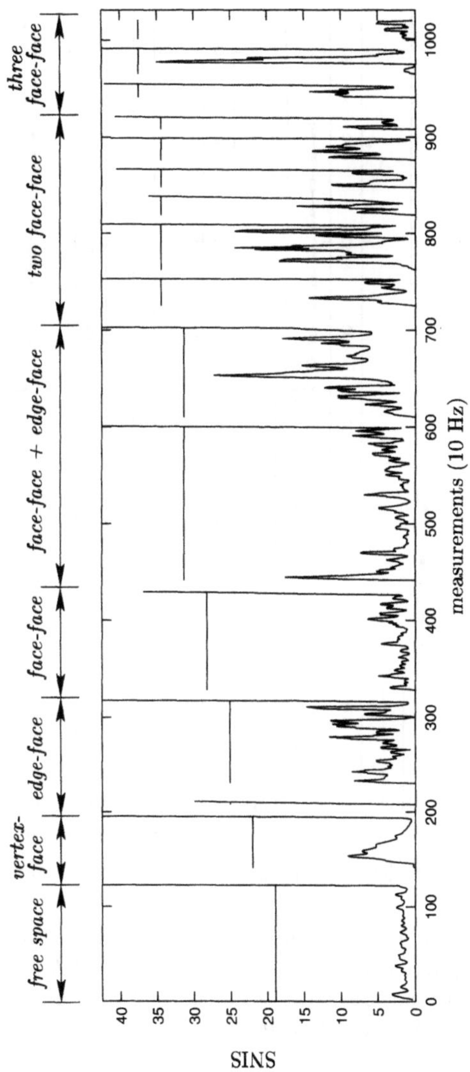

Fig. 8.3. SNIS consistency test for the *Iterated Extended Kalman Filter* when dealing with *small* uncertainties on the geometrical parameters. The confidence boundary (horizontal lines) is chosen at 60% which causes several false alarms

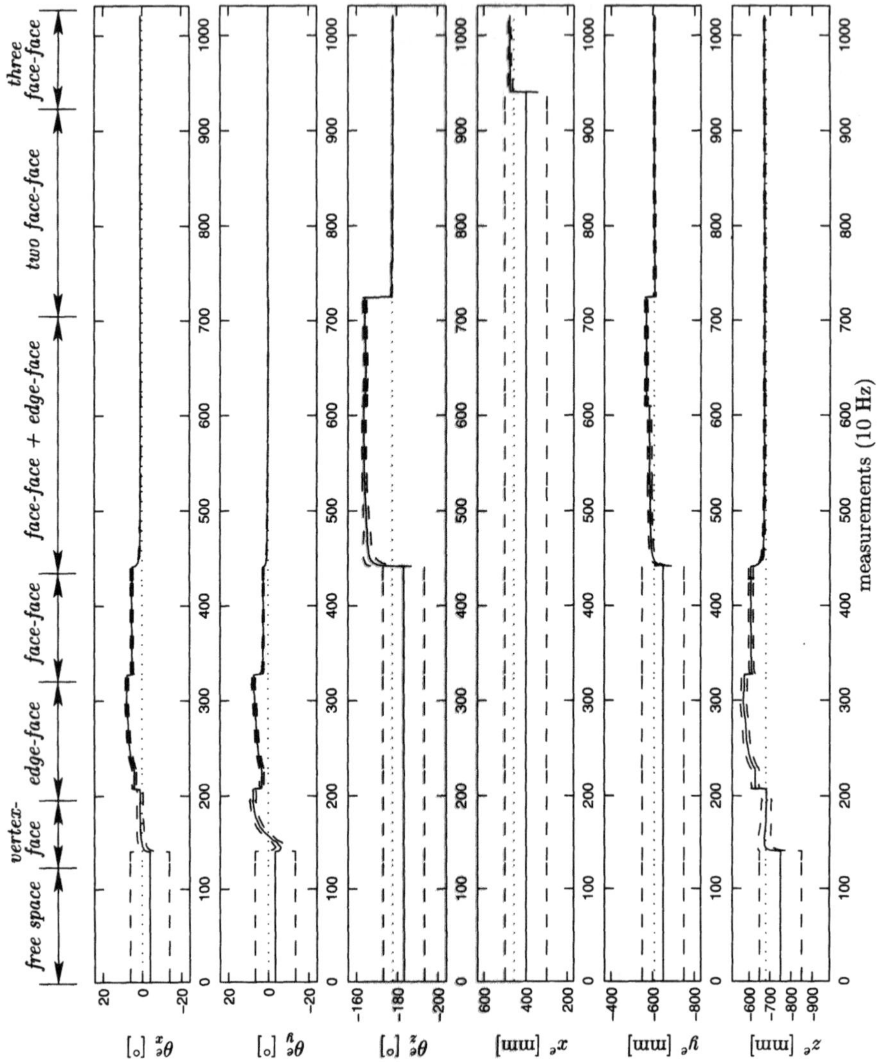

Fig. 8.4. Results of the *Iterated Extended Kalman Filter* when dealing with *small* uncertainties on the geometrical parameters: estimates (*full lines*) for the *environment parameters*, their true values (*dotted lines*) and their uncertainty visualised by twice the standard deviation (*dashed lines*)

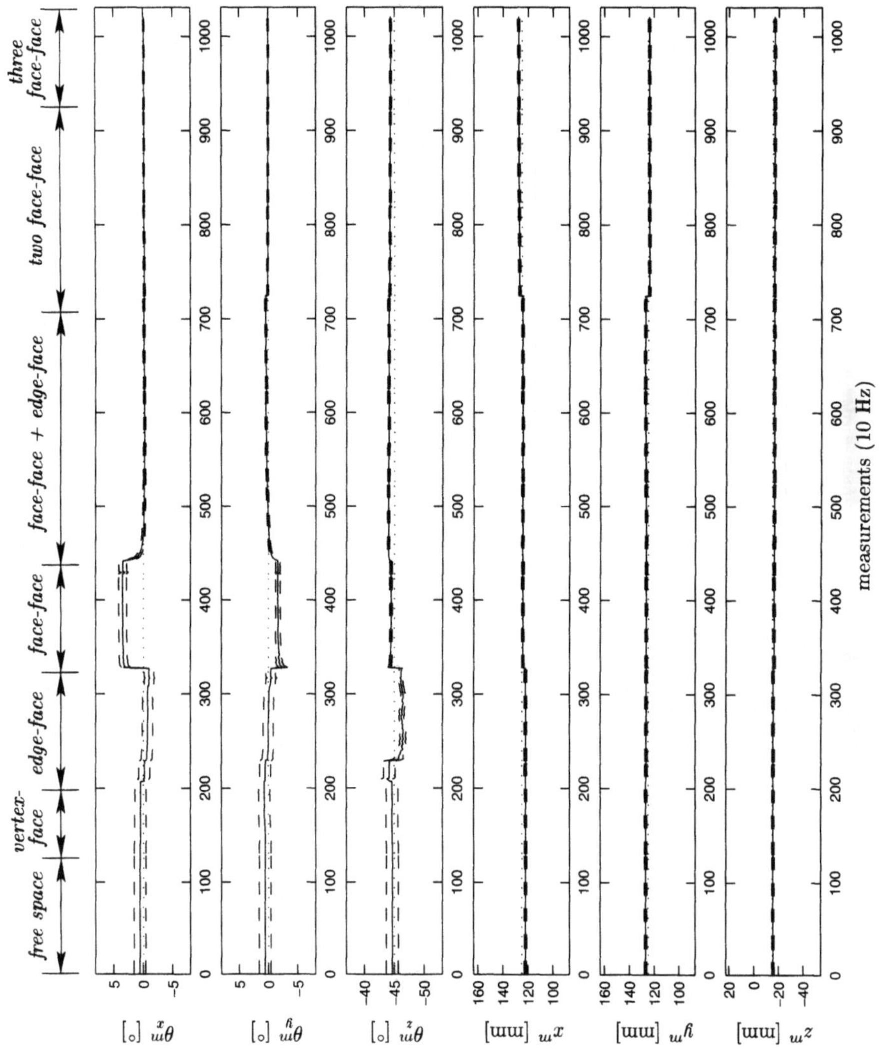

Fig. 8.5. Results of the *Iterated Extended Kalman Filter* when dealing with *small* uncertainties on the geometrical parameters: estimates (*full lines*) for the *grasping parameters*, their true values (*dotted lines*) and their uncertainty visualised by twice the standard deviation (*dashed lines*)

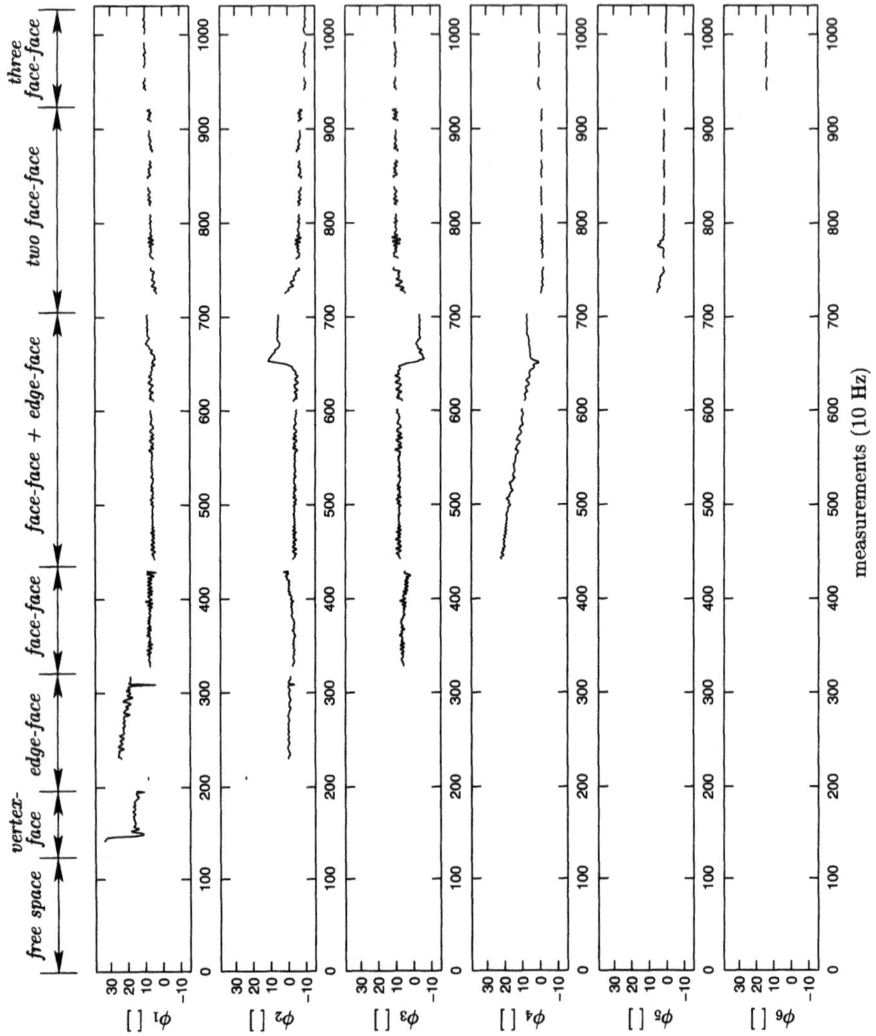

Fig. 8.6. Results of the *Iterated Extended Kalman Filter* when dealing with *small* uncertainties on the geometrical parameters: estimates for the ϕ *variables.* These variables represent the wrench decomposition in the contact wrench directions of the chosen elementary contacts

8.4 IEKF Results for Large Uncertainties

This section shows the results of the IEKF when the uncertainties on the initial parameters are larger.

Initial Uncertainty

The estimation is repeated with the following larger covariances:

$$\begin{aligned}
\sigma^2_{\theta^e_x} &= \sigma^2_{\theta^e_y} = \sigma^2_{\theta^e_z} = (20°)^2; \\
\sigma^2_{x^e} &= \sigma^2_{y^e} = \sigma^2_{z^e} = (500 \text{ mm})^2; \\
\sigma^2_{\theta^m_x} &= \sigma^2_{\theta^m_y} = \sigma^2_{\theta^m_z} = (20°)^2; \\
\sigma^2_{x^m} &= \sigma^2_{y^m} = \sigma^2_{z^m} = (500 \text{ mm})^2;
\end{aligned}$$

and the less accurate initial estimates:

$$\begin{aligned}
&\hat{\theta}^e_x = 15.5°; && \hat{x}^e = 600 \text{ mm}; && \hat{\theta}^m_x = -15.0°; && \hat{x}^m = 400 \text{ mm}; \\
&\hat{\theta}^e_y = 9.2°; && \hat{y}^e = -200 \text{ mm}; && \hat{\theta}^m_y = 16.0°; && \hat{y}^m = 400 \text{ mm}; \\
&\hat{\theta}^e_z = -194.8°; && \hat{z}^e = -1050 \text{ mm}; && \hat{\theta}^m_z = -23.5°; && \hat{z}^m = -400 \text{ mm}.
\end{aligned}$$

CF Transition Monitoring and CF Recognition

The CF transitions are monitored with the same SNIS consistency test as used for the small uncertainties ($l = 3$ and a 60% confidence boundary). Figure 8.7 shows the results, these are interpreted below together with the discussion on the geometrical parameter estimates. The CF *recognition* after an inconsistency detection is done manually (see Sect. 7.4.3).

Geometrical Parameter Estimation and Wrench Decomposition

Figures 8.8 and 8.9 show the estimates of the environment and the grasping uncertainties.

During the motion in free space (first 124 measurements), none of the parameter estimates changes because the motion in free space does not provide any information about them.

At measurement 124, an inconsistency is detected (Fig. 8.7). The contact is unstable until measurement 141, after which the vertex-face filter corresponding the new CF is started. The geometrical parameters (Figs. 8.8 and 8.9) are not fully observed during this CF. Since the initial uncertainties are large and we use an IEKF estimator, the results are inconsistent (Chap. 4). E.g., this is clearly the case for the estimates of the parameters θ^m_x and z^m.

The CF transition at measurement 196 results in an inconsistency detection (see Fig. 8.7). After stabilisation (measurement 141), the estimation is proceeded in an edge-face CF. The estimates of the geometrical parameters are changed into values which are consistent with the previous estimates and the new measurements (Figs. 8.8–8.9). In this CF also the estimates of z^e, θ_y^m and x^m become clearly inconsistent. The filter reports several inconsistencies (see Fig. 8.7) which were not present in the first experiment for small initial uncertainties on the geometrical parameters. These inconsistencies are already an indication that the parameter estimates are inconsistent.

After the transition to a face-face CF, no further filtering results are presented because all parameter estimates are inconsistent (except for the totally unobserved θ_z^e, x^e and y^e, see Fig. 8.2). The SNIS test indicates inconsistency of the measurements with the CF model.

Figure 8.10 shows the estimates of the wrench decomposition variables ϕ during the vertex-face and the edge-face CF.

8.5 NMSKF Results for Large Uncertainties

Initial Uncertainty

The NMSKF makes the identification of the parameters under large uncertainties possible. The results are given for the same initial uncertainties as considered in Sect. 8.4:

$$\begin{aligned}
\sigma^2_{\theta^e_x} = \sigma^2_{\theta^e_y} = \sigma^2_{\theta^e_z} &= (20°)^2;\\
\sigma^2_{x^e} = \sigma^2_{y^e} = \sigma^2_{z^e} &= (500\text{ mm})^2;\\
\sigma^2_{\theta^m_x} = \sigma^2_{\theta^m_y} = \sigma^2_{\theta^m_z} &= (20°)^2;\\
\sigma^2_{x^m} = \sigma^2_{y^m} = \sigma^2_{z^m} &= (500\text{ mm})^2;
\end{aligned}$$

$$\begin{array}{llll}
\hat{\theta}^e_x = 15.5° & \hat{x}^e = 600\text{ mm} & \hat{\theta}^m_x = -15.0° & \hat{x}^m = 400\text{ mm}\\
\hat{\theta}^e_y = 9.2° & \hat{y}^e = -200\text{ mm} & \hat{\theta}^m_y = 16.0° & \hat{y}^m = 400\text{ mm}\\
\hat{\theta}^e_z = -194.8° & \hat{z}^e = -1050\text{ mm} & \hat{\theta}^m_z = -23.5° & \hat{z}^m = -400\text{ mm}
\end{array}$$

The formerly used IEKF was not able to solve this estimation problem.

CF Transition Monitoring and CF Recognition

A SNIS consistency test ($l = 3$ and confidence boundary 10%) detects the CF transitions. The confidence boundary is chosen low because the SNIS test checks the consistency of the measurements with the more flexible model in $\mathbf{x}'$. The 10% boundary is experimentally determined and results in quick detection of all CF transitions.

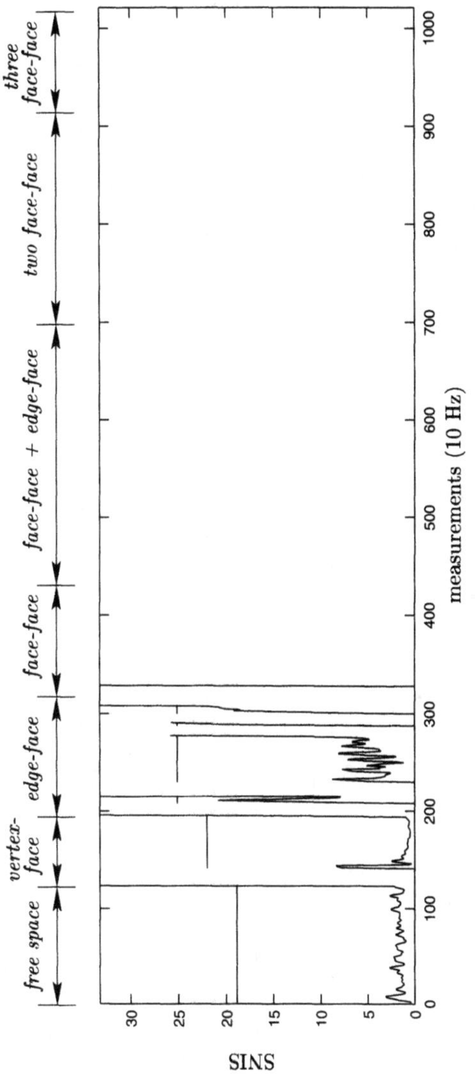

Fig. 8.7. SNIS consistency test for the *Iterated Extended Kalman Filter* when dealing with *large* uncertainties on the geometrical parameters. The confidence boundary (horizontal lines) is chosen at 60% which causes several false alarms

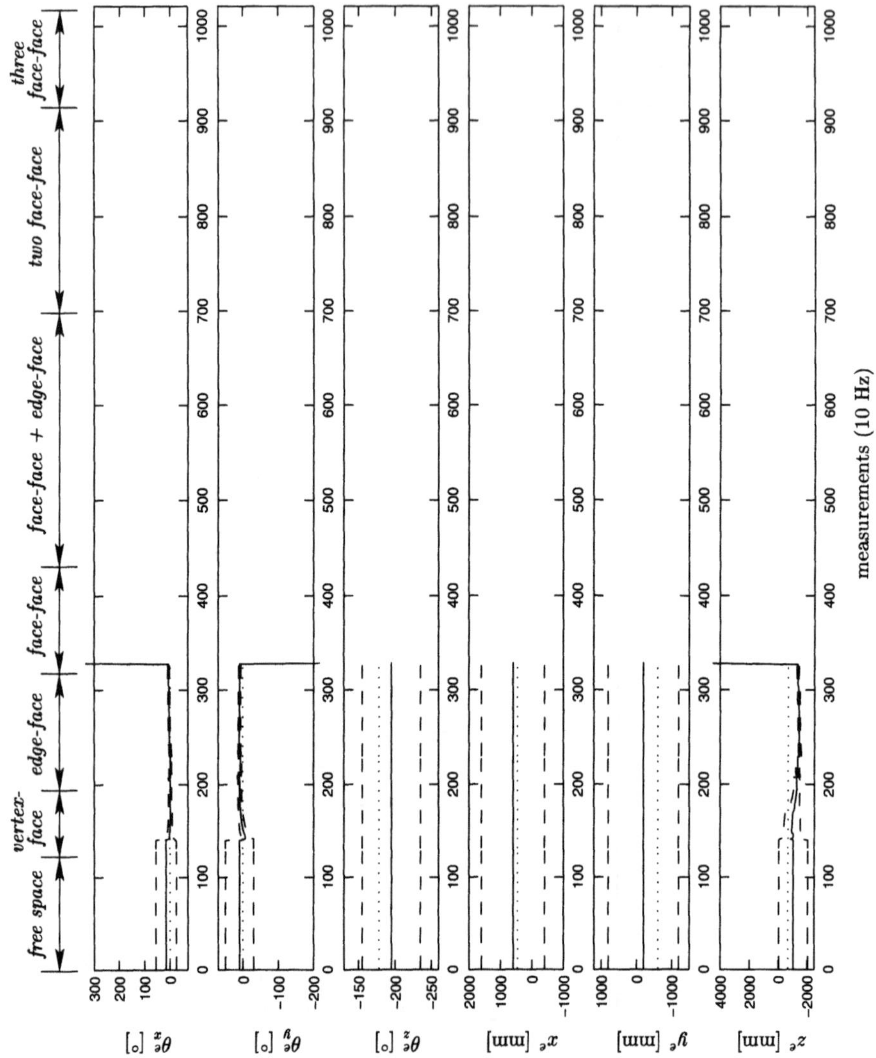

Fig. 8.8. Results of the *Iterated Extended Kalman Filter* when dealing with *large* uncertainties on the geometrical parameters: estimates (*full lines*) for the *environment parameters*, their true values (*dotted lines*) and their uncertainty visualised by twice the standard deviation (*dashed lines*)

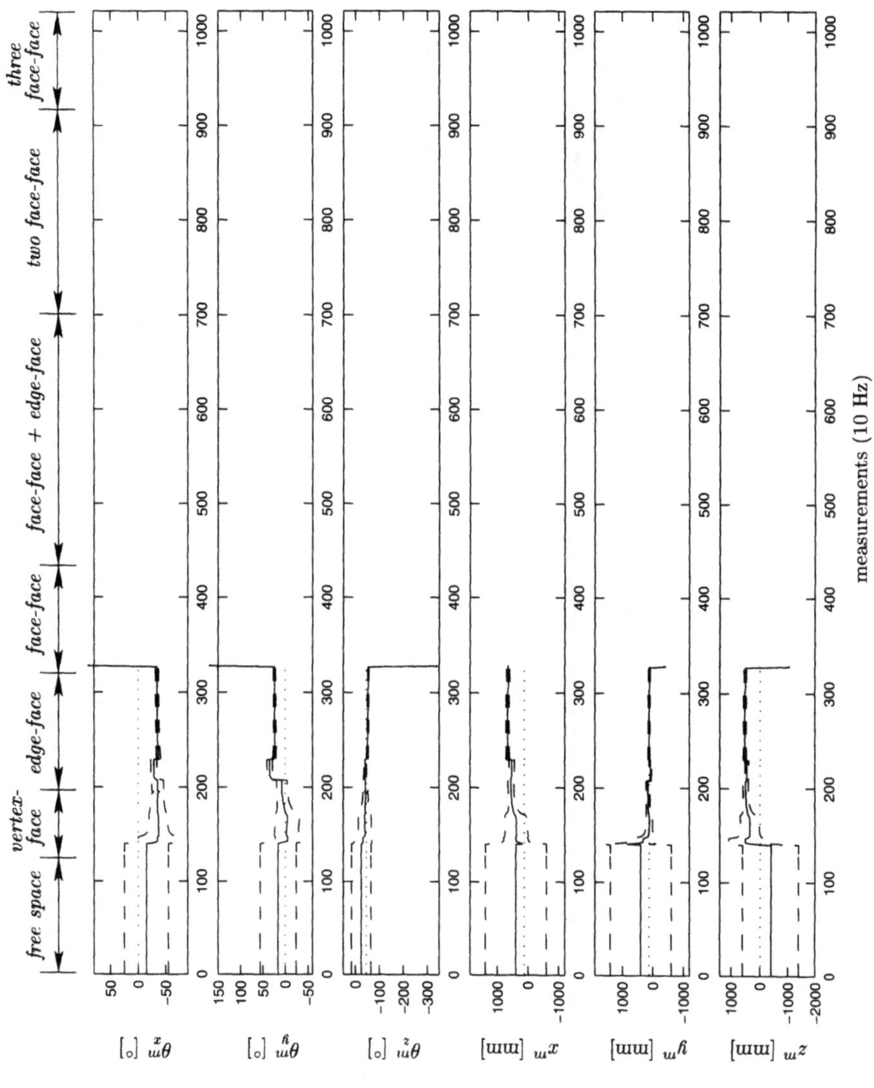

Fig. 8.9. Results of the *Iterated Extended Kalman Filter* when dealing with *large* uncertainties on the geometrical parameters: estimates (*full lines*) for the *grasping parameters*, their true values (*dotted lines*) and their uncertainty visualised by twice the standard deviation (*dashed lines*)

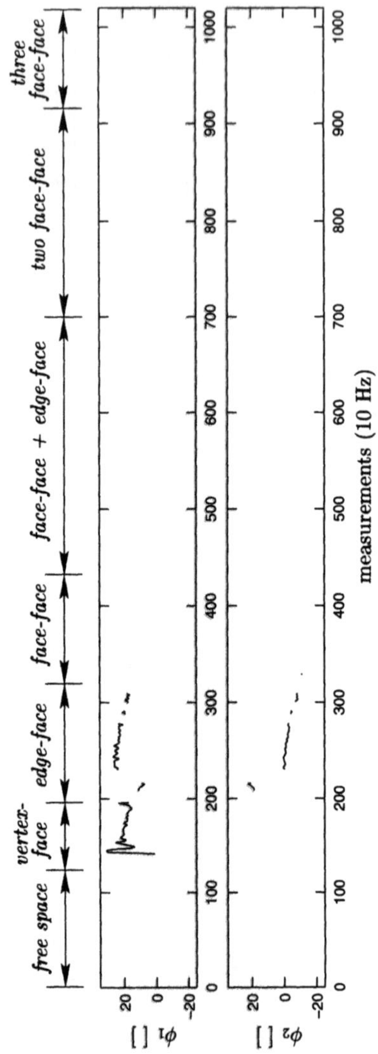

Fig. 8.10. Results of the *Iterated Extended Kalman Filter* when dealing with *large* uncertainties on the geometrical parameters: estimates for the ϕ *variables.* These variables represent the wrench decomposition in the contact wrench directions of the chosen elementary contacts

Figure 8.11 shows that the test has less false alarms than the SNIS test of the IEKF in Fig. 8.3. E.g., none of the false alarms in the last two CFs is present with the new filter. Hence, these false alarms were due to linearization errors of the IEKF (and the corresponding inconsistency of the estimates), disturbing the filter even for small uncertainties on the parameters.

The CF *recognition* after an inconsistency is done manually (see Sect. 7.4.3).

Geometrical Parameter Estimation and Wrench Decomposition

The NMSKF is computationally more expensive than the IEKF. The processing of the measurements in the two face-face CF takes approximately 0.05 sec on a Pentium III 800 MHz, 128 Mb RAM laptop (C++ program). The two face-face CF is the computationally most expensive CF. This means that online processing of 10 Hz twist and wrench measurements is feasible.

Equation (5.9)

$$p(\mathbf{x}_k|\hat{\mathbf{Z}}_k) \sim \exp\Big(-\frac{1}{2}\big(\mathbf{g}(\mathbf{x}_k) - \hat{\mathbf{x}}'_{k|k}\big)^T \mathbf{P}'^{-1}_{k|k}\big(\mathbf{g}(\mathbf{x}_k) - \hat{\mathbf{x}}'_{k|k}\big)\Big);$$

describes the (non-Gaussian) analytical posterior pdf over $\mathbf{x}$. We compare two methods to obtain the mean and variance for the geometrical parameters.

1. *A Monte Carlo sampler* returns the posterior sample mean and covariance. Figure 8.12 shows the sample histograms of the geometrical parameter estimates, after having processed 100, 200 and 300 measurements. The sample mean and covariance are a good approximation for the expected value and covariance of the analytical pdf (the deviations between the true values and the estimates of θ_x^e and θ_y^e are due to unmodelled friction). Unfortunately, this method is computationally too expensive to be used online.
2. An *Iterated Extended Kalman Filter* (IEKF) returns a good approximation of the expected value and covariance *if* the measurements fully observe the system (Chap. 4). The combination of the NMSKF and the IEKF is called the Cascaded Kalman Filter (CKF, see Sect. 5.2.2).
   Figures 8.13 and 8.14 show the results of the CKF for the environment and grasping uncertainties. Comparing these results with the results of the Monte Carlo sampler (Fig. 8.12) shows that the IEKF returns estimates in the peak of the posterior pdf plotted in Fig. 8.12. However, the IEKF sometimes returns underestimated covariances, i.e., inconsistent estimates. These inconsistent results however are not used further on to make future estimates (see Chap. 5).
   The advantage of the CKF is that it is computationally not very expensive and therefore can be used *online*. The IEKF filtering during the (computationally most expensive) face-face CF takes about 0.30 sec per estimate. Hence, both the NMSKF and the IEKF can be run on this same computer

by processing the measurements at 10 Hz and producing new estimates for the geometrical parameters every 6 measurements (0.6 seconds). Figure 8.15 presents the estimates for the wrench decomposition variables ϕ.

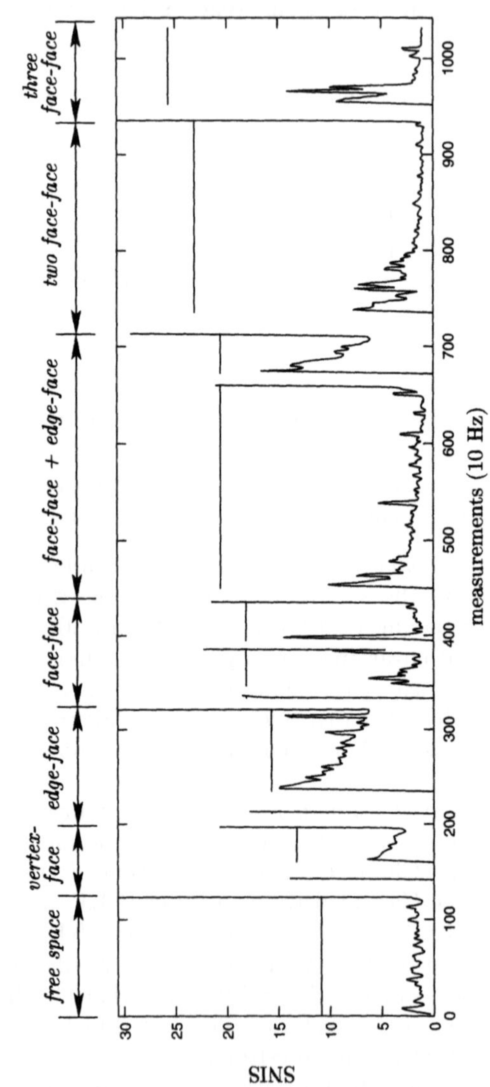

Fig. 8.11. SNIS consistency test for the *Non-minimal State Kalman Filter* when dealing with *large* uncertainties on the geometrical parameters. The confidence boundary (horizontal lines) is chosen at 10%

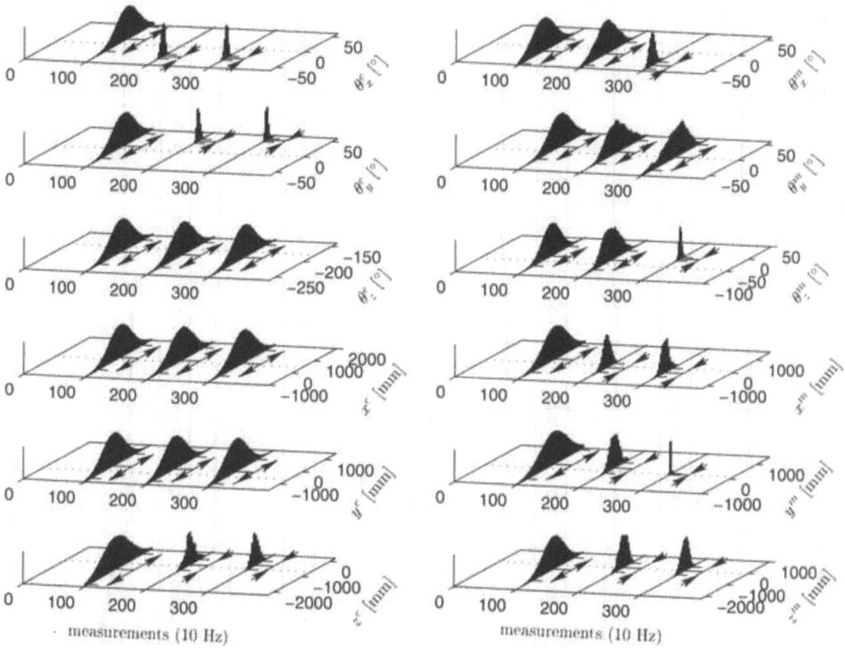

Fig. 8.12. Results of the *Non-minimal State Kalman Filter* when dealing with *large* uncertainties on the geometrical parameters. The posterior pdfs over the *environment and grasping parameters* are shown. The pdfs are calculated by Monte Carlo sampling of the analytical posterior pdf after having processed 100 measurements (vertex-face CF), 200 measurements (edge-face CF) and 300 measurements (face-face CF). The arrows indicate the sample means plus/minus twice the sample standard deviation

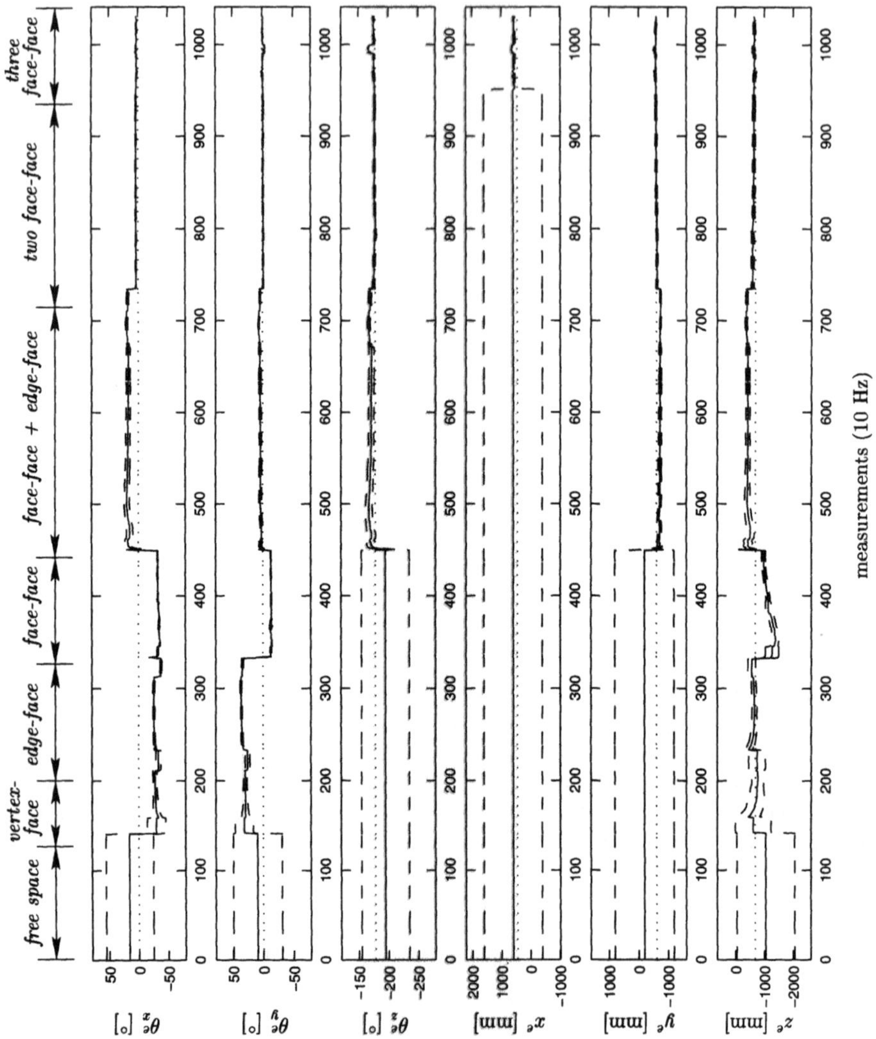

Fig. 8.13. Results of the *Cascaded Kalman Filter* when dealing with *large* uncertainties on the geometrical parameters: estimates (*full lines*) for the *environment parameters*, their true values (*dotted lines*) and their uncertainty visualised by twice the standard deviation (*dashed lines*)

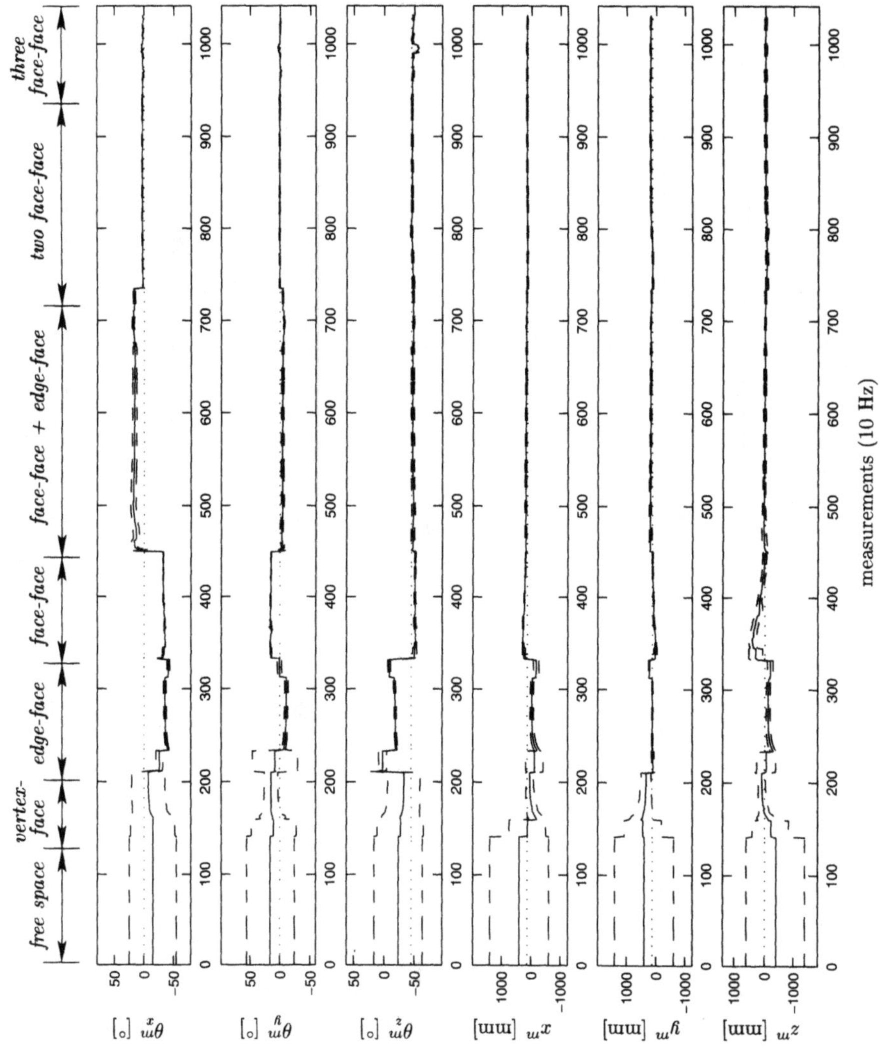

Fig. 8.14. Results of the *Cascaded Kalman Filter* when dealing with *large* uncertainties on the geometrical parameters: estimates (*full lines*) for the *grasping parameters*, their true values (*dotted lines*) and their uncertainty visualised by twice the standard deviation (*dashed lines*)

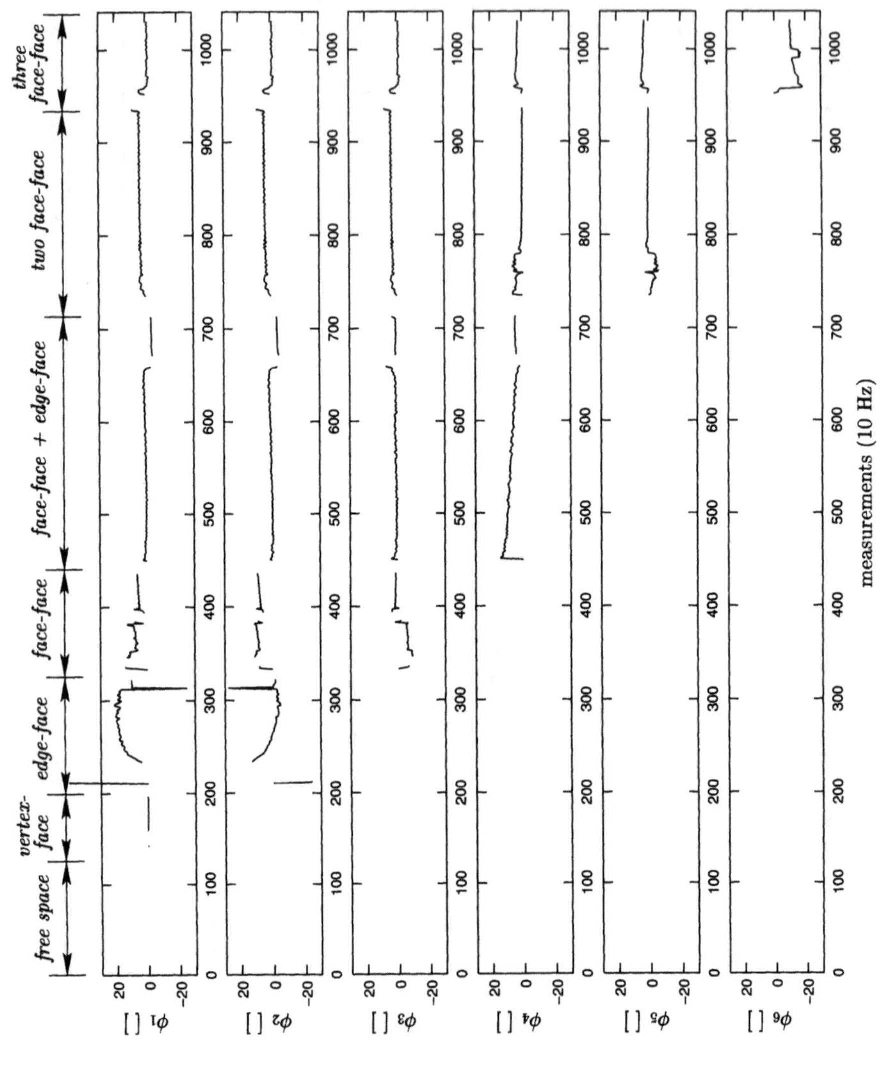

Fig. 8.15. Results of the *Cascaded Kalman Filter* when dealing with *large* uncertainties on the geometrical parameters: estimates for the ϕ *variables*. These variables represent the wrench decomposition in the contact wrench directions of the chosen elementary contacts

Accuracy of the Estimates

The estimates of Figs. 8.13 and 8.14 during the last two CFs are zoomed in to the same scale as used on Fig. 8.4 for the estimates of the environment parameters under small initial uncertainty. The result is plotted in Fig. 8.16. The estimates are significantly less accurate than those obtained under small uncertainty. This means that the information in the initial estimates used to obtain the results of Fig. 8.4 is "complementary" to the information in the processed measurements. In Fig. 8.16 we notice that, even for the large initial uncertainties assumed, the angle estimates at the end of the experiment are rather accurate. The most uncertain angle estimates are those of θ_z^e and θ_z^m with standard deviations of approximately 0.6°. The most uncertain position estimates (i.e., z^e and z^m), on the other hand, have standard deviations of approximately 20 mm.

Closer Look at the Estimates of z^e and z^m

Figure 8.17 presents the uncertainty ellipse over the estimates z^e and z^m at the end of the experiment. The ellipse is drawn at the distance of twice the standard deviation and hence can be compared to the uncertainty boundaries of Fig. 8.16. The ellipse shows that the estimates are strongly correlated: their sum is accurately known (standard deviation 4.4 mm) and their difference inaccurately known (standard deviation 27.9 mm).

Figure 8.18 shows the physical distances to which the parameters z^e and z^m correspond. The parameter z^e corresponds to the Z-coordinate of the world frame in the environment frame (see Fig. 1.1), i.e., minus the physical height of the table. The parameter z^m corresponds to the Z-coordinate of the gripper frame in the manipulated object frame (see Fig. 1.1), i.e., minus the height of the position of the gripper frame above the cube.

When a face-face contact is established between the cube and the table, i.e., during the last four CFs of the experiment, the distances expressed by z^e and z^m are aligned. Knowing their sum but not their difference means that the height of the gripper frame above the world frame is known, but the height of the table is not. The difference is an unobserved parameter combination because the same twists (translational motions parallel to the face and a rotational motion about the normal to the face) and wrenches (moments parallel to the face and a force about the normal to the face) are allowed for whatever table height.

The parameters z^e and z^m would be estimated more accurately if the measurements *during the earlier vertex-face and edge-face CFs* are more informative. The CF-compliant motions for these CFs in the experiment, however, are only constant rotational velocities, chosen to assure the CF transitions.

The next chapter describes the calculation of task plans which allow the observation of the parameters to a user-defined accuracy tolerance.

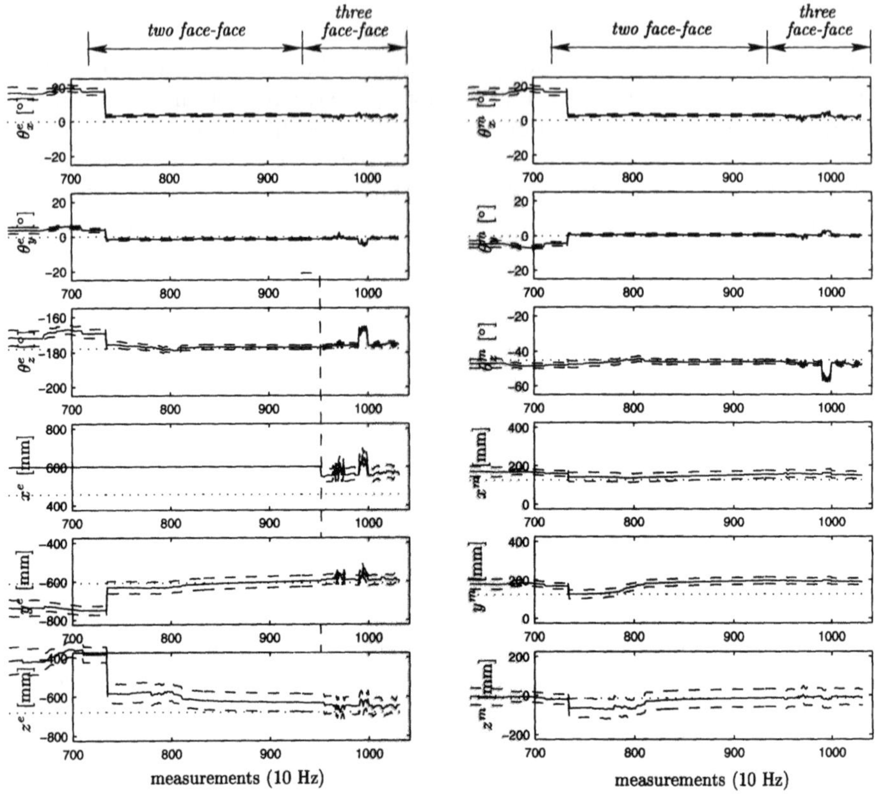

Fig. 8.16. Results of the *Cascaded Kalman Filter* when dealing with *large* uncertainties on the geometrical parameters. The full lines show the estimates of the environment and grasping parameters during the last 2 CFs. The dashed lines indicate their uncertainty visualised by twice the standard deviation. The estimates of the orientations of the objects are accurate, the estimates of their positions, however, are rather inaccurate

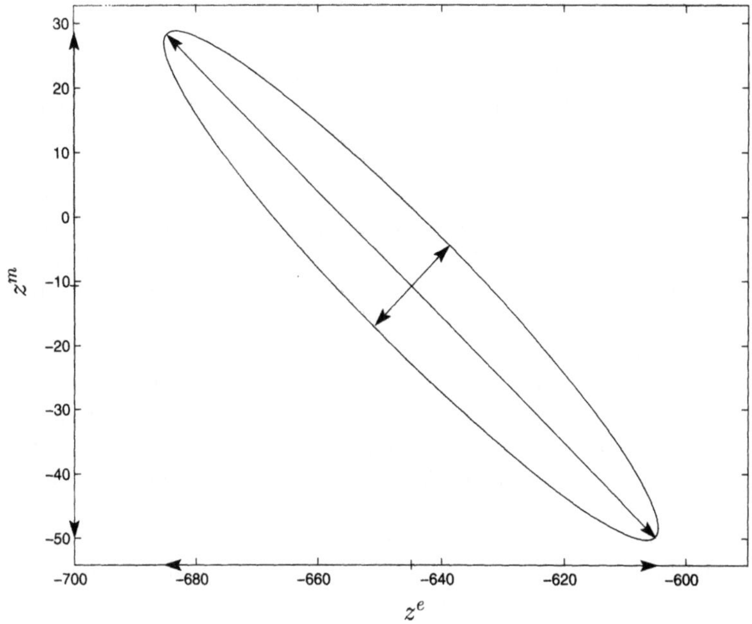

Fig. 8.17. 2σ-uncertainty ellipse of the estimates z^e and z^m at the end of the experiment. The estimates are strongly correlated: the sum of both parameters is accurately known, their difference is inaccurately known

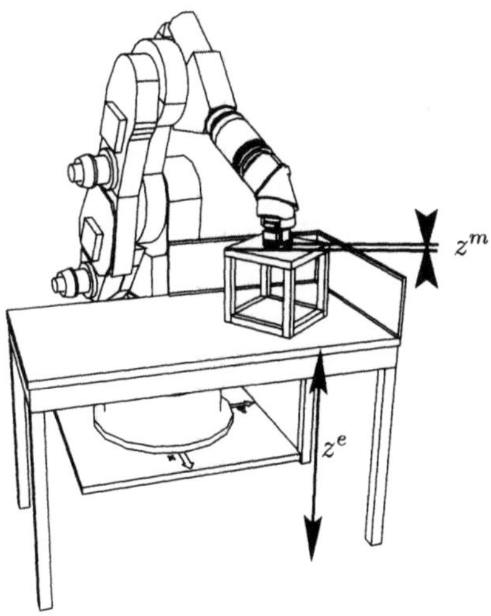

Fig. 8.18. The parameter z^e corresponds to minus the physical height of the table. The parameter z^m corresponds to minus the height of the position of the gripper frame above the cube

8.6 Conclusions

This chapter validates the theory of the previous chapters with an experiment. The experiment consists in the estimation of the geometrical parameters and the recognition of CFs during a cube-in-corner assembly, Fig. 1.1.

The *estimation of the geometrical parameters* is performed by an IEKF when the initial uncertainties are small. However, this estimator cannot handle large uncertainties. When the initial uncertainties are large, the estimation problem is solved by an NMSKF. The NMSKF is computationally more expensive, but it can still process the measurements online.

The performed CF-compliant motions in the experiment are constant translational and rotational velocities which trigger the CF transitions. The estimation results under large uncertainty show that the measurements are not sufficiently exciting to identify all parameters. The next chapter discusses how to design a task plan which guarantees informative measurements.

The *CF transition detection* is performed with a SNIS consistency test. All transitions are detected by all filters. Several false alarms are reported. The *CF recognition* is only possible when dealing with small uncertainties on the geometrical parameters. For large uncertainties, several CFs are probable. The general solution of calculating a discrete probability distribution over the CFs at all time is a topic of future research.

9

Task Planning with Active Sensing

Measure what is measurable, and
make measurable what is not so.

Galileo Galilei, (1564–1642).

9.1 Introduction

This chapter describes the calculation of a compliant motion task plan which improves the observation of the inaccurately known geometrical parameters (i.e., the positions, orientations and dimensions of the contacting objects). This is called *active sensing*. The experiment of the previous chapter shows

1. that active sensing is *necessary* in order to obtain accurate parameter estimates. The motions executed during the experiment are constant translational and rotational velocities which trigger the CF transitions, see Fig. 8.1. However, these motions are not sufficiently exciting to identify all geometrical parameters.
2. that the *wrench measurements are less reliable* than the pose and twist measurements due to the unmodelled friction. Therefore, the active-sensing task plans in this chapter are designed based on pose and twist information only. I.e., the pose and twist measurements of the active sensing task plan provide enough information to estimate the geometrical parameters to the desired accuracy.
   Because the task is planned based on only the pose and twist information, the symbol $\mathbf{x}$ in this chapter refers to the vector of the geometrical parameters without the wrench decomposition variables, which are only needed in the wrench equations.

Our approach to active sensing is to *optimise* the task plan (i) by *minimising* an objective function, such as the expected execution time, which is an important criterion in industrial applications; and (ii) by *constraining* the task plans to plans which observe the geometrical parameter estimates to the required accuracy.

The *task plan* consists of (i) the desired *sequence of CFs* and (ii) the *motions* to be executed in each CF of the sequence: *guarded motions* in free space and *CF-compliant motions* in contact. Figure 9.1 shows a CF sequence consisting of free space, a vertex-face CF and a double vertex-face CF. Figures 9.2 and 9.3 show two possible CF-compliant paths during the vertex-face CF. A CF-compliant path is defined by the full rigid body motion of the manipulated object. The figures show only the path of the contacting vertex. A CF-compliant path comprises (i) an *active sensing motion* which results in informative measurements for the estimation of the geometrical parameters; and (ii) a *transition motion* which enhances the desired contact transition while avoiding unwanted transitions.

At planning time, the true parameter values are not known, hence, the calculated plan is –at its best– only the *expected* optimum given the prior uncertainty on the parameters and the measurement model uncertainty. It does *not guarantee* the optimality of the objective function nor the estimation of the parameters to the required accuracy. Hence, online *re-planning* (based on the more accurate parameter estimates at execution time) is necessary. Consequently, we focus on *approximate* optimisation solutions for online use. The used approximations are (i) the assumption that the *estimates* of the geometrical parameters are the *true* values; (ii) the decoupling of the optimisation problem of the task plan into the optimisation of the CF sequence and the optimisation of the active sensing motions for each CF of the optimal sequence; and (iii) the use of a local optimisation routine or even heuristics for the calculation of the active sensing motions.

Model-based active sensing for compliant motion is a new research area. Hence, this chapter primarily aims at formulating the active sensing problem and decoupling it into smaller optimisation problems. This work does *not* analyse which optimisation *algorithms* are computationally most efficient for online use.

Section 9.2 presents the amount of information about the geometrical parameters which is available in the pose and twist measurements. Section 9.3 discusses the task plan optimisation with constraints on the uncertainty of the parameter estimates at the end of the plan. Section 9.3 additionally presents an algorithm to calculate the transition motions which allow to move from one CF to the next.

For each CF it is possible to calculate the parameter combinations which are observable in this CF. This allows us to decouple the active sensing requirement into a requirement of the CF sequence and requirements of the active sensing motions in each CF of this sequence. Sections 9.4 and 9.5 de-

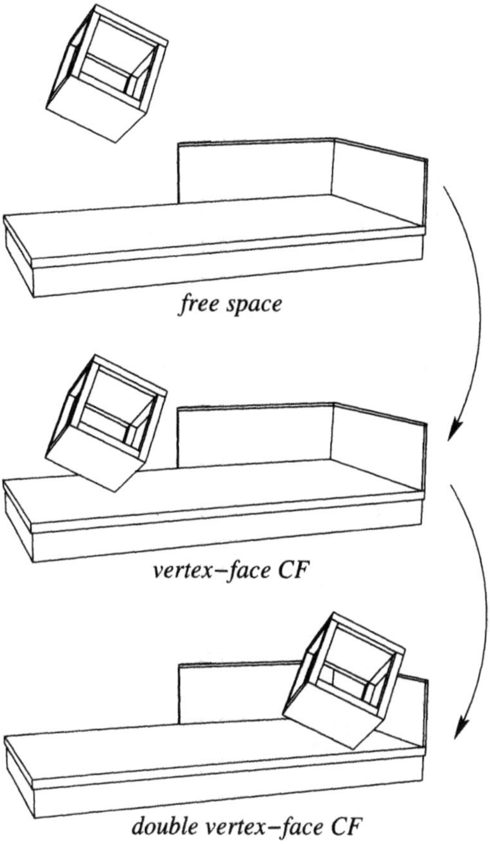

Fig. 9.1. CF sequence of free space motion, a vertex-face CF and a double vertex-face CF

scribe solutions to the decoupled optimisation problems of the CF sequence and of the active sensing motions respectively. Section 9.6 discusses online re-planning.

9.2 Measurement Information

The pose and twist measurement equations (6.3) and (6.4) are of the form $\mathbf{h}_k(\mathbf{x}, \mathbf{z}_k) = \mathbf{0}_{n_h \times 1}$. The information in a measurement $\hat{\mathbf{z}}_k$ about the geometrical parameters $\mathbf{x}$ during compliant motion depends on both the type of the CF and the executed CF-compliant path:

- different CFs correspond to different measurement functions $\mathbf{h}_k$. E.g., during a vertex-face CF it is impossible to observe the parameter combinations corresponding to the translation in the face and the orientation about the

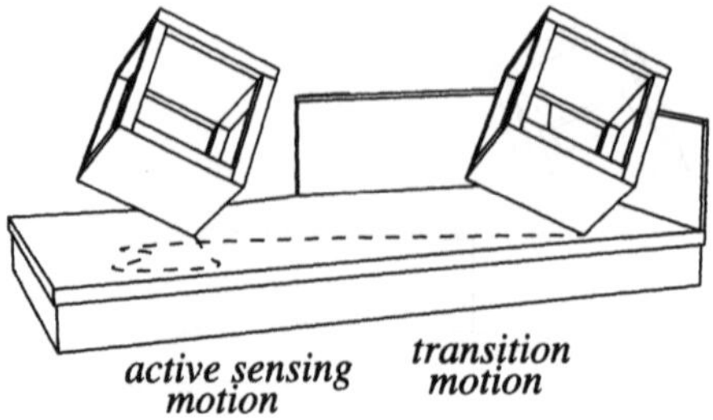

Fig. 9.2. CF-compliant motion during the vertex-face CF. The first part of the motion is designed to sufficiently excite all parameters (active sensing motion), the second part of the motion is designed to enhance the CF transition to the double vertex-face CF (transition motion). A CF-compliant path is defined by the full rigid body motion of the manipulated object. The figure shows only the path of the contacting vertex

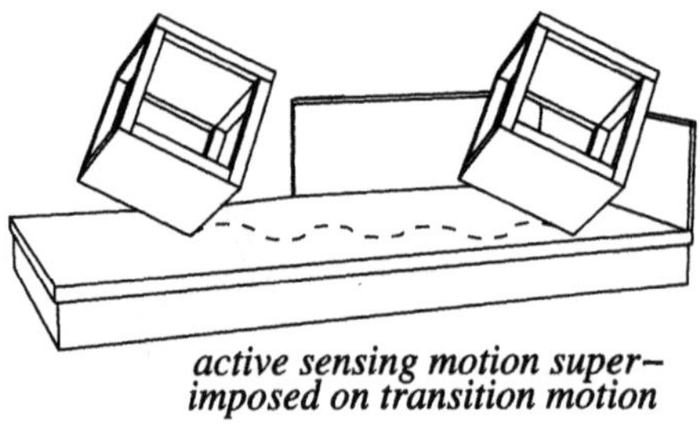

Fig. 9.3. CF-compliant motion during the vertex-face CF. The active sensing motion is designed as a deviation of the transition motion. A CF-compliant path is defined by the full rigid body motion of the manipulated object. The figures shows only the path of the contacting vertex

face normal, see Fig. 8.2. By making contact with other parts of the environment object, these parameters can become observable.

- different CF-compliant paths correspond to different measurements $\hat{\mathbf{z}}_k$. E.g., Fig. 9.4 indicates the unobserved orientation of the face in a vertex-face CF when the vertex moves over a straight line on the face. A more arbitrary CF-compliant path of the contact point observes the full face orientation (Fig. 9.5).

Hence, active sensing is important both on the level of determining the CF sequence and the choice of a CF-compliant path in each CF of the sequence.

9.2.1 Measure of Information

Formulating the active sensing problem implies the use of a *measure of accuracy* of the measurements and state estimates or inversely a *measure of their uncertainty*. Section 3.5 contains a review of two well-known measures,

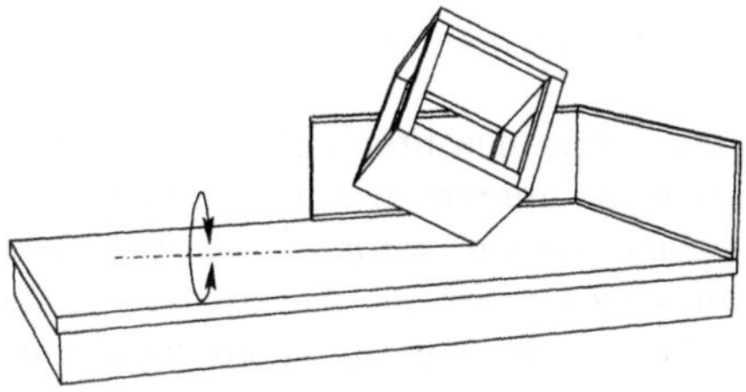

Fig. 9.4. The observation of the face orientation during a vertex-face CF depends on the executed CF-compliant path. After a straight line motion of the contact point, one of the orientation uncertainties of the face remains unobserved

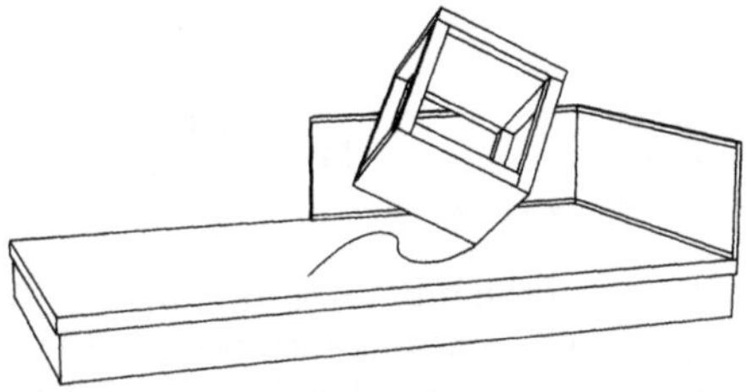

Fig. 9.5. The observation of the face orientation during a vertex-face CF depends on the executed CF-compliant path. A more arbitrary path of the contact point observes the full face orientation

i.e., the entropy of the pdf and loss functions of the covariance matrix of the estimate.

The robot controller of an autonomous robot requires a single estimate of the geometrical parameters. Hence, these parameters need to be estimated unambiguously to a certain accuracy. Therefore, the *covariance matrix* is a more appropriate measure of uncertainty than the entropy (see Sect. 3.5.3).

9.2.2 Information in a Single Measurement

Because at planning time, the true values of the geometrical parameters are unknown and only a prior distribution $p(\mathbf{x})$ is available, the *expected Fisher information* in a future measurement $\mathbf{z}$ is given by

$$\mathbf{I}_{VT,meas} = E_{p(\mathbf{x})}\left[E_{p(\mathbf{z}|\mathbf{x})}\left[-\nabla_{\mathbf{x}}\nabla_{\mathbf{x}}^T \ln p(\mathbf{z}|\mathbf{x})\right]\right]; \tag{9.1}$$

see Sect. 3.5.2. $\mathbf{I}_{VT,meas,k}$ is the *expected* information gain due to a future measurement $\hat{\mathbf{z}}_k$. This is the expected information gain for the parameter estimates when using an estimator with no information loss:

$$\mathbf{P}_{expected,k}^{-1} = \mathbf{P}_{k-1}^{-1} + \mathbf{I}_{VT,meas,k}. \tag{9.2}$$

Implicit and Nonlinear Measurement Equation

For the implicit and nonlinear measurement equation $\mathbf{h}_k(\mathbf{x}, \mathbf{z}_k) = \mathbf{0}_{n_h \times 1}$, the expectations in (9.1) have no analytical solution. Numerical approximation is computationally expensive. Hence, we *analytically approximate* the solution by working with a *linearized* measurement model obtained by a first-order Taylor series around the estimate of the geometrical parameters $\hat{\mathbf{x}}$ and the expected measurement value $\hat{\mathbf{z}}_k$. The approximate measurement equation is

$$\mathbf{h}_k(\mathbf{x}, \mathbf{z}_k) \approx \mathbf{h}_k(\hat{\mathbf{x}}, \hat{\mathbf{z}}_k) + \mathbf{H}_k(\mathbf{x} - \hat{\mathbf{x}}) + \mathbf{D}_k(\mathbf{z}_k - \hat{\mathbf{z}}_k); \tag{9.3}$$

where

$$\mathbf{H}_k = \left.\frac{\partial \mathbf{h}_k(\mathbf{x}, \mathbf{z}_k)}{\partial \mathbf{x}}\right|_{\mathbf{x}=\hat{\mathbf{x}}, \mathbf{z}_k=\hat{\mathbf{z}}_k}; \quad \mathbf{D}_k = \left.\frac{\partial \mathbf{h}_k(\mathbf{x}, \mathbf{z}_k)}{\partial \mathbf{z}_k}\right|_{\mathbf{x}=\hat{\mathbf{x}}, \mathbf{z}_k=\hat{\mathbf{z}}_k}.$$

This linearization is similar to the one used to apply the Kalman Filter equations to implicit measurement equations (see Sect. 7.3.1). The differences are (i) that the Kalman Filter approximation in Sect. 7.3.1 used the measured value $\hat{\mathbf{z}}_k$ instead of the expected measurement value $\hat{\mathbf{z}}_k$ used here. The expected measurement value $\hat{\mathbf{z}}_k$ is calculated based on the parameter estimate $\hat{\mathbf{x}}$ and the applied CF-compliant path; and (ii) that we do not consider any linearization errors here ($\boldsymbol{\rho}^*_{m,k} \equiv \mathbf{0}_{m \times 1}$, $\mathbf{R}^*_k \equiv \mathbf{0}_{m \times m}$).

The expected Fisher information (9.1) of a measurement $\hat{\mathbf{z}}_k$ with measurement equation (9.3) is

$$\mathbf{I}_{VT,meas,k} = \mathbf{I}_{FI,k}(\hat{\mathbf{x}}) = \mathbf{H}_k^T \left(\mathbf{D}_k \mathbf{R}_k \mathbf{D}_k^T\right)^{-1} \mathbf{H}_k; \tag{9.4}$$

if the measurement model uncertainty $\mathbf{R}_k$ is due to uncertainty on the measurement $\mathbf{z}_k$. This value equals the Fisher information $\mathbf{I}_{FI,k}(\mathbf{x})$, evaluated in $\hat{\mathbf{x}}$ (see Sect. 3.5.2). If $\hat{\mathbf{x}}$ approaches the true value of the parameter vector, then the information matrix (9.4) approaches the Fisher information matrix. Hence, re-planning at execution time (when more accurate estimates become available) results in more optimal task plans.

Informative Estimator

The information matrix $\mathbf{I}_{VT,meas,k}$ is a characteristic of the measurement $\hat{\mathbf{z}}_k$, and not of the applied estimator. This decouples the implementation of the task planner from the estimator component. However, the information gain (9.2) can only be reached by an estimator which does not loose any information, e.g., the Non-minimal State Kalman Filter of Chap. 5. When an approximate filter is used, e.g., one of the common Kalman Filter (KF) variants of Chap. 4, its information loss needs to be taken into account at planning time.

Robustness Against Non-Minimal Contact Modelling

Equations (9.2)–(9.4) are similar to a KF measurement update of the state covariance matrix for measurement equations linearized around $\hat{\mathbf{x}}$. Appendix D shows the robustness of the KF formulas against non-minimal contact modelling. The proof of the robustness of (9.2)–(9.4) is similar.

9.2.3 CF-Observable Parameter Space

$\mathbf{I}_{VT,meas,k}$ is the expected information gain of a measurement $\hat{\mathbf{z}}_k$ about the parameter vector $\mathbf{x}$. The eigenvalue decomposition of this matrix, $\mathbf{I}_{VT,meas,k} = \mathbf{U}_k \mathbf{\Sigma}_k \mathbf{U}_k^T$, indicates the expected information gains on the diagonal of $\mathbf{\Sigma}_k$ for each of the parameter directions (linear combinations of the parameters) in the columns of $\mathbf{U}_k$; i.e. the eigenvectors. Each parameter direction with a non-zero expected information gain is a parameter direction in which the expected posterior covariance value $\mathbf{P}_{expected,k}$ is smaller than the prior covariance value $\mathbf{P}_{k-1}$.

Definition: CF-Observable Parameter Space

The *observable parameter space of a CF* or *"CF-observable parameter space"* is the space of parameter directions that *can* be observed during CF-compliant motion.
The observable parameter space of a CF is a characteristic of the CF given the type of the measurements and the allowed measurement range.

Spanning Set for the CF-Observable Parameter Space

We describe the observable parameter space of $CF = CF_i$ by a spanning set $\mathbf{O}_i$ of this space. For instance, consider the state vector of the experiment in the previous chapter:

$$\mathbf{x} = [x^m \; y^m \; z^m \; \theta_x^m \; \theta_y^m \; \theta_z^m \; x^e \; y^e \; z^e \; \theta_x^e \; \theta_y^e \; \theta_z^e]^T; \tag{9.5}$$

and consider a face-face CF for which both the Z-axis of {m} and the Z-axis of {e} are perpendicular to the contacting faces (see Fig. 6.1). A spanning set for the observable parameter space for pose and twist measurements is:

$$\mathbf{O} = \begin{bmatrix} 0\text{ mm} & 0\text{ mm} & 0\text{ mm} & 0\text{ mm} & 0\text{ mm} \\ 0\text{ mm} & 0\text{ mm} & 0\text{ mm} & 0\text{ mm} & 0\text{ mm} \\ 0\text{ mm} & 0\text{ mm} & 0\text{ mm} & 0\text{ mm} & 1\text{ mm} \\ 1\text{ rad} & 0\text{ rad} & 0\text{ rad} & 0\text{ rad} & 0\text{ rad} \\ 0\text{ rad} & 1\text{ rad} & 0\text{ rad} & 0\text{ rad} & 0\text{ rad} \\ 0\text{ rad} & 0\text{ rad} & 0\text{ rad} & 0\text{ rad} & 0\text{ rad} \\ 0\text{ mm} & 0\text{ mm} & 0\text{ mm} & 0\text{ mm} & 0\text{ mm} \\ 0\text{ mm} & 0\text{ mm} & 0\text{ mm} & 0\text{ mm} & 0\text{ mm} \\ 0\text{ mm} & 0\text{ mm} & 0\text{ mm} & 0\text{ mm} & 1\text{ mm} \\ 0\text{ rad} & 0\text{ rad} & 1\text{ rad} & 0\text{ rad} & 0\text{ rad} \\ 0\text{ rad} & 0\text{ rad} & 0\text{ rad} & 1\text{ rad} & 0\text{ rad} \\ 0\text{ rad} & 0\text{ rad} & 0\text{ rad} & 0\text{ rad} & 0\text{ rad} \end{bmatrix}.$$

The first two columns of $\mathbf{O}$ indicate that the orientation of the contacting face of the manipulated object is observable. The next two columns indicate that the orientation of the contacting face of the environment object is observable. The last column indicates that the sum of the variables z^m and z^e is observable (see also Fig. 8.18).

A spanning set of a CF-observable parameter space for pose and twist measurements can be obtained by analysing $\mathbf{H}_k$ for a limited number of poses and twists:

1. The set of eigenvectors of $\mathbf{I}_{VT,meas,k}$ corresponding to non-zero eigenvalues[1] is a spanning set of the space of parameter directions observed by a twist or pose measurement. Appendix G shows that if $\mathbf{I}_{VT,meas,k}$ is calculated by (9.4), the set of *right singular directions of matrix* $\mathbf{H}_k$ *corresponding to non-zero singular values* is equally a spanning set of this space.
2. Appendix G proves that the calculation of the spanning set of the CF-observable parameter space can be limited to the calculation of a spanning set of the space of parameter directions observed by *one pose measurement* and spanning sets of the spaces of parameter directions observed by twists equal to the *twist spanning set directions, i.e., the columns of* $\mathbf{J}_k(\mathbf{x}, \mathbf{t}_{d,k})$, *at several poses.*
   The number of poses at which the observable parameter directions for twists need to be calculated, depends on the CF. E.g., for a vertex-face CF at a given pose $\mathbf{t}_{d,1}$, the full orientation of the face in the world frame is observed by motions of the contacting vertex in the two allowed translational velocity degrees of freedom. The rotational velocities allow to identify two of the three position coordinates of the position of the contacting vertex with respect to the gripper frame. The unobservable position coordinate is the position of the vertex along a normal to the face (see Fig. 9.6). The analysis at *two* poses of the manipulated object which are rotated about an axis which is not the face normal, is sufficient to determine the CF-observable parameter directions of a vertex-face CF.

[1] Because the matrix $\mathbf{I}_{VT,meas,k}$ is a symmetric positive semi-definite matrix, its eigenvalue decomposition equals its singular value decomposition.

A CF-independent strategy is to calculate spanning sets for the observable parameter directions based on $\mathbf{H}_k$ for one pose measurement and for twists equal to the twist spanning set directions *at a large number of poses*. The calculation of the spanning sets of the CF-observable parameter spaces is performed offline, hence, its computational cost is of lower importance.

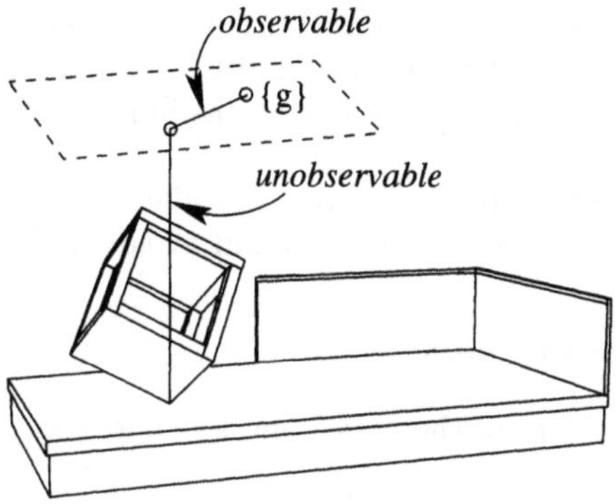

Fig. 9.6. The instantaneously observable and unobservable coordinates of the contacting vertex during a vertex-face CF

Observable Parameter Directions of a CF Sequence

The space of observable parameter directions of a *CF sequence* $CF_1, \ldots, CF_{n_{Seq}}$ is the union of the CF-observable parameter spaces of the different CFs in the sequence. A spanning set for this union is $\mathbf{O}_{Seq} = \left[\mathbf{O}_1 \ldots \mathbf{O}_{n_{Seq}}\right]$.

9.3 General Active Sensing Task Plan

A general task plan consists of (i) the CF sequence $\boldsymbol{\pi}$, see Fig. 9.1, and (ii) the CF-compliant paths $\mathbf{U}_i$ to execute in each of the n_{Seq} CFs of the sequence, see Figs. 9.2 and 9.3.

The task plan is optimised with respect to an objective function with the accurate observation of the parameters as constraint (Sect. 9.3.1). I.e., the covariance matrix of the parameter estimates after the task execution $\mathbf{P}_{end,Seq}(\boldsymbol{\pi}, \mathbf{U}_1, \ldots, \mathbf{U}_{n_{Seq}})$ should be smaller than a user-defined uncertainty tolerance matrix $\mathbf{P}_{toleranced}$.

Section 9.3.2 presents how we decouple the optimisation problem into $n_{Seq} + 1$ optimisation problems: first, the optimal CF sequence $\boldsymbol{\pi}$ is calculated (Sect. 9.4), then, the CF-compliant paths $\mathbf{U}_i$ for each of the CFs in the optimal sequence are defined ($1 \leq i \leq n_{Seq}$) (Sect. 9.5).

Definitions: Active Sensing Motion, Transition Motion.

The CF-compliant path $\mathbf{U}_i$ has two components: (i) the *active sensing motion* $\mathbf{U}_{a,i}$, which is designed to observe the geometrical parameters to the required accuracy; and (ii) the *transition motion* $\mathbf{U}_{t,i}$, which is designed to trigger the transition to the next CF. Note that $\mathbf{U}_{a,i}$ and $\mathbf{U}_{t,i}$ do not need to be consecutive paths in time, the active sensing motion $\mathbf{U}_{a,i}$ can, for instance, be a description of a motion superimposed on $\mathbf{U}_{t,i}$ (see Fig. 9.3). In this case the active sensing motion and the transition motion do not need to be CF-compliant; only the *resultant* path $\mathbf{U}_i$ is CF-compliant.

This chapter describes the calculation of CF-compliant active sensing and transition motions for sequential execution (as illustrated in Fig. 9.2).

Section 9.3.3 describes an algorithm to calculate a CF-compliant transition motion which closes the desired contacts and avoids undesired transitions.

9.3.1 Optimisation Problem

The optimisation problem of the task plan is denoted as:

$$\begin{aligned}
\min_{\boldsymbol{\pi},\mathbf{U}_1,\ldots,\mathbf{U}_{n_{Seq}}} \quad & \mathbf{f}_{obj,Seq}(\boldsymbol{\pi},\mathbf{U}_1,\ldots,\mathbf{U}_{n_{Seq}}); \\
\text{subject to} \quad & \mathbf{P}_{end,Seq}(\boldsymbol{\pi},\mathbf{U}_1,\ldots,\mathbf{U}_{n_{Seq}}) \leq \mathbf{P}_{toleranced}; \\
& \mathbf{c}_{ineq,Seq}(\boldsymbol{\pi},\mathbf{U}_1,\ldots,\mathbf{U}_{n_{Seq}}) < \mathbf{0}_{n_i\times 1}; \\
& \mathbf{c}_{eq,Seq}(\boldsymbol{\pi},\mathbf{U}_1,\ldots,\mathbf{U}_{n_{Seq}}) = \mathbf{0}_{n_e\times 1}; \\
& CF_1 \in \textsc{ListOfInitialCFs}; \\
& CF_{n_{Seq}} \in \textsc{ListOfGoalCFs}.
\end{aligned}$$

where $\mathbf{f}_{obj,Seq}(\boldsymbol{\pi},\mathbf{U}_1,\ldots,\mathbf{U}_{n_{Seq}})$ is the objective function to minimise. E.g., this can be the cost of the plan in terms of time, energy, number of CF transitions, etc. The n_i inequality constraints $\mathbf{c}_{ineq,Seq}(\boldsymbol{\pi},\mathbf{U}_1,\ldots,\mathbf{U}_{n_{Seq}}) < \mathbf{0}_{n_i\times 1}$ and the n_e equality constraints $\mathbf{c}_{eq,Seq}(\boldsymbol{\pi},\mathbf{U}_1,\ldots,\mathbf{U}_{n_{Seq}}) = \mathbf{0}_{n_e\times 1}$ denote constraints on the executed CF sequence and CF-compliant paths such as the robot work-space limits, the maximum number of allowed constraint directions, maximum velocities, accelerations, etc.

9.3.2 Decoupling of the Optimisation Problem

The notion of the CF-observable parameter space allows us to define CF-sequences for which CF-compliant motions *can be found* which observe the geometrical parameter estimates to the required accuracy. This allows the *decoupling of the active sensing requirement* into a requirement of the CF sequence on the one hand and requirements of the CF-compliant paths in each of the CFs on the other hand. The *decoupling of the optimisation problem* is, of course, only possible if also the objective function and the other constraints can be decoupled. This is problem dependent.

This chapter focuses on the following two optimisation problems:

Problem 9.1 *The optimisation of the task plan to a minimum number of CF transitions* $n_{Seq}(\boldsymbol{\pi})$, *i.e.,*

$$\begin{aligned} \min_{\boldsymbol{\pi}} \quad & n_{Seq}(\boldsymbol{\pi}); \\ \textit{subject to} \quad & \mathbf{P}_{end,Seq}(\boldsymbol{\pi}, \mathbf{U}_1, \ldots, \mathbf{U}_{n_{Seq}}) \leq \mathbf{P}_{toleranced}; \\ & E_{kin,Seq}(\mathbf{U}_1, \ldots, \mathbf{U}_{n_{Seq}}) - E_{kin,max} < 0; \\ & CF_1 \in \text{LISTOFINITIALCFS}; \\ & CF_{n_{Seq}} \in \text{LISTOFGOALCFS}. \end{aligned}$$

Problem 9.2 *The optimisation of the task plan to the minimum execution time* $t_{Seq}(\boldsymbol{\pi}, \mathbf{U}_1, \ldots, \mathbf{U}_{n_{Seq}})$, *i.e.,*

$$\begin{aligned} \min_{\boldsymbol{\pi}, \mathbf{U}_1, \ldots, \mathbf{U}_{n_{Seq}}} \quad & t_{Seq}(\boldsymbol{\pi}, \mathbf{U}_1, \ldots, \mathbf{U}_{n_{Seq}}); \\ \textit{subject to} \quad & \mathbf{P}_{end,Seq}(\boldsymbol{\pi}, \mathbf{U}_1, \ldots, \mathbf{U}_{n_{Seq}}) \leq \mathbf{P}_{toleranced}; \\ & E_{kin,Seq}(\mathbf{U}_1, \ldots, \mathbf{U}_{n_{Seq}}) - E_{kin,max} < 0; \\ & CF_1 \in \text{LISTOFINITIALCFS}; \\ & CF_{n_{Seq}} \in \text{LISTOFGOALCFS}. \end{aligned}$$

The second constraint of both problems indicates that the kinetic energy of the CF-compliant motions is constrained to $E_{kin,max}$.

Decoupling Problem 9.1

In optimisation problem 9.1, the *CF sequence* can be optimised independently of the CF-compliant paths. The active sensing constraint of the CF sequence[2] expresses that the parameter directions which need to be observed better, i.e., directions for which $\mathbf{P}_{init}$ is larger than $\mathbf{P}_{toleranced}$, are in the observable parameter space of the sequence:

[2] DIRECTIONSTOOBSERVE($\mathbf{P}_1, \mathbf{P}_2$) refers to all parameter directions for which the uncertainty is larger under covariance matrix $\mathbf{P}_1$ than under covariance matrix $\mathbf{P}_2$. OBSERVABLESPACE($\mathbf{O}$) denotes the space of observable parameter directions given its spanning set $\mathbf{O}$. The constraint DIRECTIONSTOOBSERVE($\mathbf{P}_1, \mathbf{P}_2$) $\in$ OBSERVABLESPACE($\mathbf{O}$) is satisfied if the eigenvectors $\mathbf{V}_i$ of the matrix $\mathbf{P}_1 - \mathbf{P}_2$ for which the eigenvalues are larger than zero, are vectors in OBSERVABLESPACE($\mathbf{O}$). These vectors $\mathbf{V}_i$ are vectors in OBSERVABLESPACE($\mathbf{O}$) if the vectors can be written as linear combinations of the spanning vectors, i.e., $\exists \boldsymbol{\psi}_i : \mathbf{V}_i = \mathbf{O}\, \boldsymbol{\psi}_i$.

$$
\begin{aligned}
\min_{\boldsymbol{\pi}} \quad & n_{Seq}(\boldsymbol{\pi}); && (9.6)\\
\text{subject to} \quad & \textsc{DirectionsToObserve}(\mathbf{P}_{init}, \mathbf{P}_{toleranced}) \\
& \qquad \in \textsc{ObservableSpace}(\mathbf{O}_{Seq}(\boldsymbol{\pi})); && (9.7)\\
& CF_1 \in \textsc{ListOfInitialCFs}; && (9.8)\\
& CF_{n_{Seq}} \in \textsc{ListOfGoalCFs}. && (9.9)
\end{aligned}
$$

The *CF-compliant motion trajectories* do not need to be optimised, they are only subject to the following constraints:

$$\mathbf{P}_{end,i}(\mathbf{U}_i) \leq \mathbf{P}_{toleranced,i}; \tag{9.10}$$

$$E_{kin,i}(\mathbf{U}_i) - E_{kin,max} < 0; \tag{9.11}$$

where $\mathbf{P}_{toleranced,i}$ is the toleranced uncertainty at the end of $\mathbf{U}_i$. At the end of the CF sequence the toleranced uncertainty is

$$\mathbf{P}_{toleranced,n_{Seq}} = \mathbf{P}_{toleranced}. \tag{9.12}$$

The other tolerances $\mathbf{P}_{toleranced,i}$ are defined such that $(i = n_{Seq}, \ldots, 1)$

$$
\begin{aligned}
&\textsc{DirectionsToObserve}(\mathbf{P}_{toleranced,i-1}, \mathbf{P}_{toleranced,i}) \\
&\qquad \in \textsc{ObservableSpace}(\mathbf{O}_i); && (9.13)
\end{aligned}
$$

where $\mathbf{P}_{toleranced,0}$ equals $\mathbf{P}_{init}$. There is some freedom in the choice of the matrices $\mathbf{P}_{toleranced,i}$ because some parameter directions are observable in several CFs. Our *choice* is to observe the parameter directions to the required accuracy *during the first CF in which they are observable.*

Decoupling Problem 9.2

Optimisation problem 9.2 *can not be decoupled* because the execution time of the active sensing motion $\mathbf{U}_{a,i}$ influences the objective function. The combined optimisation problem of the CF sequence and the CF-compliant paths is computationally very expensive. Our approach is to decouple the optimisation problem by *approximating the "extra execution time" $t_{a,i}$ of the CF-compliant path due to active sensing with a user-defined value* during the optimisation of the CF-sequence. For an active sensing motion $\mathbf{U}_{a,i}$ which is executed separately of the transition motion $\mathbf{U}_{t,i}$ (Fig. 9.2), this "extra time" $t_{a,i}$ is the execution time of the active sensing path $\mathbf{U}_{a,i}$. For an active sensing motion $\mathbf{U}_{a,i}$ which is superimposed on the transition motion $\mathbf{U}_{t,i}$ (Fig. 9.3), this "extra time" is the difference in execution time of the combined motion $\mathbf{U}_i$ and of the transition motion $\mathbf{U}_{t,i}$ calculated when there is no need for active sensing.

By approximating the "extra execution time" of the CF-compliant path due to the active sensing with a user-defined value, the optimisation problem of the *CF sequence* is again independent of the active sensing paths $\mathbf{U}_{a,i}$:

$$\min_{\boldsymbol{\pi}} \quad t_{Seq}(\boldsymbol{\pi}); \tag{9.14}$$

$$\text{subject to} \quad \text{DIRECTIONSTOOBSERVE}(\mathbf{P}_{init}, \mathbf{P}_{toleranced}) \in \text{OBSERVABLESPACE}(\mathbf{O}_{Seq}(\boldsymbol{\pi})); \tag{9.15}$$

$$E_{kin,Seq}(\mathbf{U}_{t,1}, \ldots, \mathbf{U}_{t,n_{Seq}}) - E_{kin,max} < 0; \tag{9.16}$$

$$CF_1 \in \text{LISTOFINITIALCFS}; \tag{9.17}$$

$$CF_{n_{Seq}} \in \text{LISTOFGOALCFS}. \tag{9.18}$$

The constraint on the kinetic energy of the applied motion constrains the transition motions $\mathbf{U}_{t,i}$ which enhance the CF transitions.

The total execution time $t_{Seq}(\boldsymbol{\pi})$ is the sum of the execution times of the CF-compliant paths $\mathbf{U}_i$ plus the *stabilisation time* $t_{s,i}$ (the duration of transition effects at CF transition, due to dynamics and elasticity of the controller and the robot-environment interaction) at each transition. The execution time of a CF-compliant path $\mathbf{U}_i$ is the active sensing time $t_{a,i}$ plus the time $t_{t,i}$ of the transition motion.

$$t_{Seq}(\boldsymbol{\pi}) = \sum_{i=1}^{i=n_{Seq}-1} (t_{a,i} + t_{t,i} + t_{s,i}) + t_{a,n_{Seq}}.$$

The active sensing time $t_{a,i}$ and the stabilisation time $t_{s,i}$ are approximated by user-defined values. The transition motion $\mathbf{U}_{t,i}$ is simulated with the algorithm described in Sect. 9.3.3. For a given pose, the algorithm returns a twist which reduces the distances between the features corresponding to the desired elementary contacts to close, while avoiding undesired contacts by keeping the corresponding distances above a user-defined value.

After the optimisation of the CF sequence, the *active sensing paths* $\mathbf{U}_{a,i}$ in each of the CFs of the optimal sequence ($1 \leq i \leq n_{Seq}$) are optimised:

$$\min_{\mathbf{U}_{a,i}} \quad t_{a,i}(\mathbf{U}_i); \tag{9.19}$$

$$\text{subject to} \quad \mathbf{P}_{end,i}(\mathbf{U}_i) \leq \mathbf{P}_{toleranced,i}; \tag{9.20}$$

$$E_{kin,i}(\mathbf{U}_i) - E_{kin,max} < 0. \tag{9.21}$$

The matrices $\mathbf{P}_{toleranced,i}$ are defined by (9.12)–(9.13).

9.3.3 Transition Motion

This section describes the generation of transition motions to move from a CF to the next one. The presented solution works well in the absence of obstacles. More sophisticated (but computationally more expensive) algorithms need to be considered in other cases.

Given the current pose $\mathbf{t}_{d,k}$, the algorithm calculates a desired twist $\mathbf{t}_k = \mathbf{J}_{i,k}(\mathbf{x}, \mathbf{t}_{d,k})\ \chi_k$ in the twist space of the current $CF = CF_i$. The twist (i) performs a position closure at the constraints of the next CF while (ii) avoiding undesired contacts.

The idea is similar to the work of McCarragher [183, 184] on polyhedral CFs. However, the presented algorithm considers more general CFs and does not need any "maintaining conditions" because the desired twist is calculated in the contact's twist space.

Calculation of a CF-Compliant Transition Motion

Given a pose $\mathbf{t}_{d,k}$, the calculated twist $\mathbf{t}_k = \mathbf{J}_{i,k}(\mathbf{x}, \mathbf{t}_{d,k})\, \boldsymbol{\chi}_k$

1. performs a position closure at the constraints of the next CF. $\mathbf{h}_{d,next,k}(\mathbf{x}, \mathbf{t}_{d,k})$ is a vector of which each element represents the distance[3] to contact closure for an additional elementary contact of the next CF.
   The twist of the manipulated object is chosen such that these contacts are closed simultaneously:

$$\mathbf{h}_{d,next,k}(\mathbf{x}, \mathbf{t}_{d,k}) \sim \mathbf{A}_{next,k}\, \boldsymbol{\chi}_k;$$

$$\mathbf{A}_{next,k} = \begin{bmatrix} [\mathbf{n}_{1,k}^T \ \mathbf{0}_{1\times3}]\ {}^{a}_{c1}\mathbf{S}_t \\ [\mathbf{n}_{2,k}^T \ \mathbf{0}_{1\times3}]\ {}^{a}_{c2}\mathbf{S}_t \\ \ldots \qquad \ldots \\ [\mathbf{n}_{m,k}^T \ \mathbf{0}_{1\times3}]\ {}^{a}_{cm}\mathbf{S}_t \end{bmatrix} \mathbf{J}_{i,k}(\mathbf{x}, \mathbf{t}_{d,k});$$

   where $\mathbf{n}_{j,k}, 1 \leq j \leq m$ is a unit vector normal[3] to the jth elementary contact to close, expressed in a frame {cj} in the contacting point[3] of the manipulated object. {a} represents the frame in which $\mathbf{J}_{i,k}(\mathbf{x}, \mathbf{t}_{d,k})$ is expressed.
2. avoids undesired contacts. I.e., if the distance[3] between a pair of features of the manipulated object and environment object is smaller than a user-defined distance $h_{d,avoid,desired}$ and the corresponding contact is undesired, then the twist is chosen to maintain the distance $h_{d,avoid,desired}$ between the features:

$$\mathbf{h}_{d,avoid,desired} - \mathbf{h}_{d,avoid,k}(\mathbf{x}, \mathbf{t}_{d,k}) \sim \mathbf{A}_{avoid,k}\, \boldsymbol{\chi}_k;$$

$$\mathbf{A}_{avoid,k} = \begin{bmatrix} [\mathbf{n}_{1,k}^T \ \mathbf{0}_{1\times3}]\ {}^{a}_{c1}\mathbf{S}_t \\ [\mathbf{n}_{2,k}^T \ \mathbf{0}_{1\times3}]\ {}^{a}_{c2}\mathbf{S}_t \\ \ldots \qquad \ldots \\ [\mathbf{n}_{l,k}^T \ \mathbf{0}_{1\times3}]\ {}^{a}_{cl}\mathbf{S}_t \end{bmatrix} \mathbf{J}_{i,k}(\mathbf{x}, \mathbf{t}_{d,k});$$

---

[3] The *distance* between two features of an elementary contact which is not closed is defined as the shortest distance between the features. The *contact normal* of an elementary contact which is not closed is defined as the direction of the shortest distance between the two features which define the elementary contact. The *contact points* on the environment and manipulated object for an elementary contact which is not closed are defined as the closest points on the two features which define the elementary contact.

where $\mathbf{n}_{j,k}, 1 \leq j \leq l$ is a unit vector normal[3] to the jth elementary contact to avoid, expressed in a frame {cj} in the contacting point[3] of the manipulated object. {a} represents the frame in which $\mathbf{J}_{i,k}(\mathbf{x}, \mathbf{t}_{d,k})$ is expressed.

Fortunately, not all distances between all pairs of features of the manipulated object and environment object need to be verified. It is sufficient to *restrict* the verification to the elementary contacts of the CFs neighbouring to the current CF which are not in contact during the next, desired, CF.

If the next CF is *less constrained* than the current CF, the transition motion arises fully from contact avoidance. The previously established, undesired, elementary contacts are broken.

Using the symbols

$$\boldsymbol{\Delta}\mathbf{h}_{d,k}(\mathbf{x}, \mathbf{t}_{d,k}) = \begin{bmatrix} \mathbf{h}_{d,next,k}(\mathbf{x}, \mathbf{t}_{d,k}) \\ \mathbf{h}_{d,avoid,desired} - \mathbf{h}_{d,avoid,k}(\mathbf{x}, \mathbf{t}_{d,k}) \end{bmatrix};$$

$$\mathbf{A}_k(\mathbf{x}, \mathbf{t}_{d,k}) = \begin{bmatrix} \mathbf{A}_{next,k}(\mathbf{x}, \mathbf{t}_{d,k}) \\ \mathbf{A}_{avoid,k}(\mathbf{x}, \mathbf{t}_{d,k}) \end{bmatrix};$$

the constraints on the desired twist $\mathbf{J}_{i,k}(\mathbf{x}, \mathbf{t}_{d,k})\, \chi_k$ are described by

$$\boldsymbol{\Delta}\mathbf{h}_{d,k}(\mathbf{x}, \mathbf{t}_{d,k}) \sim \mathbf{A}_k(\mathbf{x}, \mathbf{t}_{d,k})\, \chi_k. \tag{9.22}$$

Generally, (9.22) leaves a freedom in the choice of χ_k, i.e., several χ_k result in the same $\mathbf{A}_k(\mathbf{x}, \mathbf{t}_{d,k})\chi_k$. We choose the χ_k which performs the motion with minimal required kinetic energy of the manipulated object:

$$\min_{\chi_k} \frac{1}{2} \mathbf{t}_k^T \mathbf{M} \mathbf{t}_k;$$

i.e.,

$$\chi_k \sim \mathbf{A}^{\#}_{W^{-1},k}(\mathbf{x}, \mathbf{t}_{d,k})\, \boldsymbol{\Delta}\mathbf{h}_{d,k}(\mathbf{x}, \mathbf{t}_{d,k}); \tag{9.23}$$

$$\mathbf{A}^{\#}_{W^{-1},k}(\mathbf{x}, \mathbf{t}_{d,k}) = \mathbf{W}_k^{-1}(\mathbf{x}, \mathbf{t}_{d,k})\, \mathbf{A}_k^T(\mathbf{x}, \mathbf{t}_{d,k}) \times$$
$$\left(\mathbf{A}_k(\mathbf{x}, \mathbf{t}_{d,k})\, \mathbf{W}_k^{-1}(\mathbf{x}, \mathbf{t}_{d,k})\, \mathbf{A}_k^T(\mathbf{x}, \mathbf{t}_{d,k})\right)^{-1};$$
$$\mathbf{W}_k(\mathbf{x}, \mathbf{t}_{d,k}) = \mathbf{J}_{i,k}^T(\mathbf{x}, \mathbf{t}_{d,k})\, \mathbf{M}\, \mathbf{J}_{i,k}(\mathbf{x}, \mathbf{t}_{d,k});$$

where $\mathbf{M}$ is the inertia matrix of the manipulated object, expressed in the same frame as the twist spanning set $\mathbf{J}_{i,k}(\mathbf{x}, \mathbf{t}_{d,k})$. Minimising the kinetic energy of the manipulated object is only one possible choice; other possible criteria are the minimisation of the kinetic energy of the whole robot system, the maximisation of the distance to singularities or collisions, etc.

Equation (9.23) defines χ_k up to a scaling factor. The obtained value is rescaled such that the kinetic energy of the manipulated object equals the maximum allowed value $E_{kin,max}$.

Computational Cost

The algorithm only uses simple linear algebraic formulas, hence, *online* use is not a problem.

The calculation of a transition motion *in simulation*, however, is slow because at each time step the pose needs to be calculated based on the previous pose and the generated twist command. This numerical "integration" of the twists needs to be performed using very small time increments.

Automatic Generation of the Equations

The equations for the calculation of the transition motion are automatically generated based on the elementary contact library of Chap. 6. The contact modelling is easier if non-minimal models are allowed. This is the case because non-minimal modelling just adds constraints on the transition twist which are combinations of already existing constraints. This does not influence the resulting twist.

Note that the numerically calculated twist spanning set $\mathbf{J}_{i,k}(\mathbf{x}, \mathbf{t}_{d,k})$ (Sect. 6.4.1) is minimal, even for non-minimal contact models. This implies that $\mathbf{W}_k(\mathbf{x}, \mathbf{t}_{d,k})$ is a regular matrix and that the calculation of $\mathbf{A}^{\#}_{W^{-1},k}(\mathbf{x}, \mathbf{t}_{d,k})$ can use a regular inverse of this matrix.

Wrench Control Command

Austin and McCarragher [15, 16, 17] describe the generation of a wrench instead of a twist control command. The wrench command is based on pulling forces for elementary contacts to close and pushing forces for elementary contacts to avoid. By doing so, non-minimal contact models result in larger pulling or pushing forces of the non-minimally modelled contacts. The "transition wrench" is the *resultant* of the imposed wrenches on each of the contacts to avoid or to close. Hence, the result is different for different decompositions of the same CF. Therefore, we prefer a twist control command over a wrench control command.

9.4 CF Sequence

This section presents the calculation of (i) the *shortest* CF sequence, i.e., optimisation problem 9.1, (9.6)–(9.9); and (ii) the *fastest* CF sequence, i.e., optimisation problem 9.2, (9.14)–(9.18).

A CF sequence is denoted by a vector $\boldsymbol{\pi}$. This vector includes the *total number* of CFs n_{Seq} and the *values* $CF_1, \ldots, CF_{n_{Seq}}$ of the CFs in the sequence. For every CF, the CF-observable parameter space for pose and twist measurements is calculated beforehand (see Sect. 9.2.3 and Appendix G).

Problems 9.1 and 9.2 are *discrete* optimisation problems under *uncertainty*:

1. The CF sequence $\boldsymbol{\pi}$ only takes *discrete* values. Not all CF sequences are possible either, only transitions between neighbouring CFs are allowed. These transitions are summarised in a CF graph. The allowed CFs and CF transitions can be further reduced depending on the task.
2. At execution time, the task is subject to *uncertainty*: (i) unexpected CF transitions can occur because the geometrical parameters are inaccurately known. This implies that a control command intended to make a specific CF transition can cause another CF transition (*process uncertainty*); and (ii) the measurements do not always uniquely identify the CFs, i.e., during the task execution only a probability distribution over the CFs can be calculated (*measurement uncertainty*).

Section 3.6 describes the existing algorithms for discrete optimisation under uncertainty. These algorithms (i) provide only the *expected* optimum and (ii) are computationally intractable for all but the smallest problems.

9.4.1 Optimisation Solution

In our approach, *the initial estimates of the geometrical parameters are assumed to be the true values.* This assumption eliminates both the process and the measurement uncertainty and the optimisation problem becomes *a deterministic graph search.*

Algorithm 9.1 presents the implemented graph search algorithm. This algorithm assumes that the value of the objective function increases with each added CF transition. This is the case for both the number of transitions (problem 9.1) and the execution time (problem 9.2). The assumption allows us to discard sequences which have a larger objective function than the previously found solution.

The algorithm preferably expands the CF sequences which are close to the goal. In the context of the above described optimisation problem, this means "close to the toleranced accuracy" and "close to the goal CF" (see Algorithm 9.1, lines 4 and 5). The applied distance measures are:

1. The distance between the toleranced accuracy and the accuracy of the parameter estimates after executing a CF sequence, is defined as the rank of the space of parameter directions which are not yet observed and for which the initial covariance is larger than the toleranced covariance.
2. The distance between the last CF of a sequence and the goal CF is defined as the minimum number of elementary CFs to break, increased with the minimum number of elementary CFs to close in order to move from the CF to the goal CF. If there is a list of goal CFs, the measure is determined by the "closest" goal CF.

Algorithm 9.1 Graph search algorithm for optimal CF sequence

```
Initialise the BestFinishedSequence to be empty
Initialise the ListOfActiveSequences to the list of sequences containing all
possible initial CFs as first elements
while the ListOfActiveSequences is not empty do
  Copy from the ListOfActiveSequences the list of the sequences which are
  "closest to the toleranced accuracy"
  Take from the above list the list of the sequences which are "closest to the
  goal CF"
  Take from the above list the ActiveSequence with the smallest objective
  function
  Remove the ActiveSequence from the ListOfActiveSequences
  Find the NumberOfNeighbouring neighbouring CFs of the last CF of the
  ActiveSequence
  for j = 1 to NumberOfNeighbouring do
    Create a sequence CurrentActiveSequence by adding the jth neigh-
    bouring CF to the ActiveSequence
    Calculate the new objective function value for the CurrentActive-
    Sequence
    if the CurrentActiveSequence has a smaller objective function value
    than the BestFinishedSequence then
      Add a spanning set of the CF-observable space of the jth neighbour-
      ing CF to the spanning set of the observable parameter space of the
      CurrentActiveSequence
      if the CurrentActiveSequence ends with a CF which is a possible
      goal CF and if the observable parameter space allows the observation to
      the required accuracy then
        Save the CurrentActiveSequence as the BestFinishedSequence
        Remove the sequences with higher objective function value than the
        BestFinishedSequence from the ListOfActiveSequences
      else
        Add the CurrentActiveSequence to the ListOfActive-
        Sequences
      end if
    end if
  end for
end while
Return the BestFinishedSequence
```

9.4.2 Example

The theory is illustrated by the planning of several tasks with a cube as manipulated object and a corner object as environment object. The position and orientation of both objects are inaccurately known, i.e., we want to estimate twelve geometrical parameters. The space of allowed active sensing motions is the full twist space of the manipulated object in every CF. All task executions start in free space. The side length of the cube is 250 mm, its mass is 1 kg and is located on the edges of the cube. The maximum allowed kinetic energy is 0.0002 Nm. E.g for a cube with mass 1 kg this corresponds to a translational velocity of 20 mm/s. Note that this velocity is low because the task plan is to be executed under uncertainty. High speeds would cause brutal collisions and contact loss during compliant motion. The user-defined parameters of the task planning are the distance to undesired contacts $h_{d,avoid,desired} = 100$ mm, the stabilisation time $t_{s,i} = 1$ second and the active sensing time $t_{a,i} = 2$ seconds, for all CFs $(1 \leq i \leq n_{Seq})$.

The simulation programs are written in Matlab and run on a Pentium III 800 MHz, 128 Mb RAM laptop. The examples show

1. that the optimisation of the shortest CF sequence (Problem 9.1) is fast. The time to solve the optimisation problem is between 0.15 seconds and 12.5 seconds for the presented problems. The calculation of the fastest CF sequence (Problem 9.2), on the other hand, is computationally much more expensive due to the numerical simulation of the transition motions (Sect. 9.3.3). The calculation time to solve the presented optimisation problems is between 13 minutes and two weeks. The indicated time laps comprise the time to find the solution *and* to verify its optimality.
2. that the extra computation time for the calculation of the fastest CF sequence results in sequences which are considerably faster; in all examples we note execution times of two thirds or less than that of the shortest CF sequence. The gain in execution time is due to making contacts between features which are already close to each other.

Observe the Orientation of the Horizontal Face

As a first example, the program calculates a CF sequence with a *minimum number of CF transitions.* The sequences are constrained to sequences which observe the parameters corresponding to the orientation of the horizontal face of the corner object. No goal CF is specified. The time to solve the optimisation problem is about 0.15 seconds. The solution (Fig. 9.7) has only one CF transition, i.e., a transition to a vertex-face CF with the horizontal face. Its execution time[4] is 34.0 seconds.

Figure 9.8 presents the solution for the minimisation of *the execution time.* The calculation time to solve the optimisation problem is about 13 minutes. The execution time of the solution is 21.8 seconds. This is faster than the solution of Fig. 9.7 because the considered vertex is closer to the face.

[4] The long execution times are due to the constraint on the kinetic energy.

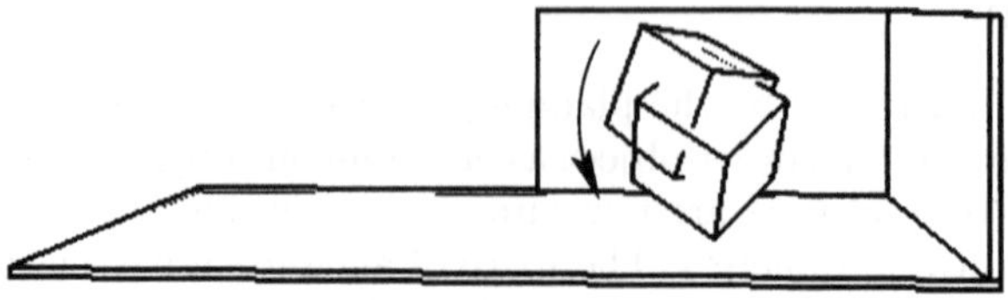

Fig. 9.7. The *shortest* CF sequence which observes the orientation of a face is the transition to a vertex-face CF

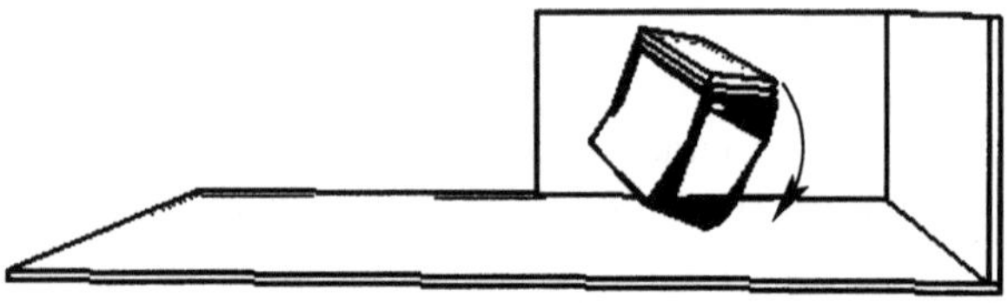

Fig. 9.8. The *fastest* CF sequence which observes the orientation of a face is the transition to a vertex-face CF by making contact with the vertex closest to the face

Observe All Geometrical Parameters Using Only Vertex-Face CFs

The next optimisation problem searches a CF sequence which observes all the geometrical parameters, being the twelve parameters corresponding to the positions and orientations of the cube and the corner object. By using a reduced CF graph, the sequence is restricted to only guarded motions (motion in free space) and single *vertex-face CFs*. I.e., the geometrical parameters are identified by "hopping" between different vertex-face CFs. Note that during guarded motions none of the geometrical parameters are observable, hence, all active sensing motions are compliant motions (in this example motions during *vertex-face* contact).

Figure 9.9 shows the result of the optimal CF sequence when the *number of CF transitions* is minimised. Vertex-face CFs are made with the horizontal face, the right face and the rear face respectively. This optimal CF sequence is obtained in about 10.7 seconds. Its execution time is 3 minutes 9 seconds. Especially the last transition motion is time consuming (1 minute 40 seconds) because a vertex not so close to the face is moved to contact the face while other (closer) vertices need to stay at distance $h_{d,avoid,desired} = 100$ mm of the face.

Figure 9.10 shows the result of the optimal CF sequence when the *execution time* is minimised. Vertex-face CFs are made with the rear face, the horizontal face and the right face respectively. This optimal CF sequence is obtained in about 3 hours and 30 minutes. The execution time of this sequence is 1 minute 22 seconds.

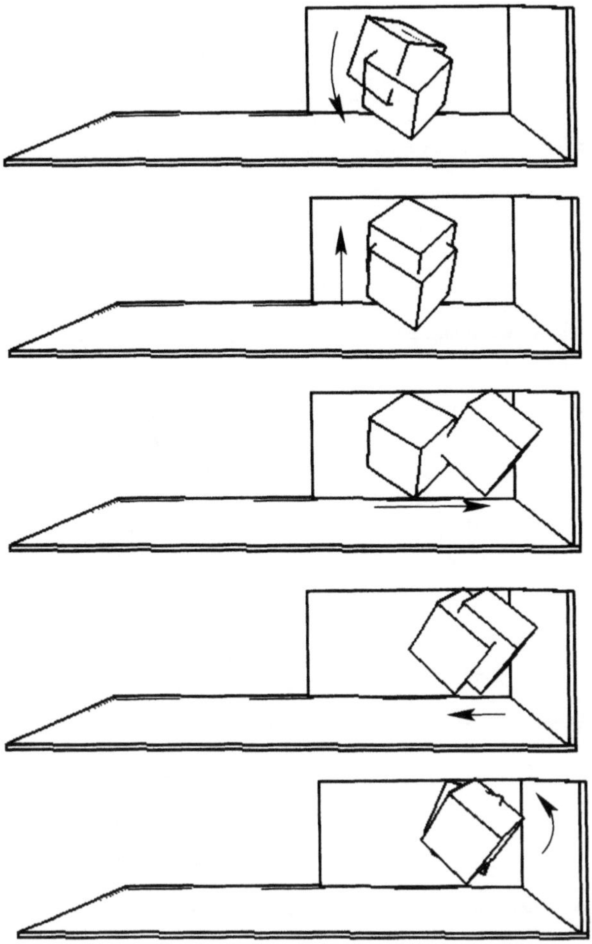

Fig. 9.9. The *shortest* CF sequence observing all positions and orientations of the cube and the corner object. Only free space motions and vertex-face CFs are allowed. Vertex-face CFs are made with the horizontal face, the right face and the rear face respectively

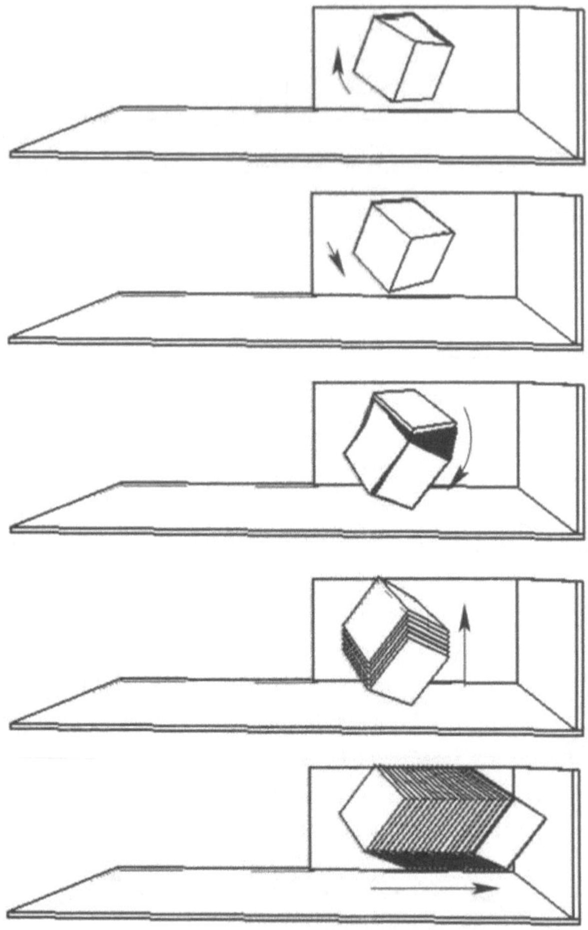

Fig. 9.10. The *fastest* CF sequence observing all positions and orientations of the cube and the corner object. Only free space motions and vertex-face CFs are allowed. Vertex-face CFs are made with the rear face, the horizontal face and the right face respectively

Observe All Geometrical Parameters and Perform a Cube-In-Corner Assembly

Next, a task plan is optimised to observe all geometrical parameters and perform a cube-in-corner assembly.

Figure 9.11 shows the *shortest* CF sequence for this cube-in-corner assembly under uncertainty. The CF sequence is one which gradually adds contacts until the cube is in the corner. First, the cube makes a vertex-face CF, an edge-face CF and a face-face CF with the horizontal face. Next, an edge-face CF with the rear face is added, after which a two face-face CF with the horizontal and rear faces is established. The cube is finally slid into the corner to the three face-face CF. The optimal solution is obtained in 10.7 seconds. Its execution time is 3 minutes 35 seconds.

Figure 9.12 shows the *fastest* CF sequence for this cube-in-corner assembly under uncertainty. The cube first makes a vertex-face CF, an edge-face CF and a face-face CF with the right face. Next, an edge-face contact with the horizontal face is added. This is followed by a rotation to a two face-face CF. The cube is finally slid into the corner. The optimisation procedure takes about two weeks. The execution time of this sequence is 2 minutes and 26 seconds.

Conclusion

The optimisation of the shortest CF sequence (Problem 9.1) is fast, the calculation of the *fastest* CF sequence (Problem 9.2), on the other hand, is computationally expensive. The computational cost originates from the simulation of the guarded motions and CF-compliant paths which trigger the transitions (Sect. 9.3.3).

The extra computational cost results in sequences which are considerably faster; in all examples we note *execution times* of two thirds or less than that of the shortest CF sequence. The gain in execution is due to making contacts between features which are already close to each other.

In all cases the fastest CF sequence has the *same number of CF transitions* as the shortest sequence. This invites us to approximate the calculation of the fastest CF sequence by combining both optimisation methods: first a list of shortest sequences is calculated, the simulation of the execution times of this limited list then shows which of these sequences is the fastest. This is subject to future research.

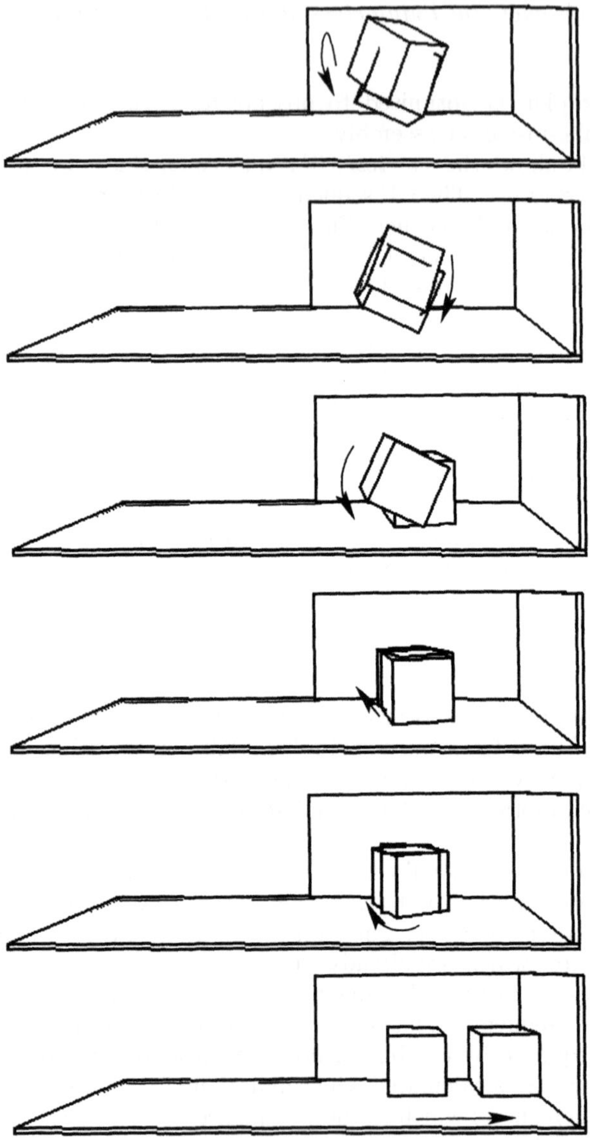

Fig. 9.11. The *shortest* CF sequence observing all geometrical parameters and performing a cube-in-corner assembly. First, the cube makes a vertex-face CF, an edge-face CF and a face-face CF with the horizontal face. Next, an edge-face CF with the rear face is added, followed by the rotation to a two face-face CF with the horizontal and rear faces. The cube is finally slid into the corner

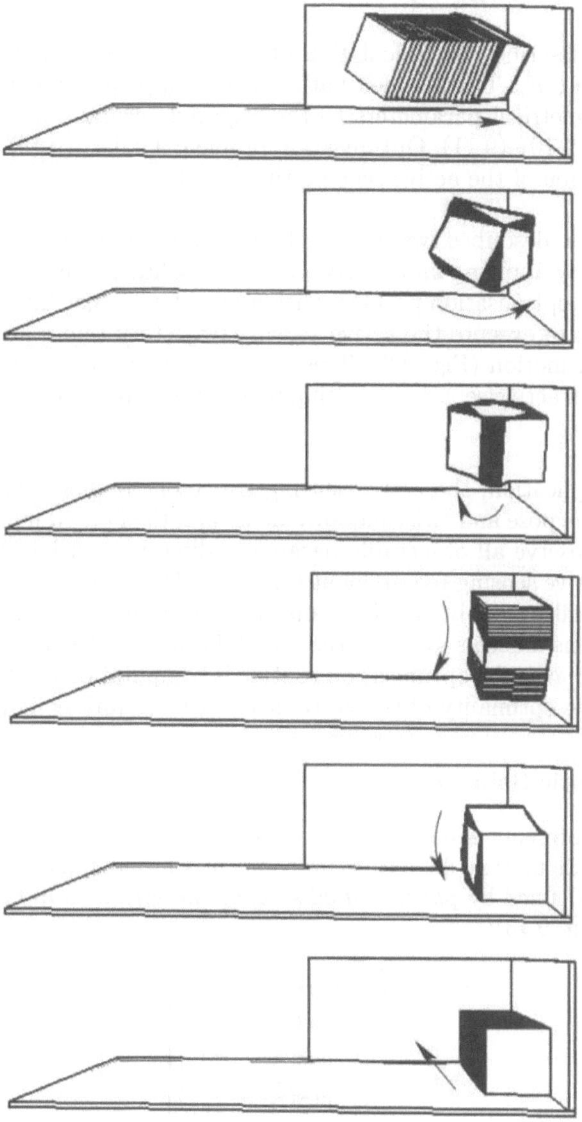

Fig. 9.12. The *fastest* CF sequence observing all geometrical parameters and performing a cube-in-corner assembly. First, the cube makes a vertex-face CF, an edge-face CF and a face-face CF with the right face. Next, an edge-face contact with the horizontal face is added. This is followed by a rotation to a two face-face CF. The cube is finally slid into the corner

9.5 Active Sensing Motion

This section describes the calculation of an active sensing motion $\mathbf{U}_{a,i}$ *in a specific* $CF = CF_i$. The planned path is constrained to the paths which observe the geometrical parameters to the required accuracy, see optimisation problem 9.1, (9.10)–(9.11). Optimisation problem 9.2, i.e., (9.19)–(9.21), adds the minimisation of the active sensing time $t_{a,i}(\mathbf{U}_i)$.

This section describes two ways to find an appropriate active sensing motion $\mathbf{U}_{a,i}$: (i) by applying an intuitive, heuristic solution (Sect. 9.5.1); and (ii) by solving the optimisation problem with a line search algorithm (Sect. 9.5.2).

We choose to execute the active sensing trajectory *before* and not during the transition motion (Fig. 9.2). This allows us to spend the limited kinetic energy fully on active sensing, resulting in a fast estimation of the geometrical parameters.

In our application, the active sensing trajectory is short due to the high accuracy of the pose and twist measurements and the limited number of poses required to observe all observable parameter directions. A lot of trajectories satisfy the active sensing requirement (9.20)-(9.21), hence, optimally designed trajectories will probably not differ much from intuitive, heuristic solutions. The simulations indeed show that the search for more optimal solutions is not interesting because the optimisation routine is computationally expensive and the gain in the optimality of the active sensing trajectory is limited.

9.5.1 Heuristic Solution

This section presents a heuristic active sensing solution. The chosen trajectory excites all motion degrees of freedom of the manipulated object. The trajectory is defined by *phase-shifted sine/cosine velocities* in the six motion directions (Fig. 9.13):

$$\mathbf{t} = \begin{bmatrix} a_{1,i}\cos(\frac{2\pi t}{t_{a,i}}) \\ a_{2,i}\cos(\frac{\pi}{3} - \frac{2\pi t}{t_{a,i}}) \\ a_{3,i}\cos(\frac{2\pi}{3} - \frac{2\pi t}{t_{a,i}}) \\ a_{4,i}\cos(\frac{3\pi}{3} - \frac{2\pi t}{t_{a,i}}) \\ a_{5,i}\cos(\frac{4\pi}{3} - \frac{2\pi t}{t_{a,i}}) \\ a_{6,i}\cos(\frac{5\pi}{3} - \frac{2\pi t}{t_{a,i}}) \end{bmatrix}.$$

The velocities are specified in a frame in the manipulated object's centre of mass with axes parallel to the object's principal axes. The amplitude of the different motion components is determined by the inertia in the corresponding motion direction ($\mathbf{M}$ is the inertia matrix of the manipulated object):

$$a_{d,i} \sim \sqrt{\frac{1}{\mathbf{M}(d,d)}};$$

the scaling factor is such that the resulting motion attains the maximum allowed kinetic energy $E_{kin,max}$.

The trajectory length $t_{a,i}$ can be made shorter or longer depending on the application, i.e., depending on the measurement covariances, the required gain in parameter accuracy, the allowed dynamic contents of the path, the maximum allowed kinetic energy, etc. Figure 9.14 shows the paths when applied to a cube object for three different lengths $t_{a,i}$.

The obtained motion at any time k is *projected onto the user-defined space of allowed active sensing twists.* This space is represented by the spanning set $\mathbf{J}_{actsens,i,k}(\mathbf{x}, \mathbf{t}_{d,k})$. This space of allowed active sensing twists equals the twist space of the current CF or is a subspace of it. For instance, consider a face-face CF with the Z-axis of {m} perpendicular to the contacting faces. A possible spanning set for the twist space, expressed in {m}, is:

$$\mathbf{J}_{i,k} = \begin{bmatrix} 1 \text{ mm/s} & 0 \text{ mm/s} & 0 \text{ mm/s} \\ 0 \text{ mm/s} & 1 \text{ mm/s} & 0 \text{ mm/s} \\ 0 \text{ mm/s} & 0 \text{ mm/s} & 0 \text{ mm/s} \\ 0 \text{ rad/s} & 0 \text{ rad/s} & 0 \text{ rad/s} \\ 0 \text{ rad/s} & 0 \text{ rad/s} & 0 \text{ rad/s} \\ 0 \text{ rad/s} & 0 \text{ rad/s} & 1 \text{ rad/s} \end{bmatrix}.$$

If the active sensing motions are limited to translational velocities, then for this CF a spanning set of the allowed active sensing twists is

$$\mathbf{J}_{actsens,i,k} = \begin{bmatrix} 1 \text{ mm/s} & 0 \text{ mm/s} \\ 0 \text{ mm/s} & 1 \text{ mm/s} \\ 0 \text{ mm/s} & 0 \text{ mm/s} \\ 0 \text{ rad/s} & 0 \text{ rad/s} \\ 0 \text{ rad/s} & 0 \text{ rad/s} \\ 0 \text{ rad/s} & 0 \text{ rad/s} \end{bmatrix}.$$

The projection of the twist $\mathbf{t}_k$ onto the space of active sensing twists is chosen to minimise the loss in kinetic energy:

$$\mathbf{t}_{proj} = \mathbf{J}_{actsens,i} \, \mathbf{J}^{\#}_{M,actsens,i} \, \mathbf{t};$$

$$\mathbf{J}^{\#}_{M,actsens,i} = \left(\mathbf{J}^{T}_{actsens,i} \, \mathbf{M} \, \mathbf{J}_{actsens,i}\right)^{-1} \mathbf{J}^{T}_{actsens,i} \, \mathbf{M}.$$

9.5.2 Solution by Optimisation

This section presents an optimised solution to the active sensing problem. The optimisation problem depends on the chosen *representation* of possible active sensing paths $\mathbf{U}_{a,i}$:

- The evolution of $\mathbf{U}_{a,i}$ can be restricted to a trajectory, described by a nominal trajectory and a *parameterised* deviation of this trajectory. In this way, the optimisation problem is reduced to a finite-dimensional, parameterised optimisation problem. An example is the parameterisation of the deviation as a finite sine/cosine series.

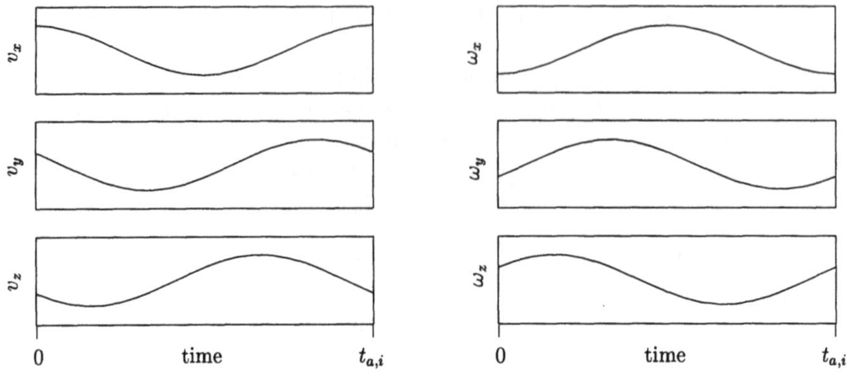

Fig. 9.13. Heuristic active sensing motion determined by phase-shifted sine/cosine velocities in the six motion directions. The velocities are defined in a frame in the manipulated object's centre of mass with axes parallel to the object's principal axes

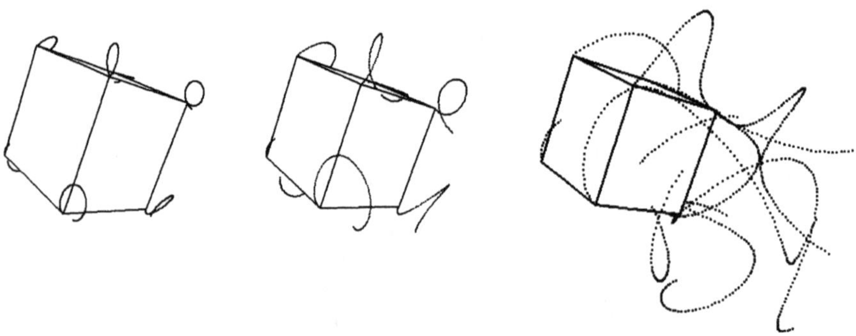

Fig. 9.14. Heuristic active sensing motions drawn as the trajectories of consecutive corner positions of a cube. The trajectory is drawn for different lengths $t_{a,i}$: the second and third trajectories are two respectively five times longer than the first trajectory

- A more general way to describe the trajectory is a (discrete) *sequence* of actions which are not restricted to a certain form of trajectory. This optimisation problem is computationally very expensive.

This chapter presents the optimisation of a *parameterised* active sensing path $\mathbf{U}_{a,i}$. The path is defined by the initial pose and an L-dimensional sine/cosine parameterised trajectory of the manipulated object twist in the D-dimensional space of allowed active sensing motions (D is the number of columns in a minimal spanning set $\mathbf{J}_{actsens,i}$ of the active sensing twist space):

$$\mathbf{t} = \mathbf{J}_{actsens,i}\, \chi_i; \tag{9.24}$$

$$\chi_i = \begin{bmatrix} \sum_{l=1}^{l=L}\left(a_{1,i,l}\cos(\frac{2\pi l t}{t_{a,i}}) + b_{1,i,l}\sin(\frac{2\pi l t}{t_{a,i}})\right) \\ \dots \\ \sum_{l=1}^{l=L}\left(a_{D,i,l}\cos(\frac{2\pi l t}{t_{a,i}}) + b_{D,i,l}\sin(\frac{2\pi l t}{t_{a,i}})\right) \end{bmatrix}. \tag{9.25}$$

This choice (i) limits the optimisation time, and (ii) allows us to impose constraints, such as the dynamic contents of the trajectory, into the path representation. These constraints are automatically satisfied without complicating the optimisation algorithm.

The parameter vector $\mathbf{p}_i$, representing the manipulated object twist, contains the amplitudes of the different sine/cosine terms and the total time $t_{a,i}$ of the active sensing motion:

$$\begin{aligned} \mathbf{p}_i^T = \big[\, & a_{1,i,1}, \ \dots, \ a_{1,i,L}, \\ & b_{1,i,1}, \ \dots, \ b_{1,i,L}, \\ & \quad\ \dots, \\ & a_{D,i,1}, \dots, \ a_{D,i,L}, \\ & b_{D,i,1}, \ \dots, \ b_{D,i,L}, \\ & t_{a,i} \qquad\qquad\quad \big]. \end{aligned} \tag{9.26}$$

Looking for the global optimum $\hat{\mathbf{p}}_i$ of optimisation problem 9.2, (9.19)–(9.21), is very expensive and the "optimal" solution to the optimisation problem is only an *expected* optimal solution anyway, because it is based on *estimates* of the geometrical parameters. Hence, we restrict ourselves to the use of a *local line search optimisation algorithm* [199].

9.5.3 Example

The active sensing motions are calculated for all CFs in the sequences determined in Sect. 9.4.2, i.e., the CFs of Figs. 9.7–9.12. Note that free space motion does not provide any information about the geometrical parameters. Hence, all active sensing motions are compliant motions.

The position and orientation of the cube and the corner objects are initially inaccurately known with covariance $(20°)^2$ for all angles and $(500\text{ mm})^2$ for all distances. The toleranced uncertainties of the estimates at the end of the task are $(1°)^2$ for all angles and $(1\text{ mm})^2$ for all distances. The pose measurement covariance is $(1°)^2$ for angles and $(1\text{ mm})^2$ for distances. Poses are measured at 10 Hz. Twist measurements are derived from the pose measurements.

Heuristic Trajectory

The heuristic trajectory (Sect. 9.5.1) is applied with a trajectory length $t_{a,i}$ of *2 seconds*. For every CF of Figs. 9.7–9.12, this active sensing motion is sufficient to observe all parameters to the required accuracy[5]. The motion is even sufficient to observe the CF-observable parameters in every CF to the desired accuracy if *no* information from previous CFs is passed on, i.e., if the initial uncertainty on the geometrical parameters in every CF is reset to $(20°)^2$ for the angles and $(500\text{ mm})^2$ for the distances.

Figure 9.15 shows the applied twist for the face-face plus edge-face CF of Fig. 9.11. The twist is expressed in the gripper frame. The heuristic trajectory of Fig. 9.13 is projected onto the 2D twist space of the CF.

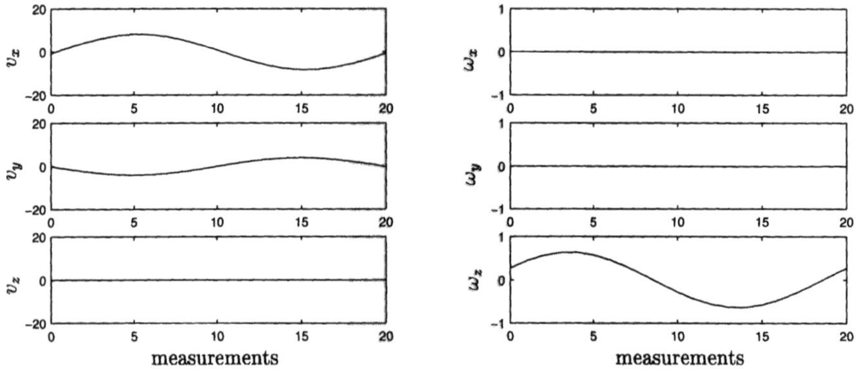

Fig. 9.15. Heuristic active sensing motion in the face-face plus edge-face CF of Fig. 9.11. The heuristic trajectory of Fig. 9.13 is projected onto the 2D twist space of the CF. The twist is expressed in the gripper frame

Line Search Optimisation

The line search algorithm (Sect. 9.5.2) is tested with a twist parameterisation which has one sine/cosine term per motion direction, i.e., (9.24)–(9.26) with $L = 1$. The optimisation is started with the heuristic trajectory as initial value:

$$a_{d,i,1} \sim \cos(\frac{2d\pi}{D})\sqrt{\frac{1}{\mathbf{J}^{dT}_{actsens,i}\ \mathbf{M}\ \mathbf{J}^{d}_{actsens,i}}};$$

$$b_{d,i,1} \sim \sin(\frac{2d\pi}{D})\sqrt{\frac{1}{\mathbf{J}^{dT}_{actsens,i}\ \mathbf{M}\ \mathbf{J}^{d}_{actsens,i}}};$$

$$t_{a,i} = 2 \text{ seconds};$$

[5] The heuristic trajectory with a trajectory length $t_{a,i}$ of *1 second* is a solution for *some* of the CFs.

where D is the number of columns of $\mathbf{J}_{actsens,i}$ and $\mathbf{J}^d_{actsens,i}$ represents its dth column. The values $a_{d,i,1}$, $b_{d,i,1}$ are scaled such that the resulting motion attains the maximum allowed kinetic energy E_{max}.

To limit the calculation time of the optimisation routine, a lower bound of 1 second is put on the trajectory time $t_{a,i}$. This also limits the dynamic contents of the trajectory.

The optimisation routine results in somewhat faster active sensing motions. However, the routine is computationally expensive because at each optimisation iteration the (changed) active sensing motion trajectory needs to be re-simulated in order to calculate the new $\mathbf{P}_{end,i}(\mathbf{U}_i)$, (9.2)–(9.4). The optimisation of the active sensing motions takes between 30 seconds and 9 minutes 20 seconds.

Figure 9.16 presents the optimised active sensing motion for the face-face plus edge-face CF of Fig. 9.11. For this CF, the optimisation routine reached the user-defined lower bound of 1 second on the active sensing time.

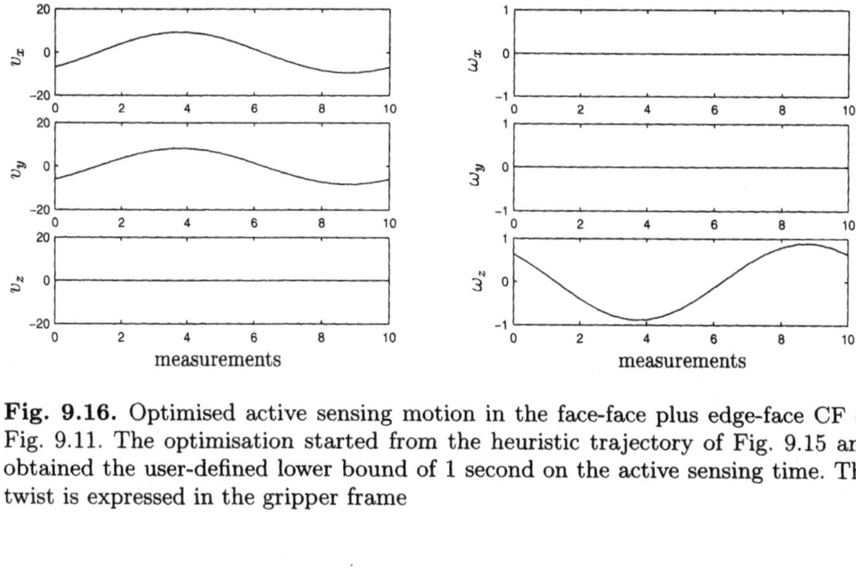

Fig. 9.16. Optimised active sensing motion in the face-face plus edge-face CF of Fig. 9.11. The optimisation started from the heuristic trajectory of Fig. 9.15 and obtained the user-defined lower bound of 1 second on the active sensing time. The twist is expressed in the gripper frame

Conclusion

The simulations show that executing a short heuristic path is sufficient to obtain accurate estimates for all geometrical parameters. The search for more optimal solutions is not interesting because the optimisation routine is computationally expensive and the gain in the optimality of the active sensing trajectory is limited. Even more, due to the uncertainty in the geometrical parameters, the "optimal solution" is in fact only an *expected* optimal solution.

9.6 Task Execution and Online Re-planning

CF-Compliant Motion

The CF-compliant motion comprises an active sensing motion (Sect. 9.5) and a transition motion (Sect. 9.3.3). During the CF-compliant motion the estimator continuously updates the geometrical parameter estimates.

The calculated active sensing motion is preceded by a smooth trajectory to arrive at the non-zero initial velocities. The active sensing motion is interrupted when the desired accuracy of the geometrical parameter estimates is obtained. If at the end of the calculated active sensing trajectory, the desired accuracy is not yet obtained, an extra active sensing motion is executed. An online *re-planning* of this motion is computationally expensive. However, the simulations show that heuristic solutions, which are developed independently of the manipulated object's pose and CF, are able to observe all observable parameter directions in a short time.

When the desired accuracy is obtained, the transition motion is executed. Also the transition between the active sensing motion and the transition motion is smoothed. The transition motion is executed until a CF transition is detected.

CF Transition

If the new CF is the planned one, the task execution continues as planned. If, however, the CF after transition is *not* the next CF of the planned sequence, *re-planning* is necessary. First, an active sensing motion is executed in order to recognise the CF and estimate the geometrical parameters. Force control is needed in those directions which are not yet recognised as motion degree of freedom directions.

After active sensing, the CF sequence needs to be *re-planned. One possible solution* is to make a transition to the previous CF and continue the task plan execution. The previously made erroneous transition is now avoided because the corresponding geometrical parameters are accurately known.

Another solution is to re-plan the CF sequence starting from the current CF and the goal CF. If the new plan is chosen to minimise the number of CF transitions (optimisation problem 9.1), the result is obtained quickly. Re-planning for the minimisation of the execution time is not relevant because it is computationally expensive and, hence, results in lost execution time. Returning to the previous CF is fast. Hence, the execution of the original task plan is a better choice in this case.

A CF transition can also occur during an active sensing motion. This undesired transition can be prevented during the task *planning* by calculating the transition motion with a large distance to undesired contacts $h_{d,avoid,desired}$,

Sect. 9.3.3. Of course, an unwanted transition can not be excluded during the task *execution* under inaccurate knowledge of the geometrical parameters. The CF-transition is dealt with in the way as before: i.e., an active sensing motion is performed in the new CF and the CF sequence is re-planned.

Under large uncertainty of the geometrical parameters, it is not always possible to uniquely identify the CF (Sect. 7.4.3). In this case the probability distribution over the CFs is calculated [105]. Calculating an adapted task plan for the easy recognition of the CF, e.g., by transitions to neighbouring CFs which are uniquely identifiable, is a topic of future research.

9.7 Conclusions

This chapter presents the optimisation of compliant motion task plans which are constrained by the need for accurate observation of the geometrical parameters. This is a new research area. Hence, we primarily focus on formulating the active sensing problem and decoupling it into smaller optimisation problems. We do *not* look for the computationally most efficient optimisation algorithms for online use. The main topics of this chapter are:

1. the definition of the *CF-observable parameter space*. The notion of the CF-observable parameter space allows us to *decouple* the active sensing requirement of the task plan into a requirement of the CF sequence and requirements of the active sensing motions in each CF.
2. the presentation of an algorithm to calculate CF-compliant *transition motions*. The algorithm is restricted to motions in the absence of obstacles between the elements (surfaces, edges, etc) to join.
3. the description of *practical (sub-optimal) solutions* to the optimisation problems, making online *re-planning* feasible.

10

General Conclusions

A journey of a thousand miles begins with a single step.

Confucius, (551–479 B.C.).

This work deals with force-controlled robot tasks under uncertainty. The aim is to increase the robot's autonomy in order to perform compliant motion tasks in unstructured environments. A general autonomous compliant motion system consists of three components, i.e., the task planner, the force controller and the estimator. This book focuses on solutions where these components are based on the same geometrical contact models. The models specify the constraints, the motion degree of freedom directions and the ideal contact force directions for every contact formation (CF). The models are a function of the inaccurately known geometrical parameters, i.e., the positions, orientations and dimensions of the contacting objects. During the compliant motion, the parameters are estimated more accurately based on pose, twist and wrench measurements.

This work describes the latest contributions in the areas of fast recursive estimation, contact modelling and task planning with active sensing. The next section presents a detailed overview of those contributions.

10.1 Main Topics

Analysis of Well-Known Fast Recursive Estimators

The autonomous robot needs a fast recursive estimator able to deal with highly nonlinear measurement models. The commonly used technique for fast recursive estimation is the Kalman Filter (KF) algorithm. This algorithm, however, requires a linear system description, i.e., a linear process and a linear measurement model. Such systems are rarely found in the real world.

Fast recursive filtering for nonlinear models is still obtained by the KF algorithm by using *linearizations* of the system description. Different ways of linearization, i.e., different KF variants, yield other results. Chapter 4 presents a comparison of the best known KF variants, i.e., the Extended Kalman Filter, the Iterated Extended Kalman Filter and the Unscented Kalman Filter.

The main topics in this chapter are:

1. The Unscented Kalman Filter is shown to be a filter which linearizes the models by statistical linear regression. This allows a better understanding of the filter performance and creates the possibility to compare the filter to other KF variants. Also the less known Central Difference Filter and first-order Divided Difference Filter fit in the framework of this "Linear Regression Kalman Filter".
2. The comparison of the different KF variants is based on two criteria, namely the *consistency* and the *information content* of the estimates.
   This work relates the consistency and information content of the estimates to (i) how the linearization is performed and (ii) how the linearization errors are taken into account. This understanding of the linearization processes allows us to make a well-founded choice of filter for a specific application.
3. The filters' performances are compared separately for process and measurement updates. The performances in both updates can be very different.
   The separate analysis of the process and measurement updates yields an insight in the exact nature and performance of these updates. This insight can be exploited to construct the most optimal filter for a particular application. This filter can use different types of model linearization in the process and measurement updates.

A New Approach to Fast Recursive Estimation

Chapter 5 describes *a new exact Bayesian filter* able to deal with static systems (parameter estimation) with any kind of nonlinear *measurement model* subject to additive Gaussian measurement uncertainty. The filter is also applicable to a limited class of dynamic systems. The new filter is appropriate for real-time use, because its core consists of a simple KF algorithm.

Experimental results of this new filter are given (i) for the (dynamic) estimation of the contact point position along a known contour based on curvature measurements (Chap. 5), and (ii) for the (static) estimation of the geometrical parameters for autonomous compliant motion (Chap. 8). In both experiments the filter is able to process the measurements at the sampling rate and the estimation results are excellent.

These estimation problems cannot be solved by the classical KF variants without extensive application-dependent configuration and tuning; i.e., choosing the most appropriate number of KFs to use in parallel, each starting from appropriately chosen initial conditions. We solve these problems with a single non-linear filter, which in addition needs *much* less application-dependent tuning.

An advantage of the new filter is its theoretical simplicity for people already familiar with the KF algorithm. The filter can be interpreted as a KF linearizing the process and measurement models in a non-minimal state space. Therefore, the filter is called the *"Non-minimal State Kalman Filter"*. The "mean vector" and "covariance matrix" of the Gaussian in the non-minimal state space are sufficient statistics of the (non-Gaussian) posterior pdf over the original (minimal) state.

This filter is not an approximation, but an *exact* result of Bayesian probability theory. Hence, it *does not loose any information* during measurement processing. The filter is general and not at all limited to the applications considered in this work. It is a major contribution to a wide area of applications where real-time nonlinear parameter estimation is important.

Contact Modelling

Our research group focuses on autonomous compliant motion based on *explicit* contact models. The parameters of the models correspond to the inaccurately known positions, orientations and dimensions of the contacting objects. The shapes of the contacting objects are supposed to be known (up to the dimensions).

Previous work modelled the contacts based on the "Virtual Contact Manipulator" describing the *relative* degrees of freedom between the manipulated object and the environment objects.

Chapter 6 presents a contact modelling technique

1. which does not need the description of the Virtual Contact Manipulator. This reduces the modelling effort considerably, and results in a minimal set of closure equations.
2. which generates the contact models for multiple-contact CFs automatically based on a *library of elementary CFs*. Contacts between generally curved surfaces can be added to the library if analytical surface descriptions are available which allow an analytical description of the contact point(s) and normal(s).
   This automatic generation of the contact models is indispensable for use outside the research community.

The library of elementary contacts is used in the described estimator (Chap. 7) and planner (Chap. 9) of the autonomous compliant motion system. The implementation of both components is shown to be invariant to non-minimal contact modelling, i.e., modelling a contact with more elementary contacts than necessary.

Estimator Component of an Autonomous Compliant Motion System

The estimator component of our autonomous compliant motion system (i) estimates the geometrical parameters and (ii) recognises the CF. The geometrical parameter estimation and the CF recognition are based on measurements of the pose and twist of the manipulated object and of the contact wrench. The (CF-dependent) measurement equations relate these measurements to the geometrical parameters. A recursive filter estimates the geometrical parameters given the CF. Inconsistency of the measurements with the measurement equations indicates that the corresponding CF is not valid.

The nonlinear parameter estimation problem under *large uncertainty* is solved by the use of the Non-minimal State Kalman Filter. Previously, none of the existing Kalman Filter variants was able to deal with large uncertainties.

The possibility of estimation under large uncertainty creates the opportunity to model objects without initial knowledge, i.e., the objects are represented by an unknown number of features (vertices, surfaces, etc) with inaccurately known positions, orientations and dimensions.

The filter equations are *automatically generated* based on the elementary contact library of Chap. 6.

The *propagation of the estimates* from one CF to the next is straightforward because we always consider all parameters, even the instantaneously unobservable ones, in the estimator. This does not increase the computational load of the estimator.

The presented techniques are experimentally validated in Chap. 8. The results correspond to the theoretical expectations.

The experimental validation required solutions to the practical problems of (i) the instability of the contact after a transition, and (ii) the numerical instability of the KF algorithm when a part of the parameter values are observed very accurately while others are still inaccurately known.

The Bayesian approach makes the estimation results consistent with the available information in the measurements and initial estimates. No information is created, nor lost. This implies that the results are invariant to changes in units or frames.

Planner of an Autonomous Compliant Motion System

The planner of the autonomous compliant motion system calculates the control set-points for the force controller. The task planner determines the task plan to be executed, i.e., the CF sequence and the CF-compliant motions. A specific concern of this work is the development of task plans in view of the accurate estimation of the geometrical parameters (active sensing, Chap. 9). Active sensing is a new research area in fine-motion planning. The experiment of Chap. 8 showed its necessity when accurate parameter estimation is important.

The main focus points of Chap. 9 are:

1. the definition of the *CF-observable parameter space*. The notion of the CF-observable parameter space allows us to *decouple* the active sensing requirement of the task plan into a requirement of the CF sequence and requirements of the active sensing motions in each CF.
2. the description of *practical solutions* to the optimisation problems making online *re-planning* feasible.
3. the *automatic* generation of all equations based on the elementary contact library of Chap. 6.
4. the description of an algorithm to calculate CF-compliant transition motions, i.e., motions which enhance the desired CF transitions while avoiding undesired contacts. The algorithm is restricted to motions in the absence of obstacles between the features (surfaces, edges, etc) to join.

10.2 Limitations and Future Work

This work makes a step forward in the development of a fully autonomous compliant motion system. The latest research results in the areas of contact modelling, estimation and planning are described. Some aspects remain open for future research:

CF Recognition in the Estimator

The correct *CF recognition* under large geometrical uncertainties is still an open problem. For small uncertainties on the geometrical parameters, the SNIS test eliminates the wrong CF hypotheses. For large uncertainties, however, several CFs are probable. The general solution to CF recognition is the calculation of the discrete probability *distribution* over the CFs [105].

Active Sensing for CF Recognition

The unique CF recognition under large geometrical uncertainty, i.e., the reduction of this discrete probability distribution to one "peak", is determined by the executed CF sequence. Active sensing for quick CF recognition is a topic of future research.

Software

For general use by non-expert programmers, the software needs to be improved to obtain high-quality "plug-and-play" modules.

Experimental Validation of the Planning Results

The online task execution and re-planning needs to be experimentally validated. This requires the online use of the contact library, the estimator and the planning module.

Efficient Optimisation Routines for Active Sensing

Task execution under uncertainty makes online re-planning indispensable. The re-planning of active sensing tasks involves both discrete (CF-sequence) and continuous (CF-compliant paths) optimisation problems. The efficiency of the implemented optimisation algorithms is crucial.

Non-Rigid Contacts

Finally, we notice that this whole work is based on the assumption of *rigid* contacting objects. The extension to flexible objects is important for applications in surgery on tissues and organs. Flexibilities influence the controller, the estimator and the planner.

10.3 Fast Model Building: A New Research Direction

The newly developed Non-minimal State Kalman Filter allows parameter estimation with very uncertain (or even absent) initial estimates. This creates the opportunity to actually *build* models of objects. To this aim, the objects are represented by an unknown number of features (vertices, faces, cylinders, etc) with inaccurately known positions, orientations and dimensions.

Possible application areas include

1. the demonstration of robot tasks involving "unknown", i.e., previously unmodelled, objects. The advantage of explicit geometrical modelling versus implicit modelling (e.g., neural networks) is that explicit modelling allows us to separate the identification of the object model parameters from the task characteristics. Hence, even after one learning cycle, the task can be replayed under uncertainty (e.g., other positions and orientations of the contacting objects).
2. model extraction (reverse engineering) based on stereo vision data, distance information from laser scanning, contact data of a Coordinate Measuring Machine, etc. All these measurement models are easily transformed to and interpreted as "closure equations".

A

The Linear Regression Kalman Filter

The Linear Regression Kalman Filter (LRKF, Sect. 4.2) has the following properties:

1. it linearizes the process and measurement functions by statistical linear regression of the functions through a number of regression points in state space;
2. it defines the uncertainty due to linearization errors on the linearized process or measurement function as the sample covariance matrix of the deviations between the function values of the nonlinear and the linearized function in the regression points.

This appendix contains the derivation of the process and measurement update of this LRKF (Sects. A.2 and A.3). First, Sect. A.1 describes the linear regression formulas.

A.1 Statistical Linear Regression

Consider a nonlinear function $\mathbf{y} = \mathbf{g}(\mathbf{x})$ evaluated in r points $(\boldsymbol{\mathcal{X}}^j, \boldsymbol{\mathcal{Y}}^j)$ where $\boldsymbol{\mathcal{Y}}^j = \mathbf{g}(\boldsymbol{\mathcal{X}}^j)$. Define

$$\bar{\boldsymbol{\mathcal{X}}} = \frac{1}{r}\sum_{j=1}^{r} \boldsymbol{\mathcal{X}}^j; \qquad \bar{\boldsymbol{\mathcal{Y}}} = \frac{1}{r}\sum_{j=1}^{r} \boldsymbol{\mathcal{Y}}^j;$$

$$\mathbf{P}_{xx} = \frac{1}{r}\sum_{j=1}^{r} (\boldsymbol{\mathcal{X}}^j - \bar{\boldsymbol{\mathcal{X}}})(\boldsymbol{\mathcal{X}}^j - \bar{\boldsymbol{\mathcal{X}}})^T;$$

$$\mathbf{P}_{xy} = \frac{1}{r}\sum_{j=1}^{r} (\boldsymbol{\mathcal{X}}^j - \bar{\boldsymbol{\mathcal{X}}})(\boldsymbol{\mathcal{Y}}^j - \bar{\boldsymbol{\mathcal{Y}}})^T;$$

$$\mathbf{P}_{yy} = \frac{1}{r}\sum_{j=1}^{r} (\boldsymbol{\mathcal{Y}}^j - \bar{\boldsymbol{\mathcal{Y}}})(\boldsymbol{\mathcal{Y}}^j - \bar{\boldsymbol{\mathcal{Y}}})^T.$$

The *y-regression* is the *linear regression* $\mathbf{y} = \mathbf{A}_{lr}\mathbf{x} + \mathbf{b}_{lr}$ that minimises the sum of the squared errors between the function values of the nonlinear and linearized functions in the regression points:

$$\mathbf{e}_j = \mathcal{Y}^j - \left(\mathbf{A}\mathcal{X}^j + \mathbf{b}\right); \tag{A.1}$$

$$(\mathbf{A}_{lr}, \mathbf{b}_{lr}) = \arg \min_{(\mathbf{A},\mathbf{b})} \sum_{j=1}^{r} \mathbf{e}_j^T \mathbf{e}_j. \tag{A.2}$$

The solution to (A.2) is [8]:

$$\mathbf{A}_{lr} = \mathbf{P}_{xy}^T \, \mathbf{P}_{xx}^{-1} = \mathbf{P}_{yx} \, \mathbf{P}_{xx}^{-1}; \tag{A.3}$$

$$\mathbf{b}_{lr} = \bar{\mathcal{Y}} - \mathbf{A}_{lr}\bar{\mathcal{X}}. \tag{A.4}$$

The sample covariance matrix of the deviations $\mathbf{e}_j$ is

$$\begin{aligned}
\mathbf{P}_{ee} &= \frac{1}{r}\sum_{j=1}^{r} \mathbf{e}_j \mathbf{e}_j^T; \\
&= \frac{1}{r}\sum_{j=1}^{r} \left((\mathcal{Y}^j - \bar{\mathcal{Y}}) - \mathbf{A}_{lr}\left(\mathcal{X}^j - \bar{\mathcal{X}}\right)\right)\left((\mathcal{Y}^j - \bar{\mathcal{Y}}) - \mathbf{A}_{lr}\left(\mathcal{X}^j - \bar{\mathcal{X}}\right)\right)^T; \\
&= \mathbf{P}_{yy} - \mathbf{A}_{lr}\,\mathbf{P}_{xy} - \mathbf{P}_{yx}\,\mathbf{A}_{lr}^T + \mathbf{A}_{lr}\,\mathbf{P}_{xx}\,\mathbf{A}_{lr}^T; \\
&= \mathbf{P}_{yy} - \mathbf{A}_{lr}\,\mathbf{P}_{xx}\,\mathbf{A}_{lr}^T.
\end{aligned} \tag{A.5}$$

A.2 The LRKF Process Update

The Linear Regression Kalman Filter (LRKF) uses the function values of r regression points $\mathcal{X}^j_{k-1|k-1}$ in state space to model the behaviour of the process function in the "uncertainty region" around the updated state estimate $\hat{\mathbf{x}}_{k-1|k-1}$. The regression points are chosen such that their mean and covariance matrix equal the state estimate $\hat{\mathbf{x}}_{k-1|k-1}$ and its covariance matrix $\mathbf{P}_{k-1|k-1}$:

$$\hat{\mathbf{x}}_{k-1|k-1} = \bar{\mathcal{X}}_{k-1|k-1} = \frac{1}{r}\sum_{j=1}^{r} \mathcal{X}^j_{k-1|k-1};$$

$$\begin{aligned}
\mathbf{P}_{k-1|k-1} &= \frac{1}{r}\sum_{j=1}^{r} \left(\mathcal{X}^j_{k-1|k-1} - \bar{\mathcal{X}}_{k-1|k-1}\right)\left(\mathcal{X}^j_{k-1|k-1} - \bar{\mathcal{X}}_{k-1|k-1}\right)^T; \\
&= \frac{1}{r}\sum_{j=1}^{r} \left(\mathcal{X}^j_{k-1|k-1} - \hat{\mathbf{x}}_{k-1|k-1}\right)\left(\mathcal{X}^j_{k-1|k-1} - \hat{\mathbf{x}}_{k-1|k-1}\right)^T.
\end{aligned}$$

The function values of the regression points are

$$\mathcal{X}^j_{k|k-1} = \mathbf{f}_{k-1}(\mathcal{X}^j_{k-1|k-1}).$$

Table A.1. Equivalence of the symbols in Sects. A.1, A.2 and A.3

Sect. A.1	Sect. A.2	Sect. A.3
$\mathbf{x}$	$\mathbf{x}_{k-1}$	$\mathbf{x}_k$
$\mathbf{y}$	$\mathbf{x}_k$	$\mathbf{z}_k$
$\mathbf{g}$	$\mathbf{f}_{k-1}$	$\mathbf{h}_k$
$\mathbf{A}_{lr}$ (A.3)	$\mathbf{F}_{k-1}$	$\mathbf{H}_k$
$\mathbf{b}_{lr}$ (A.4)	$\mathbf{b}_{k-1}$	$\mathbf{d}_k$
$\mathcal{X}^j$	$\mathcal{X}^j_{k-1\|k-1}$	$\mathcal{X}^j_{k\|k-1}$
$\mathcal{Y}^j$	$\mathcal{X}^j_{k\|k-1}$	$\mathcal{Z}^j_k$
$\mathbf{P}_{ee}$ (A.5)	$\mathbf{Q}^*_{k-1}$	$\mathbf{R}^*_k$

The LRKF algorithm uses a linearized process function (4.1) where $\mathbf{F}_{k-1}$, $\mathbf{b}_{k-1}$ and $\mathbf{Q}^*_{k-1}$ are obtained by statistical linear regression through the $(\mathcal{X}^j_{k-1|k-1}, \mathcal{X}^j_{k|k-1})$ points, $j = 1, \ldots, r$. The statistical regression is described in Sect. A.1. Replacing the symbols according to Table A.1 results in

$$\begin{aligned}
\mathbf{F}_{k-1} &= \left(\frac{1}{r}\sum_{j=1}^{r}\left(\mathcal{X}^j_{k|k-1} - \bar{\mathcal{X}}_{k|k-1}\right)\left(\mathcal{X}^j_{k-1|k-1} - \bar{\mathcal{X}}_{k-1|k-1}\right)^T\right) \times \\
&\quad \left(\frac{1}{r}\sum_{j=1}^{r}\left(\mathcal{X}^j_{k-1|k-1} - \bar{\mathcal{X}}_{k-1|k-1}\right)\left(\mathcal{X}^j_{k-1|k-1} - \bar{\mathcal{X}}_{k-1|k-1}\right)^T\right)^{-1}; \\
&= \left(\frac{1}{r}\sum_{j=1}^{r}\left(\mathcal{X}^j_{k|k-1} - \bar{\mathcal{X}}_{k|k-1}\right)\left(\mathcal{X}^j_{k-1|k-1} - \hat{\mathbf{x}}_{k-1|k-1}\right)^T\right) \mathbf{P}^{-1}_{k-1|k-1}; \\
\mathbf{b}_{k-1} &= \bar{\mathcal{X}}_{k|k-1} - \mathbf{F}_{k-1}\bar{\mathcal{X}}_{k-1|k-1}; \\
&= \frac{1}{r}\sum_{j=1}^{r}\mathcal{X}^j_{k|k-1} - \mathbf{F}_{k-1}\,\hat{\mathbf{x}}_{k-1|k-1}; \\
\mathbf{Q}^*_{k-1} &= \frac{1}{r}\sum_{j=1}^{r}\left(\mathcal{X}^j_{k|k-1} - \bar{\mathcal{X}}_{k|k-1}\right)\left(\mathcal{X}^j_{k|k-1} - \bar{\mathcal{X}}_{k|k-1}\right)^T - \mathbf{F}_{k-1}\times \\
&\quad \left(\frac{1}{r}\sum_{j=1}^{r}\left(\mathcal{X}^j_{k-1|k-1} - \bar{\mathcal{X}}_{k-1|k-1}\right)\left(\mathcal{X}^j_{k-1|k-1} - \bar{\mathcal{X}}_{k-1|k-1}\right)^T\right)\mathbf{F}^T_{k-1}; \\
&= \frac{1}{r}\sum_{j=1}^{r}\left(\mathcal{X}^j_{k|k-1} - \bar{\mathcal{X}}_{k|k-1}\right)\left(\mathcal{X}^j_{k|k-1} - \bar{\mathcal{X}}_{k|k-1}\right)^T \\
&\quad - \mathbf{F}_{k-1}\mathbf{P}_{k-1|k-1}\mathbf{F}^T_{k-1}.
\end{aligned}$$

The LRKF *process update equations* follow from the well-known linear Kalman Filter equations (4.3)–(4.4):

$$\begin{aligned}
\hat{\mathbf{x}}_{k|k-1} &= \mathbf{F}_{k-1}\hat{\mathbf{x}}_{k-1|k-1} + \mathbf{b}_{k-1};\\
&= \mathbf{F}_{k-1}\hat{\mathbf{x}}_{k-1|k-1} + \frac{1}{r}\sum_{j=1}^{r}\boldsymbol{\mathcal{X}}^j_{k|k-1} - \mathbf{F}_{k-1}\hat{\mathbf{x}}_{k-1|k-1};\\
&= \frac{1}{r}\sum_{j=1}^{r}\boldsymbol{\mathcal{X}}^j_{k|k-1};\\
\mathbf{P}_{k|k-1} &= \mathbf{F}_{k-1}\mathbf{P}_{k-1|k-1}\mathbf{F}^T_{k-1} + \mathbf{Q}^*_{k-1} + \mathbf{C}_{k-1}\mathbf{Q}_{k-1}\mathbf{C}^T_{k-1};\\
&= \mathbf{F}_{k-1}\mathbf{P}_{k-1|k-1}\mathbf{F}^T_{k-1}\\
&\quad + \frac{1}{r}\sum_{j=1}^{r}\left(\boldsymbol{\mathcal{X}}^j_{k|k-1} - \bar{\boldsymbol{\mathcal{X}}}_{k|k-1}\right)\left(\boldsymbol{\mathcal{X}}^j_{k|k-1} - \bar{\boldsymbol{\mathcal{X}}}_{k|k-1}\right)^T\\
&\quad - \mathbf{F}_{k-1}\mathbf{P}_{k-1|k-1}\mathbf{F}^T_{k-1} + \mathbf{C}_{k-1}\mathbf{Q}_{k-1}\mathbf{C}^T_{k-1};\\
&= \frac{1}{r}\sum_{j=1}^{r}\left(\boldsymbol{\mathcal{X}}^j_{k|k-1} - \hat{\mathbf{x}}_{k|k-1}\right)\left(\boldsymbol{\mathcal{X}}^j_{k|k-1} - \hat{\mathbf{x}}_{k|k-1}\right)^T\\
&\quad + \mathbf{C}_{k-1}\mathbf{Q}_{k-1}\mathbf{C}^T_{k-1}.
\end{aligned}$$

A.3 The LRKF Measurement Update

The LRKF evaluates the measurement function in r regression points $\boldsymbol{\mathcal{X}}^j_{k|k-1}$ in the "uncertainty region" around the predicted state estimate $\hat{\mathbf{x}}_{k|k-1}$. The $\boldsymbol{\mathcal{X}}^j_{k|k-1}$ are chosen such that their mean and covariance matrix are equal to the predicted state estimate and its covariance:

$$\begin{aligned}
\hat{\mathbf{x}}_{k|k-1} &= \bar{\boldsymbol{\mathcal{X}}}_{k|k-1} = \frac{1}{r}\sum_{j=1}^{r}\boldsymbol{\mathcal{X}}^j_{k|k-1};\\
\mathbf{P}_{k|k-1} &= \frac{1}{r}\sum_{j=1}^{r}\left(\boldsymbol{\mathcal{X}}^j_{k|k-1} - \bar{\boldsymbol{\mathcal{X}}}_{k|k-1}\right)\left(\boldsymbol{\mathcal{X}}^j_{k|k-1} - \bar{\boldsymbol{\mathcal{X}}}_{k|k-1}\right)^T;\\
&= \frac{1}{r}\sum_{j=1}^{r}\left(\boldsymbol{\mathcal{X}}^j_{k|k-1} - \hat{\mathbf{x}}_{k|k-1}\right)\left(\boldsymbol{\mathcal{X}}^j_{k|k-1} - \hat{\mathbf{x}}_{k|k-1}\right)^T.
\end{aligned}$$

The function values of the regression points through the nonlinear function are

$$\boldsymbol{\mathcal{Z}}^j_k = \mathbf{h}_k(\boldsymbol{\mathcal{X}}^j_{k|k-1}).$$

The LRKF algorithm uses a linearized measurement function (4.2) where $\mathbf{H}_k$, $\mathbf{d}_k$ and $\mathbf{R}_k^*$ are obtained by statistical linear regression through the points $(\boldsymbol{\mathcal{X}}^j_{k|k-1}, \boldsymbol{\mathcal{Z}}^j_k)$, $j = 1, \ldots, r$. The statistical regression is described in Sect. A.1. Replacing the symbols according to Table A.1 results in

$$
\begin{aligned}
\mathbf{H}_k &= \left(\frac{1}{r}\sum_{j=1}^{r}\left(\boldsymbol{\mathcal{Z}}^j_k - \bar{\boldsymbol{\mathcal{Z}}}_k\right)\left(\boldsymbol{\mathcal{X}}^j_{k|k-1} - \bar{\boldsymbol{\mathcal{X}}}_{k|k-1}\right)^T\right) \times \\
&\qquad \left(\frac{1}{r}\sum_{j=1}^{r}\left(\boldsymbol{\mathcal{X}}^j_{k|k-1} - \bar{\boldsymbol{\mathcal{X}}}_{k|k-1}\right)\left(\boldsymbol{\mathcal{X}}^j_{k|k-1} - \bar{\boldsymbol{\mathcal{X}}}_{k|k-1}\right)^T\right)^{-1}; \\
&= \left(\frac{1}{r}\sum_{j=1}^{r}\left(\boldsymbol{\mathcal{Z}}^j_k - \bar{\boldsymbol{\mathcal{Z}}}_k\right)\left(\boldsymbol{\mathcal{X}}^j_{k|k-1} - \hat{\mathbf{x}}_{k|k-1}\right)^T\right) \mathbf{P}^{-1}_{k|k-1}; \\
\mathbf{d}_k &= \bar{\boldsymbol{\mathcal{Z}}}_k - \mathbf{H}_k \,\bar{\boldsymbol{\mathcal{X}}}_{k|k-1}; \\
&= \frac{1}{r}\sum_{j=1}^{r}\boldsymbol{\mathcal{Z}}^j_k - \mathbf{H}_k\,\hat{\mathbf{x}}_{k|k-1}; \\
\mathbf{R}^*_k &= \frac{1}{r}\sum_{j=1}^{r}\left(\boldsymbol{\mathcal{Z}}^j_k - \bar{\boldsymbol{\mathcal{Z}}}_k\right)\left(\boldsymbol{\mathcal{Z}}^j_k - \bar{\boldsymbol{\mathcal{Z}}}_k\right)^T \\
&\qquad - \mathbf{H}_k \left(\frac{1}{r}\sum_{j=1}^{r}\left(\boldsymbol{\mathcal{X}}^j_{k|k-1} - \bar{\boldsymbol{\mathcal{X}}}_{k|k-1}\right)\left(\boldsymbol{\mathcal{X}}^j_{k|k-1} - \bar{\boldsymbol{\mathcal{X}}}_{k|k-1}\right)^T\right)\mathbf{H}^T_k; \\
&= \frac{1}{r}\sum_{j=1}^{r}\left(\boldsymbol{\mathcal{Z}}^j_k - \bar{\boldsymbol{\mathcal{Z}}}_k\right)\left(\boldsymbol{\mathcal{Z}}^j_k - \bar{\boldsymbol{\mathcal{Z}}}_k\right)^T - \mathbf{H}_k\mathbf{P}^{-1}_{k|k-1}\mathbf{H}^T_k.
\end{aligned}
$$

The LRKF *measurement update equations* follow from the linear Kalman Filter equations (4.5)–(4.9):

$$
\begin{aligned}
\hat{\mathbf{x}}_{k|k} &= \hat{\mathbf{x}}_{k|k-1} + \mathbf{K}_k\boldsymbol{\eta}_k; \\
\mathbf{P}_{k|k} &= \left(\mathbf{I}_{n\times n} - \mathbf{K}_k\mathbf{H}_k\right)\mathbf{P}_{k|k-1}; \\
\boldsymbol{\eta}_k &= \hat{\mathbf{z}}_k - \mathbf{H}_k\hat{\mathbf{x}}_{k|k-1} - \mathbf{d}_k; \\
&= \hat{\mathbf{z}}_k - \mathbf{H}_k\hat{\mathbf{x}}_{k|k-1} - \frac{1}{r}\sum_{j=1}^{r}\boldsymbol{\mathcal{Z}}^j_k + \mathbf{H}_k\hat{\mathbf{x}}_{k|k-1}; \\
&= \hat{\mathbf{z}}_k - \frac{1}{r}\sum_{j=1}^{r}\boldsymbol{\mathcal{Z}}^j_k;
\end{aligned}
$$

$$
\begin{aligned}
\mathbf{S}_k &= \mathbf{R}_k^* + \mathbf{E}_k\mathbf{R}_k\mathbf{E}_k^T + \mathbf{H}_k\mathbf{P}_{k|k-1}\mathbf{H}_k^T; \\
&= \frac{1}{r}\sum_{j=1}^{r}\left(\boldsymbol{\mathcal{Z}}_k^j - \bar{\boldsymbol{\mathcal{Z}}}_k\right)\left(\boldsymbol{\mathcal{Z}}_k^j - \bar{\boldsymbol{\mathcal{Z}}}_k\right)^T - \mathbf{H}_k\mathbf{P}_{k|k-1}^{-1}\mathbf{H}_k^T \\
&\quad + \mathbf{E}_k\mathbf{R}_k\mathbf{E}_k^T + \mathbf{H}_k\mathbf{P}_{k|k-1}\mathbf{H}_k^T; \\
&= \frac{1}{r}\sum_{j=1}^{r}\left(\boldsymbol{\mathcal{Z}}_k^j - \bar{\boldsymbol{\mathcal{Z}}}_k\right)\left(\boldsymbol{\mathcal{Z}}_k^j - \bar{\boldsymbol{\mathcal{Z}}}_k\right)^T + \mathbf{E}_k\mathbf{R}_k\mathbf{E}_k^T; \\
\mathbf{K}_k &= \mathbf{P}_{k|k-1}\mathbf{H}_k^T\mathbf{S}_k^{-1}; \\
&= \mathbf{P}_{k|k-1}\mathbf{P}_{k|k-1}^{-1} \times \\
&\quad \left(\frac{1}{r}\sum_{j=1}^{r}\left(\mathbf{X}_{k|k-1}^j - \hat{\mathbf{x}}_{k|k-1}\right)\left(\boldsymbol{\mathcal{Z}}_k^j - \bar{\boldsymbol{\mathcal{Z}}}_k\right)^T\right)\mathbf{S}_k^{-1}; \\
&= \left(\frac{1}{r}\sum_{j=1}^{r}\left(\mathbf{X}_{k|k-1}^j - \hat{\mathbf{x}}_{k|k-1}\right)\left(\boldsymbol{\mathcal{Z}}_k^j - \bar{\boldsymbol{\mathcal{Z}}}_k\right)^T\right)\mathbf{S}_k^{-1}.
\end{aligned}
$$

B

The Non-Minimal State Kalman Filter

This appendix contains the proofs of Theorems 5.1 and 5.2 describing the Non-minimal State Kalman Filter (NMSKF). The interpretation of the filter and examples are presented in Chap. 5.

B.1 The NMSKF for Static Systems

Theorem 5.1 (Analytical posterior pdf for static systems) *The posterior pdf over the state* $\mathbf{x}$*, given* k *measurements* $\hat{\mathbf{Z}}_k = \{\hat{\mathbf{z}}_1, \hat{\mathbf{z}}_2, \ldots, \hat{\mathbf{z}}_k\}$ *and the measurement model*

$$\mathbf{z}_k = \mathbf{h}_k(\mathbf{x}) + \mathbf{E}_k \boldsymbol{\rho}_{m,k} = \mathbf{H}_k\, \mathbf{g}(\mathbf{x}) + \mathbf{d}_k + \mathbf{E}_k \boldsymbol{\rho}_{m,k};$$

with zero mean Gaussian uncertainty $\boldsymbol{\rho}_{m,k}$ *which is uncorrelated over time and has covariance matrix* $\mathbf{R}_k$*, can be expressed in analytical form as*

$$p(\mathbf{x}|\hat{\mathbf{Z}}_k) \sim p(\mathbf{x}) \exp\left(-\frac{1}{2}\left(\mathbf{g}(\mathbf{x}) - \hat{\mathbf{x}}'_k\right)^T \mathbf{P}_k^{'-1} \left(\mathbf{g}(\mathbf{x}) - \hat{\mathbf{x}}'_k\right)\right). \tag{B.1}$$

$p(\mathbf{x})$ *is the prior pdf over* $\mathbf{x}$*. The sufficient statistics* $\hat{\mathbf{x}}'_k$ *and* $\mathbf{P}'_k$ *of pdf (B.1) are defined by the Kalman Filter equations for the state estimation of the* n'*-dimensional state* $\mathbf{x}' = \mathbf{g}(\mathbf{x})$ *(the iteration starts with* $\mathbf{P}_0^{'-1} = \mathbf{0}_{n' \times n'}$*, i.e., this filter has no prior information):*

$$\hat{\mathbf{x}}'_k = \hat{\mathbf{x}}'_{k-1} + \mathbf{K}_k \boldsymbol{\eta}_k; \tag{B.2}$$

$$\mathbf{P}'_k = \left(\mathbf{I}_{n' \times n'} - \mathbf{K}_k \mathbf{H}_k\right) \mathbf{P}'_{k-1}; \tag{B.3}$$

where

$$\boldsymbol{\eta}_k = \hat{\mathbf{z}}_k - \mathbf{H}_k \hat{\mathbf{x}}'_{k-1} - \mathbf{d}_k; \tag{B.4}$$

$$\mathbf{S}_k = \mathbf{E}_k \mathbf{R}_k \mathbf{E}_k^T + \mathbf{H}_k \mathbf{P}'_{k-1} \mathbf{H}_k^T; \tag{B.5}$$

$$\mathbf{K}_k = \mathbf{P}'_{k-1} \mathbf{H}_k^T \mathbf{S}_k^{-1}. \tag{B.6}$$

Proof Theorem 5.1
The prior pdf $p(\mathbf{x})$ is of the form (B.1) because $\mathbf{P}_0^{'-1} = \mathbf{0}_{n' \times n'}$. We now prove (B.1)–(B.6) by induction: given that pdf $p(\mathbf{x}|\hat{\mathbf{Z}}_{k-1})$ is of the form (B.1), we prove that $p(\mathbf{x}|\hat{\mathbf{Z}}_k)$ has also this form.

$$p(\mathbf{x}|\hat{\mathbf{Z}}_k) \sim p(\mathbf{x}|\hat{\mathbf{Z}}_{k-1}) p(\hat{\mathbf{z}}_k|\mathbf{x}, \hat{\mathbf{Z}}_{k-1}) = p(\mathbf{x}|\hat{\mathbf{Z}}_{k-1}) p(\hat{\mathbf{z}}_k|\mathbf{x});$$

where

$$p(\mathbf{x}|\hat{\mathbf{Z}}_{k-1}) \sim p(\mathbf{x}) \exp\left(-\frac{1}{2}\left(\mathbf{g}(\mathbf{x}) - \hat{\mathbf{x}}'_{k-1}\right)^T \mathbf{P}_{k-1}^{'-1} \left(\mathbf{g}(\mathbf{x}) - \hat{\mathbf{x}}'_{k-1}\right)\right);$$

$$p(\hat{\mathbf{z}}_k|\mathbf{x}) \sim \exp\left(-\frac{1}{2}\left(\mathbf{H}_k\, \mathbf{g}(\mathbf{x}) + \mathbf{d}_k - \hat{\mathbf{z}}_k\right)^T \times \left(\mathbf{E}_k \mathbf{R}_k \mathbf{E}_k^T\right)^{-1} \left(\mathbf{H}_k\, \mathbf{g}(\mathbf{x}) + \mathbf{d}_k - \hat{\mathbf{z}}_k\right)\right).$$

We simplified $p(\hat{\mathbf{z}}_k|\mathbf{x}, \hat{\mathbf{Z}}_{k-1})$ to $p(\hat{\mathbf{z}}_k|\mathbf{x})$ because the measurement model uncertainty is uncorrelated over time. Matrix calculations lead to a posterior pdf of the form (B.1):

$$p(\mathbf{x}|\hat{\mathbf{Z}}_k) \sim p(\mathbf{x}) \exp\left(-\frac{1}{2}\left(\mathbf{g}(\mathbf{x}) - \hat{\mathbf{x}}'_k\right)^T \mathbf{P}_k^{'-1} \left(\mathbf{g}(\mathbf{x}) - \hat{\mathbf{x}}'_k\right)\right);$$

where

$$\mathbf{P}_k^{'-1} = \mathbf{P}_{k-1}^{'-1} + \mathbf{H}_k^T \left(\mathbf{E}_k \mathbf{R}_k \mathbf{E}_k^T\right)^{-1} \mathbf{H}_k; \tag{B.7}$$

$$\hat{\mathbf{x}}'_k = \mathbf{P}'_k \left(\mathbf{P}_{k-1}^{'-1} \hat{\mathbf{x}}'_{k-1} + \mathbf{H}_k^T \left(\mathbf{E}_k \mathbf{R}_k \mathbf{E}_k^T\right)^{-1} \left(\hat{\mathbf{z}}_k - \mathbf{d}_k\right)\right). \tag{B.8}$$

These equations are obtained using the fact that covariance matrices are symmetric: $\mathbf{P}'_{k-1} = \mathbf{P}_{k-1}^{'T}$, $\mathbf{R}_k = \mathbf{R}_k^T$ and $\mathbf{P}'_k = \mathbf{P}_k^{'T}$.

Applying the matrix inversion lemma

$$\left(\mathbf{A} + \mathbf{B}\mathbf{C}\mathbf{B}^T\right)^{-1} = \mathbf{A}^{-1} - \mathbf{A}^{-1}\mathbf{B}\left(\mathbf{B}^T\mathbf{A}^{-1}\mathbf{B} + \mathbf{C}^{-1}\right)^{-1}\mathbf{B}^T\mathbf{A}^{-1}; \tag{B.9}$$

1. to (B.7) shows that $\mathbf{P}'_k$ is defined by (B.3)–(B.6):

$$\begin{aligned}
\mathbf{P}_k^{'-1} &= \mathbf{P}_{k-1}^{'-1} + \mathbf{H}_k^T \left(\mathbf{E}_k \mathbf{R}_k \mathbf{E}_k^T\right)^{-1} \mathbf{H}_k; \\
\mathbf{P}'_k &= \mathbf{P}'_{k-1} - \mathbf{P}'_{k-1}\mathbf{H}_k^T \left(\mathbf{H}_k \mathbf{P}'_{k-1} \mathbf{H}_k^T + \mathbf{E}_k \mathbf{R}_k \mathbf{E}_k^T\right)^{-1} \mathbf{H}_k \mathbf{P}'_{k-1}; \\
&= \mathbf{P}'_{k-1} - \mathbf{K}_k \mathbf{H}_k \mathbf{P}'_{k-1}.
\end{aligned}$$

2. to (B.8) shows that $\hat{\mathbf{x}}'_k$ is defined by (B.2), (B.4)–(B.6):

$$
\begin{aligned}
\hat{\mathbf{x}}'_k &= \mathbf{P}'_k \left(\mathbf{P}'^{-1}_{k-1} \hat{\mathbf{x}}'_{k-1} + \mathbf{H}_k^T \left(\mathbf{E}_k \mathbf{R}_k \mathbf{E}_k^T \right)^{-1} (\hat{\mathbf{z}}_k - \mathbf{d}_k) \right); \\
&= \left(\mathbf{I}_{n' \times n'} - \mathbf{K}_k \mathbf{H}_k \right) \mathbf{P}'_{k-1} \times \\
&\quad \left(\mathbf{P}'^{-1}_{k-1} \hat{\mathbf{x}}'_{k-1} + \mathbf{H}_k^T \left(\mathbf{E}_k \mathbf{R}_k \mathbf{E}_k^T \right)^{-1} (\hat{\mathbf{z}}_k - \mathbf{d}_k) \right); \\
&= \left(\mathbf{I}_{n' \times n'} - \mathbf{K}_k \mathbf{H}_k \right) \hat{\mathbf{x}}'_{k-1} \\
&\quad + \left(\mathbf{P}'_{k-1} - \mathbf{K}_k \mathbf{H}_k \mathbf{P}'_{k-1} \right) \mathbf{H}_k^T \left(\mathbf{E}_k \mathbf{R}_k \mathbf{E}_k^T \right)^{-1} (\hat{\mathbf{z}}_k - \mathbf{d}_k); \\
&= \left(\mathbf{I}_{n' \times n'} - \mathbf{K}_k \mathbf{H}_k \right) \hat{\mathbf{x}}'_{k-1} \\
&\quad + \left(\mathbf{P}'_{k-1} \mathbf{H}_k^T \mathbf{S}_k^{-1} \mathbf{S}_k - \mathbf{K}_k \mathbf{H}_k \mathbf{P}'_{k-1} \mathbf{H}_k^T \right) \times \\
&\quad \left(\mathbf{E}_k \mathbf{R}_k \mathbf{E}_k^T \right)^{-1} (\hat{\mathbf{z}}_k - \mathbf{d}_k); \\
&= \left(\mathbf{I}_{n' \times n'} - \mathbf{K}_k \mathbf{H}_k \right) \hat{\mathbf{x}}'_{k-1} \\
&\quad + \mathbf{K}_k \left(\mathbf{S}_k - \mathbf{H}_k \mathbf{P}'_{k-1} \mathbf{H}_k^T \right) \left(\mathbf{E}_k \mathbf{R}_k \mathbf{E}_k^T \right)^{-1} (\hat{\mathbf{z}}_k - \mathbf{d}_k); \\
&= \left(\mathbf{I}_{n' \times n'} - \mathbf{K}_k \mathbf{H}_k \right) \hat{\mathbf{x}}'_{k-1} \\
&\quad + \mathbf{K}_k \left(\mathbf{E}_k \mathbf{R}_k \mathbf{E}_k^T \right) \left(\mathbf{E}_k \mathbf{R}_k \mathbf{E}_k^T \right)^{-1} (\hat{\mathbf{z}}_k - \mathbf{d}_k); \\
&= \left(\mathbf{I}_{n' \times n'} - \mathbf{K}_k \mathbf{H}_k \right) \hat{\mathbf{x}}'_{k-1} + \mathbf{K}_k (\hat{\mathbf{z}}_k - \mathbf{d}_k).
\end{aligned}
$$

B.2 The NMSKF for Dynamic Systems

Theorem 5.2 (Analytical posterior pdf for dynamic systems) *Consider an estimation problem with measurements* $\hat{\mathbf{Z}}_i = \{\hat{\mathbf{z}}_1, \hat{\mathbf{z}}_2, \ldots, \hat{\mathbf{z}}_i\}$ *and process and measurement models*

$$
\begin{aligned}
\mathbf{x}'_k &= \mathbf{F}_{k-1} \mathbf{x}'_{k-1} + \mathbf{b}_{k-1} + \mathbf{C}_{k-1} \boldsymbol{\rho}_{p,k-1}; \\
\mathbf{z}_k &= \mathbf{H}_k \, \mathbf{x}'_k + \mathbf{d}_k + \mathbf{E}_k \boldsymbol{\rho}_{m,k};
\end{aligned}
$$

where $\mathbf{x}'_k = \mathbf{g}(\mathbf{x}_k)$ *and for which the zero mean Gaussian process and measurement model uncertainties* $\boldsymbol{\rho}_{p,k-1}$ *and* $\boldsymbol{\rho}_{m,k}$ *are mutually uncorrelated and uncorrelated over time. For this system, the posterior pdf over the state* $\mathbf{x}_k$ *at time step* k*, can be expressed in analytical form as*

$$
p(\mathbf{x}_k | \hat{\mathbf{Z}}_{k-1}) \sim \exp\left(-\frac{1}{2} \left(\mathbf{g}(\mathbf{x}_k) - \hat{\mathbf{x}}'_{k|k-1} \right)^T \mathbf{P}'^{-1}_{k|k-1} \left(\mathbf{g}(\mathbf{x}_k) - \hat{\mathbf{x}}'_{k|k-1} \right) \right); \tag{B.10}
$$

$$
p(\mathbf{x}_k | \hat{\mathbf{Z}}_k) \sim \exp\left(-\frac{1}{2} \left(\mathbf{g}(\mathbf{x}_k) - \hat{\mathbf{x}}'_{k|k} \right)^T \mathbf{P}'^{-1}_{k|k} \left(\mathbf{g}(\mathbf{x}_k) - \hat{\mathbf{x}}'_{k|k} \right) \right). \tag{B.11}
$$

The sufficient statistics $\hat{\mathbf{x}}'_{k|k-1}$, $\mathbf{P}'_{k|k-1}$, $\hat{\mathbf{x}}'_{k|k}$ *and* $\mathbf{P}'_{k|k}$ *are defined by the Kalman Filter equations for the state estimation of the* n'*-dimensional state* $\mathbf{x}'_k$*:*

$$\hat{\mathbf{x}}'_{k|k-1} = \mathbf{F}_{k-1}\hat{\mathbf{x}}'_{k-1|k-1} + \mathbf{b}_{k-1}; \quad \text{(B.12)}$$
$$\mathbf{P}'_{k|k-1} = \mathbf{F}_{k-1}\mathbf{P}'_{k-1|k-1}\mathbf{F}^T_{k-1} + \mathbf{C}_{k-1}\mathbf{Q}_{k-1}\mathbf{C}^T_{k-1}; \quad \text{(B.13)}$$
$$\hat{\mathbf{x}}'_{k|k} = \hat{\mathbf{x}}'_{k|k-1} + \mathbf{K}_k\boldsymbol{\eta}_k; \quad \text{(B.14)}$$
$$\mathbf{P}'_{k|k} = (\mathbf{I}_{n'\times n'} - \mathbf{K}_k\mathbf{H}_k)\,\mathbf{P}'_{k|k-1}; \quad \text{(B.15)}$$

where

$$\boldsymbol{\eta}_k = \hat{\mathbf{z}}_k - \mathbf{H}_k\hat{\mathbf{x}}'_{k|k-1} - \mathbf{d}_k; \quad \text{(B.16)}$$
$$\mathbf{S}_k = \mathbf{E}_k\mathbf{R}_k\mathbf{E}^T_k + \mathbf{H}_k\mathbf{P}'_{k|k-1}\mathbf{H}^T_k; \quad \text{(B.17)}$$
$$\mathbf{K}_k = \mathbf{P}'_{k|k-1}\mathbf{H}^T_k\mathbf{S}^{-1}_k. \quad \text{(B.18)}$$

$\mathbf{Q}_{k-1}$ *and* $\mathbf{R}_k$ *are the covariance matrices of the process and measurement model uncertainties. If no measurement* $\hat{\mathbf{z}}_k$ *is available at a certain time step* k*, then (B.14)–(B.18) reduce to* $\hat{\mathbf{x}}'_{k|k} = \hat{\mathbf{x}}'_{k|k-1}$*,* $\mathbf{P}'_{k|k} = \mathbf{P}'_{k|k-1}$*. The iteration starts with the prior uncertainty (see Sect. 5.2.3)*

$$\hat{\mathbf{x}}'_{0|0} = \begin{bmatrix}\mathbf{0}_{(n'-n)\times 1}\\ \hat{\mathbf{x}}_{0|0}\end{bmatrix}; \qquad \mathbf{P}'^{-1}_{0|0} = \begin{bmatrix}\mathbf{0}_{(n'-n)\times(n'-n)} & \mathbf{0}_{(n'-n)\times n}\\ \mathbf{0}_{n\times(n'-n)} & \mathbf{P}^{-1}_{0|0}\end{bmatrix}.$$

Proof Theorem 5.2

Before processing any measurements, the prior pdf $p(\mathbf{x})$ has the form (B.11). The first part of the proof derives the *process update*, i.e., (B.10), (B.12)–(B.13). The second part of the proof focuses on the *measurement update*, i.e., (B.11), (B.14)–(B.18).

Process update. We prove the process update by induction: given that $p(\mathbf{x}_{k-1}|\hat{\mathbf{Z}}_{k-1})$ is of the form (B.11), we prove that $p(\mathbf{x}_k|\hat{\mathbf{Z}}_{k-1})$ is of the form (B.10), (B.12)–(B.13).

$$\begin{aligned} p(\mathbf{x}_k|\hat{\mathbf{Z}}_{k-1}) &= \int_{\mathbf{x}'_{k-1}} p(\mathbf{x}_k, \mathbf{x}'_{k-1}|\hat{\mathbf{Z}}_{k-1})d\mathbf{x}'_{k-1};\\ &= \int_{\mathbf{x}'_{k-1}} p(\mathbf{x}'_{k-1}|\hat{\mathbf{Z}}_{k-1})p(\mathbf{x}_k|\mathbf{x}'_{k-1}, \hat{\mathbf{Z}}_{k-1})d\mathbf{x}'_{k-1};\\ &= \int_{\mathbf{x}'_{k-1}} p(\mathbf{x}'_{k-1}|\hat{\mathbf{Z}}_{k-1})p(\mathbf{x}_k|\mathbf{x}'_{k-1})d\mathbf{x}'_{k-1}; \end{aligned} \quad \text{(B.19)}$$

with

$$p(\mathbf{x}'_{k-1}|\hat{\mathbf{Z}}_{k-1}) \sim \exp\left(-\frac{1}{2}\left(\mathbf{x}'_{k-1} - \hat{\mathbf{x}}'_{k-1|k-1}\right)^T \mathbf{P}'^{-1}_{k-1|k-1} \times \right.$$
$$\left.\left(\mathbf{x}'_{k-1} - \hat{\mathbf{x}}'_{k-1|k-1}\right)\right);$$
$$p(\mathbf{x}_k|\mathbf{x}'_{k-1}) \sim \exp\left(-\frac{1}{2}\left(\mathbf{F}_{k-1}\mathbf{x}'_{k-1} + \mathbf{b}_{k-1} - \mathbf{g}(\mathbf{x}_k)\right)^T \times \right.$$
$$\left.\left(\mathbf{C}_{k-1}\mathbf{Q}_{k-1}\mathbf{C}^T_{k-1}\right)^{-1}\left(\mathbf{F}_{k-1}\mathbf{x}'_{k-1} + \mathbf{b}_{k-1} - \mathbf{g}(\mathbf{x}_k)\right)\right).$$

We simplified $p(\mathbf{x}_k|\mathbf{x}'_{k-1}, \hat{\mathbf{Z}}_{k-1})$ to $p(\mathbf{x}_k|\mathbf{x}'_{k-1})$ because the process and measurement model uncertainty are mutually uncorrelated and uncorrelated over time. After some matrix calculations on (B.19) a factor is identified which is independent of $\mathbf{x}'_{k-1}$ and hence can be brought outside the integral:

$$p(\mathbf{x}_k|\hat{\mathbf{Z}}_{k-1}) \sim \exp\left(-\frac{1}{2}\left(\mathbf{g}(\mathbf{x}_k) - \hat{\mathbf{x}}'_{k|k-1}\right)^T \mathbf{P}'^{-1}_{k|k-1}\left(\mathbf{g}(\mathbf{x}_k) - \hat{\mathbf{x}}'_{k|k-1}\right)\right) \times$$
$$\int_{\mathbf{x}'_{k-1}} \exp\left(-\frac{1}{2}\left(\mathbf{x}'_{k-1} - \mathbf{v}\right)^T \mathbf{V}^{-1}\left(\mathbf{x}'_{k-1} - \mathbf{v}\right)\right) d\mathbf{x}'_{k-1}; \tag{B.20}$$

where

$$\mathbf{V}^{-1} = \mathbf{P}'^{-1}_{k-1|k-1} + \mathbf{F}^T_{k-1}\left(\mathbf{C}_{k-1}\mathbf{Q}_{k-1}\mathbf{C}^T_{k-1}\right)^{-1}\mathbf{F}_{k-1};$$
$$\mathbf{v} = \mathbf{V}\left(\mathbf{P}'^{-1}_{k-1|k-1}\hat{\mathbf{x}}'_{k-1|k-1}\right.$$
$$\left. + \mathbf{F}^T_{k-1}\left(\mathbf{C}_{k-1}\mathbf{Q}_{k-1}\mathbf{C}^T_{k-1}\right)^{-1}\left(\mathbf{g}(\mathbf{x}_k) - \mathbf{b}_{k-1}\right)\right);$$
$$\mathbf{P}'^{-1}_{k|k-1} = \left(\mathbf{C}_{k-1}\mathbf{Q}_{k-1}\mathbf{C}^T_{k-1}\right)^{-1} \times$$
$$\left(\mathbf{I}_{n'\times n'} - \mathbf{F}_{k-1}\mathbf{V}\mathbf{F}^T_{k-1}\left(\mathbf{C}_{k-1}\mathbf{Q}_{k-1}\mathbf{C}^T_{k-1}\right)^{-1}\right); \tag{B.21}$$
$$\hat{\mathbf{x}}'_{k|k-1} = \mathbf{P}'_{k|k-1}\left(\mathbf{C}_{k-1}\mathbf{Q}_{k-1}\mathbf{C}^T_{k-1}\right)^{-1}\left(\mathbf{b}_{k-1} - \mathbf{F}_{k-1}\mathbf{V} \times \right.$$
$$\left.\left(\mathbf{F}^T_{k-1}\left(\mathbf{C}_{k-1}\mathbf{Q}_{k-1}\mathbf{C}^T_{k-1}\right)^{-1}\mathbf{b}_{k-1} - \mathbf{P}'^{-1}_{k-1|k-1}\hat{\mathbf{x}}'_{k-1|k-1}\right)\right). \tag{B.22}$$

These equations are obtained by using the fact that covariance matrices are symmetric: $\mathbf{P}'_{k-1|k-1} = \mathbf{P}'^T_{k-1|k-1}$, $\mathbf{Q}_{k-1} = \mathbf{Q}^T_{k-1}$, $\mathbf{V} = \mathbf{V}^T$ and $\mathbf{P}'_{k|k-1} = \mathbf{P}'^T_{k|k-1}$.

$$\int_{\mathbf{x}'_{k-1}} \exp\left(-\frac{1}{2}\left(\mathbf{x}'_{k-1} - \mathbf{v}\right)^T \mathbf{V}^{-1}\left(\mathbf{x}'_{k-1} - \mathbf{v}\right)\right) d\mathbf{x}'_{k-1} = \sqrt{(2\pi)^{n'}\det(\mathbf{V})};$$

where n' is the dimension of $\mathbf{x}'$. This expression is independent of $\mathbf{x}_k$ and hence (B.20) simplifies to

$$p(\mathbf{x}_k|\hat{\mathbf{Z}}_{k-1}) \sim \exp\left(-\frac{1}{2}\left(\mathbf{g}(\mathbf{x}_k) - \hat{\mathbf{x}}'_{k|k-1}\right)^T \mathbf{P}'^{-1}_{k|k-1}\left(\mathbf{g}(\mathbf{x}_k) - \hat{\mathbf{x}}'_{k|k-1}\right)\right);$$

i.e., a pdf of the form (B.10).

Applying the matrix inversion lemma (B.9)

1. to (B.21) shows that $\mathbf{P}'_{k|k-1}$ is defined by (B.13):

$$\begin{aligned}
\mathbf{P}'^{-1}_{k|k-1} &= \left(\mathbf{C}_{k-1}\mathbf{Q}_{k-1}\mathbf{C}^T_{k-1}\right)^{-1} \times \\
&\quad \left(\mathbf{I}_{n'\times n'} - \mathbf{F}_{k-1}\mathbf{V}\mathbf{F}^T_{k-1}\left(\mathbf{C}_{k-1}\mathbf{Q}_{k-1}\mathbf{C}^T_{k-1}\right)^{-1}\right); \\
&= \left(\mathbf{C}_{k-1}\mathbf{Q}_{k-1}\mathbf{C}^T_{k-1}\right)^{-1} - \left(\mathbf{C}_{k-1}\mathbf{Q}_{k-1}\mathbf{C}^T_{k-1}\right)^{-1}\mathbf{F}_{k-1}\times \\
&\quad \left(\mathbf{P}'^{-1}_{k-1|k-1} + \mathbf{F}^T_{k-1}\left(\mathbf{C}_{k-1}\mathbf{Q}_{k-1}\mathbf{C}^T_{k-1}\right)^{-1}\mathbf{F}_{k-1}\right)^{-1}\times \\
&\quad \mathbf{F}^T_{k-1}\left(\mathbf{C}_{k-1}\mathbf{Q}_{k-1}\mathbf{C}^T_{k-1}\right)^{-1}; \\
&= \left(\mathbf{C}_{k-1}\mathbf{Q}_{k-1}\mathbf{C}^T_{k-1} + \mathbf{F}_{k-1}\mathbf{P}'_{k-1|k-1}\mathbf{F}^T_{k-1}\right)^{-1};
\end{aligned}$$

2. to (B.22) shows that $\hat{\mathbf{x}}'_{k|k-1}$ is defined by (B.12):

$$\begin{aligned}
\hat{\mathbf{x}}'_{k|k-1} &= \mathbf{P}'_{k|k-1}\left(\mathbf{C}_{k-1}\mathbf{Q}_{k-1}\mathbf{C}^T_{k-1}\right)^{-1}\left(\mathbf{b}_{k-1} - \right. \\
&\quad \mathbf{F}_{k-1}\mathbf{V}\mathbf{F}^T_{k-1}\left(\mathbf{C}_{k-1}\mathbf{Q}_{k-1}\mathbf{C}^T_{k-1}\right)^{-1}\mathbf{b}_{k-1} + \\
&\quad \left.\mathbf{F}_{k-1}\mathbf{V}\mathbf{P}'^{-1}_{k-1|k-1}\hat{\mathbf{x}}'_{k-1|k-1}\right); \\
&= \mathbf{P}'_{k|k-1}\left(\left(\mathbf{C}_{k-1}\mathbf{Q}_{k-1}\mathbf{C}^T_{k-1}\right)^{-1} - \left(\mathbf{C}_{k-1}\mathbf{Q}_{k-1}\mathbf{C}^T_{k-1}\right)^{-1}\times\right. \\
&\quad \left.\mathbf{F}_{k-1}\mathbf{V}\mathbf{F}^T_{k-1}\left(\mathbf{C}_{k-1}\mathbf{Q}_{k-1}\mathbf{C}^T_{k-1}\right)^{-1}\right)\mathbf{b}_{k-1} + \\
&\quad \mathbf{P}'_{k|k-1}\left(\mathbf{C}_{k-1}\mathbf{Q}_{k-1}\mathbf{C}^T_{k-1}\right)^{-1}\mathbf{F}_{k-1}\mathbf{V}\mathbf{P}'^{-1}_{k-1|k-1}\hat{\mathbf{x}}'_{k-1|k-1}; \\
&= \mathbf{P}'_{k|k-1}\mathbf{P}'^{-1}_{k|k-1}\mathbf{b}_{k-1} + \mathbf{P}'_{k|k-1}\left(\mathbf{C}_{k-1}\mathbf{Q}_{k-1}\mathbf{C}^T_{k-1}\right)^{-1}\mathbf{F}_{k-1}\times \\
&\quad \mathbf{V}\mathbf{P}'^{-1}_{k-1|k-1}\hat{\mathbf{x}}'_{k-1|k-1}; \\
&= \mathbf{b}_{k-1} + \left(\mathbf{C}_{k-1}\mathbf{Q}_{k-1}\mathbf{C}^T_{k-1} + \mathbf{F}_{k-1}\mathbf{P}'_{k-1|k-1}\mathbf{F}^T_{k-1}\right)\times \\
&\quad \left(\mathbf{C}_{k-1}\mathbf{Q}_{k-1}\mathbf{C}^T_{k-1}\right)^{-1}\mathbf{F}_{k-1}\times \\
&\quad \left(\mathbf{P}'^{-1}_{k-1|k-1} + \mathbf{F}^T_{k-1}\left(\mathbf{C}_{k-1}\mathbf{Q}_{k-1}\mathbf{C}^T_{k-1}\right)^{-1}\mathbf{F}_{k-1}\right)^{-1}\times \\
&\quad \mathbf{P}'^{-1}_{k-1|k-1}\hat{\mathbf{x}}'_{k-1|k-1};
\end{aligned}$$

$$\begin{aligned}
\hat{\mathbf{x}}'_{k|k-1} &= \mathbf{b}_{k-1} + \mathbf{F}_{k-1} \times \\
&\quad \left(\mathbf{I}_{n' \times n'} + \mathbf{P}'_{k-1|k-1} \mathbf{F}_{k-1}^T \left(\mathbf{C}_{k-1} \mathbf{Q}_{k-1} \mathbf{C}_{k-1}^T\right)^{-1} \mathbf{F}_{k-1}\right) \times \\
&\quad \left(\mathbf{I}_{n' \times n'} + \mathbf{P}'_{k-1|k-1} \mathbf{F}_{k-1}^T \left(\mathbf{C}_{k-1} \mathbf{Q}_{k-1} \mathbf{C}_{k-1}^T\right)^{-1} \mathbf{F}_{k-1}\right)^{-1} \times \\
&\quad \hat{\mathbf{x}}'_{k-1|k-1}; \\
&= \mathbf{b}_{k-1} + \mathbf{F}_{k-1} \hat{\mathbf{x}}'_{k-1|k-1}.
\end{aligned}$$

Measurement update. Given a $p(\mathbf{x}_k|\hat{\mathbf{Z}}_{k-1})$ of the form (B.10), we now need to prove that $p(\mathbf{x}_k|\hat{\mathbf{Z}}_k)$ is determined by (B.11), (B.14)–(B.18). This part of the proof is similar to the proof of Theorem 5.1.

C

Frame Transformations

Chapter 6 describes the wrench spanning sets of the elementary contact library with respect to a frame {c} on the contact. In order to apply the measurement equations, the twist and wrench measurements and the wrench spanning set need to be expressed with respect to the same frame.

This appendix presents the analytical expressions for the wrench screw transformation matrices $\mathbf{S}_w$, the rotation matrices $\mathbf{R}$, the skew-symmetric matrices $[\mathbf{p}\times]$ and the homogeneous transformation matrices $\mathbf{T}$.

The matrices depend on the *definitions* of the relative position and orientation between the manipulated object frame {m}, the gripper frame {g}, the world frame {w} and the environment frame {e} (Fig. 1.1):

1. The *grasping uncertainties* $\mathbf{x}^m = [x^m\ y^m\ z^m\ \theta_x^m\ \theta_y^m\ \theta_z^m]^T$ are defined as the position *of* the {g} frame *in* the {m} frame ${}_m\mathbf{p}^{m,g} = [x^m\ y^m\ z^m]^T$ and the orientation $(\theta_x^m, \theta_y^m, \theta_z^m)$ of the {g} frame with respect to the {m} frame. The definition of these (Euler) angles is: rotating the frame {m} about its Z-axis by θ_z^m, followed by rotating about the *new* Y-axis by θ_y^m, followed by rotating about the *new* X-axis by θ_x^m, results in a frame with axes parallel to the {g} frame.

2. The *environment uncertainties* for the ith environment object $\mathbf{x}^i = [x^i\ y^i\ z^i\ \theta_x^i\ \theta_y^i\ \theta_z^i]^T$ are defined as the position *of* the {w} frame *in* the {i} frame ${}_ip^{i,w} = [x^i\ y^i\ z^i]^T$ and the orientation $(\theta_x^i, \theta_y^i, \theta_z^i)$ of the {w} frame with respect to the {i} frame. The angles are defined as: rotating the frame {i} about its Z-axis by θ_z^i, followed by rotating about the *new* Y-axis by θ_y^i, followed by rotating about the *new* X-axis by θ_x^i, results in a frame with axes parallel to the {w} frame.

3. The *transformation between the world and the gripper frame* is known as a function of the robot joint positions $\mathbf{q}$ (forward kinematics of the robot). ${}_w\mathbf{p}^{w,g} = [{}_wp_x^{w,g}\ {}_wp_y^{w,g}\ {}_wp_z^{w,g}]^T$ represents the origin *of* the gripper frame *in* the world frame and $(\theta_x, \theta_y, \theta_z)$ the orientation of the gripper frame

with respect to the world frame. The angles are defined as: rotating the frame {w} about its Z-axis by θ_z, followed by rotating about the *new* Y-axis by θ_y, followed by rotating about the *new* X-axis by θ_x, results in a frame with axes parallel to the {g} frame.

The 6×6 *Wrench Screw Transformation Matrices* $\mathbf{S}_w$*:*

$$
{}^{m}_{g}\mathbf{S}_w = \begin{bmatrix} {}^{m}_{g}\mathbf{R} & \mathbf{0}_{3\times3} \\ {}^{m}_{g}\mathbf{R}\,[{}_{m}\mathbf{p}^{g,m}\times] & {}^{m}_{g}\mathbf{R} \end{bmatrix};
$$

$$
{}^{i}_{w}\mathbf{S}_w = \begin{bmatrix} {}^{i}_{w}\mathbf{R} & \mathbf{0}_{3\times3} \\ {}^{i}_{w}\mathbf{R}\,[{}_{i}\mathbf{p}^{w,i}\times] & {}^{i}_{w}\mathbf{R} \end{bmatrix};
$$

$$
{}^{w}_{g}\mathbf{S}_w = \begin{bmatrix} {}^{w}_{g}\mathbf{R} & \mathbf{0}_{3\times3} \\ {}^{w}_{g}\mathbf{R}\,[{}_{w}\mathbf{p}^{g,w}\times] & {}^{w}_{g}\mathbf{R} \end{bmatrix};
$$

$$
{}^{c}_{m}\mathbf{S}_w = \begin{bmatrix} {}^{c}_{m}\mathbf{R} & \mathbf{0}_{3\times3} \\ [{}_{m}\mathbf{p}^{m,c}\times]\,{}^{c}_{m}\mathbf{R} & {}^{c}_{m}\mathbf{R} \end{bmatrix};
$$

$$
{}^{c}_{i}\mathbf{S}_w = \begin{bmatrix} {}^{c}_{i}\mathbf{R} & \mathbf{0}_{3\times3} \\ [{}_{i}\mathbf{p}^{i,c}\times]\,{}^{c}_{i}\mathbf{R} & {}^{c}_{i}\mathbf{R} \end{bmatrix}.
$$

The 3×3 *Skew-Symmetric Matrices* $[\mathbf{p}\times]$*:*

$$
[{}_{m}\mathbf{p}^{g,m}\times] = \begin{bmatrix} 0 & z^m & -y^m \\ -z^m & 0 & x^m \\ y^m & -x^m & 0 \end{bmatrix};
$$

$$
[{}_{i}\mathbf{p}^{w,i}\times] = \begin{bmatrix} 0 & z^i & -y^i \\ -z^i & 0 & x^i \\ y^i & -x^i & 0 \end{bmatrix};
$$

$$
[{}_{w}\mathbf{p}^{g,w}\times] = \begin{bmatrix} 0 & {}_{w}p_z^{w,g} & -{}_{w}p_y^{w,g} \\ -{}_{w}p_z^{w,g} & 0 & {}_{w}p_x^{w,g} \\ {}_{w}p_y^{w,g} & -{}_{w}p_x^{w,g} & 0 \end{bmatrix};
$$

$$
[{}_{m}\mathbf{p}^{m,c}\times] = \begin{bmatrix} 0 & -{}_{m}p_z^{m,c} & {}_{m}p_y^{m,c} \\ {}_{m}p_z^{m,c} & 0 & -{}_{m}p_x^{m,c} \\ -{}_{m}p_y^{m,c} & {}_{m}p_x^{m,c} & 0 \end{bmatrix};
$$

$$
[{}_{i}\mathbf{p}^{i,c}\times] = \begin{bmatrix} 0 & -{}_{i}p_z^{i,c} & {}_{i}p_y^{i,c} \\ {}_{i}p_z^{i,c} & 0 & -{}_{i}p_x^{i,c} \\ -{}_{i}p_y^{i,c} & {}_{i}p_x^{i,c} & 0 \end{bmatrix}.
$$

The 3×3 *Rotation Matrices* $\mathbf{R}$*:*

$$
{}^m_g\mathbf{R} = \begin{bmatrix} 1 & 0 & 0 \\ 0 & c\theta_x^m & s\theta_x^m \\ 0 & -s\theta_x^m & c\theta_x^m \end{bmatrix} \begin{bmatrix} c\theta_y^m & 0 & -s\theta_y^m \\ 0 & 1 & 0 \\ s\theta_y^m & 0 & c\theta_y^m \end{bmatrix} \begin{bmatrix} c\theta_z^m & s\theta_z^m & 0 \\ -s\theta_z^m & c\theta_z^m & 0 \\ 0 & 0 & 1 \end{bmatrix};
$$

$$
{}^i_w\mathbf{R} = \begin{bmatrix} 1 & 0 & 0 \\ 0 & c\theta_x^i & s\theta_x^i \\ 0 & -s\theta_x^i & c\theta_x^i \end{bmatrix} \begin{bmatrix} c\theta_y^i & 0 & -s\theta_y^i \\ 0 & 1 & 0 \\ s\theta_y^i & 0 & c\theta_y^i \end{bmatrix} \begin{bmatrix} c\theta_z^i & s\theta_z^i & 0 \\ -s\theta_z^i & c\theta_z^i & 0 \\ 0 & 0 & 1 \end{bmatrix};
$$

$$
{}^w_g\mathbf{R} = \begin{bmatrix} 1 & 0 & 0 \\ 0 & c\theta_x & s\theta_x \\ 0 & -s\theta_x & c\theta_x \end{bmatrix} \begin{bmatrix} c\theta_y & 0 & -s\theta_y \\ 0 & 1 & 0 \\ s\theta_y & 0 & c\theta_y \end{bmatrix} \begin{bmatrix} c\theta_z & s\theta_z & 0 \\ -s\theta_z & c\theta_z & 0 \\ 0 & 0 & 1 \end{bmatrix}.
$$

The 4×4 *Homogeneous Transformation Matrices* $\mathbf{T}$*:*

$$
{}^w_i\mathbf{T} = \begin{bmatrix} {}^w_i\mathbf{R} & {}_i\mathbf{p}^{i,w} \\ \mathbf{0}_{1\times 3} & 1 \end{bmatrix};
$$

$$
{}^i_w\mathbf{T} = \begin{bmatrix} {}^i_w\mathbf{R} & -\,{}^i_w\mathbf{R}\,{}_i\mathbf{p}^{i,w} \\ \mathbf{0}_{1\times 3} & 1 \end{bmatrix};
$$

$$
{}^g_w\mathbf{T} = \begin{bmatrix} {}^g_w\mathbf{R} & {}_w\mathbf{p}^{w,g} \\ \mathbf{0}_{1\times 3} & 1 \end{bmatrix};
$$

$$
{}^w_g\mathbf{T} = \begin{bmatrix} {}^w_g\mathbf{R} & -\,{}^w_g\mathbf{R}\,{}_w\mathbf{p}^{w,g} \\ \mathbf{0}_{1\times 3} & 1 \end{bmatrix};
$$

$$
{}^m_g\mathbf{T} = \begin{bmatrix} {}^m_g\mathbf{R} & -\,{}^m_g\mathbf{R}\,{}_m\mathbf{p}^{m,g} \\ \mathbf{0}_{1\times 3} & 1 \end{bmatrix};
$$

$$
{}^g_m\mathbf{T} = \begin{bmatrix} {}^g_m\mathbf{R} & {}_m\mathbf{p}^{m,g} \\ \mathbf{0}_{1\times 3} & 1 \end{bmatrix}.
$$

D

Kalman Filtering for Non-Minimal Measurement Models

This appendix contains the proof that the Kalman Filter (KF) is robust against non-minimal measurement equations. The KF algorithm uses the inverse of the innovation covariance matrix $\mathbf{S}_k$. For non-minimal measurement equations, this matrix is singular. This appendix contains the proof that the results of the Kalman Filter (KF) using non-minimal measurement equations are the same as the results of the KF using a minimal set of measurement equations, *whatever generalised inverse* $\mathbf{S}_k^{\#}$ *we choose.*

Generalised Inverse

The *generalised inverse* [24] $\mathbf{F}^{\#}$ of a matrix $\mathbf{F}$ is any inverse for which $\mathbf{F}\mathbf{F}^{\#}\mathbf{F} = \mathbf{F}$. It is obtained by a singular value decomposition. The singular value decomposition of $\mathbf{F}$ is

$$\mathbf{F} = \mathbf{U}\mathbf{\Sigma}\mathbf{V}^T; \tag{D.1}$$

where $\mathbf{U}$ and $\mathbf{V}$ are orthogonal matrices with dimensions $n_u \times n_u$ and $n_v \times n_v$ respectively, and where $\mathbf{\Sigma}$ is a $n_u \times n_v$ pseudo-diagonal matrix of rank $l = rank(\mathbf{F})$, i.e.,

$$\mathbf{U}^{-1} = \mathbf{U}^T;\ \mathbf{V}^{-1} = \mathbf{V}^T;\ \mathbf{\Sigma} = \left[\begin{array}{ccc|c} \sigma_1 & & & \\ & \ddots & & \mathbf{0}_{l\times(n_v-l)} \\ & & \sigma_l & \\ \hline & \mathbf{0}_{(n_u-l)\times l} & & \mathbf{0}_{(n_u-l)\times(n_v-l)} \end{array}\right]. \tag{D.2}$$

The generalised inverse of $\mathbf{F}$ is

$$\mathbf{F}^{\#} = \mathbf{V}\boldsymbol{\Sigma}^{\#}\mathbf{U}^T; \quad \boldsymbol{\Sigma}^{\#} = \left[\begin{array}{ccc|c} \frac{1}{\sigma_1} & & & \\ & \ddots & & \mathbf{X} \\ & & \frac{1}{\sigma_l} & \\ \hline & \mathbf{Y} & & \mathbf{Z} \end{array}\right]; \tag{D.3}$$

where $\mathbf{X}$, $\mathbf{Y}$ and $\mathbf{Z}$ are arbitrary matrices.

Problem statement

Assume that the set of measurement equations $\mathbf{h}_k(\mathbf{x}_k, \mathbf{z}_k) = \mathbf{0}_{n_h \times 1}$ is a minimal set. A non-minimal set of measurement equations, which contains additional linear combinations of these equations, can be written as:

$$\mathbf{h}_k^*(\mathbf{x}_k, \mathbf{z}_k) = \mathbf{B}\mathbf{h}_k(\mathbf{x}_k, \mathbf{z}_k) = \mathbf{0}_{n_* \times 1}; \qquad \mathbf{B} = \begin{bmatrix} \mathbf{B}^R \\ \mathbf{B}^S \end{bmatrix};$$

where $\mathbf{B}^R$ is a regular matrix. $\mathbf{S}_k$ and $\mathbf{S}_k^*$ indicate the innovation covariance matrices corresponding to the minimal and the non-minimal set of measurement equations respectively. Equation (7.7) shows that

$$\mathbf{S}_k^* = \mathbf{B}\mathbf{S}_k\mathbf{B}^T. \tag{D.4}$$

The update of the estimate $\hat{\mathbf{x}}_k$, its covariance matrix $\mathbf{P}_k$ and the NIS-values $\boldsymbol{\eta}_k^T\mathbf{S}_k^{-1}\boldsymbol{\eta}_k$ ((7.4)–(7.8)) have the same values for the minimal and the non-minimal description *if* the used inverse $(\mathbf{S}_k^*)^{\#}$ satisfies

$$\mathbf{B}^T(\mathbf{S}_k^*)^{\#}\mathbf{B} = \mathbf{S}_k^{-1}. \tag{D.5}$$

We prove that any generalised inverse of $\mathbf{S}_k^*$ satisfies condition (D.5); hence, whatever generalised inverse for $\mathbf{S}_k^*$ we choose, the KF estimator is robust against non-minimal measurement descriptions.

Proof

The eigenvalue decompositions of $\mathbf{S}_k$ and $\mathbf{S}_k^*$ are

$$\mathbf{S}_k = \mathbf{U}_k\boldsymbol{\Sigma}_k\mathbf{U}_k^T; \qquad \mathbf{S}_k^* = \mathbf{U}_k^*\boldsymbol{\Sigma}_k^*(\mathbf{U}_k^*)^T; \tag{D.6}$$

with orthogonal matrices $\mathbf{U}_k$ and $\mathbf{U}_k^*$

$$\mathbf{U}_k^{-1} = \mathbf{U}_k^T; \qquad (\mathbf{U}_k^*)^{-1} = (\mathbf{U}_k^*)^T. \tag{D.7}$$

From (D.4) and (D.6) we obtain:

$$\mathbf{S}_k^* = \mathbf{B}\mathbf{S}_k\mathbf{B}^T = \mathbf{E}_k \begin{bmatrix} \mathbf{\Sigma}_k & \mathbf{0}_{n_h \times (n_* - n_h)} \\ \mathbf{0}_{(n_* - n_h)\times n_h} & \mathbf{0}_{(n_* - n_h)\times(n_* - n_h)} \end{bmatrix} \mathbf{E}_k^T; \tag{D.8}$$

$$\text{with } \mathbf{E}_k = \begin{bmatrix} \mathbf{B}^R\mathbf{U}_k & \mathbf{0}_{n_h \times (n_* - n_h)} \\ \mathbf{B}^S\mathbf{U}_k & \mathbf{C}_k \end{bmatrix}; \tag{D.9}$$

for any arbitrary $\mathbf{C}_k$. Choose a regular $\mathbf{C}_k$. This makes $\mathbf{E}_k$ regular. Define $\mathbf{W}_k = (\mathbf{U}_k^*)^{-1}\mathbf{E}_k = (\mathbf{U}_k^*)^T\mathbf{E}_k$.
$\mathbf{W}_k$ and $\mathbf{U}_k^*$ are split up in sub-matrices $\mathbf{W}_{i,k}$ and $\mathbf{U}_{i,k}^*$ in such a way that the dimensions of $\mathbf{W}_{1,k}$ and $\mathbf{U}_{1,k}^*$ are $n_h \times n_h$, i.e., the dimensions of $\mathbf{B}^R$ and $\mathbf{S}_k$,

$$\mathbf{W}_k = \begin{bmatrix} \mathbf{W}_{1,k} & \mathbf{W}_{2,k} \\ \mathbf{W}_{3,k} & \mathbf{W}_{4,k} \end{bmatrix}; \qquad \mathbf{U}_k^* = \begin{bmatrix} \mathbf{U}_{1,k}^* & \mathbf{U}_{2,k}^* \\ \mathbf{U}_{3,k}^* & \mathbf{U}_{4,k}^* \end{bmatrix}.$$

$(\mathbf{U}_k^*)^T\mathbf{U}_k^* = \mathbf{I}_{n_* \times n_*}$, hence,

$$(\mathbf{U}_{1,k}^*)^T\mathbf{U}_{1,k}^* + (\mathbf{U}_{3,k}^*)^T\mathbf{U}_{3,k}^* = \mathbf{I}_{n_h \times n_h}; \tag{D.10}$$

$$(\mathbf{U}_{1,k}^*)^T\mathbf{U}_{2,k}^* + (\mathbf{U}_{3,k}^*)^T\mathbf{U}_{4,k}^* = \mathbf{0}_{n_h \times (n_* - n_h)}. \tag{D.11}$$

Replacing $\mathbf{E}_k = \mathbf{U}_k^*\mathbf{W}_k$ in (D.9) gives

$$\mathbf{B}^R\mathbf{U}_k = \mathbf{U}_{1,k}^*\mathbf{W}_{1,k} + \mathbf{U}_{2,k}^*\mathbf{W}_{3,k}; \tag{D.12}$$

$$\mathbf{B}^S\mathbf{U}_k = \mathbf{U}_{3,k}^*\mathbf{W}_{1,k} + \mathbf{U}_{4,k}^*\mathbf{W}_{3,k}. \tag{D.13}$$

Replacing $\mathbf{E}_k = \mathbf{U}_k^*\mathbf{W}_k$ in (D.8) gives

$$\mathbf{S}_k^* = \mathbf{U}_k^* \left(\mathbf{W}_k \begin{bmatrix} \mathbf{\Sigma}_k & \mathbf{0}_{n_h \times (n_* - n_h)} \\ \mathbf{0}_{(n_* - n_h)\times n_h} & \mathbf{0}_{(n_* - n_h)\times(n_* - n_h)} \end{bmatrix} \mathbf{W}_k^T \right) (\mathbf{U}_k^*)^T. \tag{D.14}$$

From (D.6) and (D.14) we obtain:

$$\mathbf{U}_k^* \left(\mathbf{W}_k \begin{bmatrix} \mathbf{\Sigma}_k & \mathbf{0}_{n_h \times (n_* - n_h)} \\ \mathbf{0}_{(n_* - n_h)\times n_h} & \mathbf{0}_{(n_* - n_h)\times(n_* - n_h)} \end{bmatrix} \mathbf{W}_k^T \right) (\mathbf{U}_k^*)^T = \mathbf{U}_k^*\mathbf{\Sigma}_k^*(\mathbf{U}_k^*)^T;$$

hence,

$$\mathbf{\Sigma}_k^* = \begin{bmatrix} \mathbf{W}_{1,k}\mathbf{\Sigma}_k\mathbf{W}_{1,k}^T & \mathbf{0}_{n_h \times (n_* - n_h)} \\ \mathbf{0}_{(n_* - n_h)\times n_h} & \mathbf{0}_{(n_* - n_h)\times(n_* - n_h)} \end{bmatrix}; \qquad \mathbf{W}_{3,k} = \mathbf{0}_{(n_* - n_h)\times n_h}. \tag{D.15}$$

The generalised inverse of $\mathbf{S}_k^*$ is ((D.1)–(D.3)):

$$(\mathbf{S}_k^*)^\# = \mathbf{U}_k^*(\mathbf{\Sigma}_k^*)^\#(\mathbf{U}_k^*)^T; \tag{D.16}$$

$$(\mathbf{\Sigma}_k^*)^\# = \begin{bmatrix} (\mathbf{W}_{1,k}^T)^{-1}\mathbf{\Sigma}_k^{-1}\mathbf{W}_{1,k}^{-1} & \mathbf{X}_k \\ \mathbf{Y}_k & \mathbf{Z}_k \end{bmatrix}. \tag{D.17}$$

Finally we calculate $\mathbf{B}^T(\mathbf{S}_k^*)^\#\mathbf{B}$ using (D.7), (D.10)–(D.13), and (D.15)–(D.17):

$$
\begin{aligned}
\mathbf{B}^T(\mathbf{S}_k^*)^{\#}\mathbf{B} &= (\mathbf{U}_k\mathbf{U}_k^T)\mathbf{B}^T\left(\mathbf{U}_k^*(\mathbf{\Sigma}_k^*)^{\#}(\mathbf{U}_k^*)^T\right)\mathbf{B}(\mathbf{U}_k\mathbf{U}_k^T);\\
&= \mathbf{U}_k\begin{bmatrix}\mathbf{B}^R\mathbf{U}_k\\ \mathbf{B}^S\mathbf{U}_k\end{bmatrix}^T\mathbf{U}_k^*(\mathbf{\Sigma}_k^*)^{\#}(\mathbf{U}_k^*)^T\begin{bmatrix}\mathbf{B}^R\mathbf{U}_k\\ \mathbf{B}^S\mathbf{U}_k\end{bmatrix}\mathbf{U}_k^T;\\
&= \mathbf{U}_k\begin{bmatrix}\mathbf{U}_{1,k}^*\mathbf{W}_{1,k}\\ \mathbf{U}_{3,k}^*\mathbf{W}_{1,k}\end{bmatrix}^T\mathbf{U}_k^*(\mathbf{\Sigma}_k^*)^{\#}(\mathbf{U}_k^*)^T\begin{bmatrix}\mathbf{U}_{1,k}^*\mathbf{W}_{1,k}\\ \mathbf{U}_{3,k}^*\mathbf{W}_{1,k}\end{bmatrix}\mathbf{U}_k^T;\\
&= \mathbf{U}_k\mathbf{W}_{1,k}^T\begin{bmatrix}\mathbf{U}_{1,k}^*\\ \mathbf{U}_{3,k}^*\end{bmatrix}^T\mathbf{U}_k^*(\mathbf{\Sigma}_k^*)^{\#}(\mathbf{U}_k^*)^T\begin{bmatrix}\mathbf{U}_{1,k}^*\\ \mathbf{U}_{3,k}^*\end{bmatrix}\mathbf{W}_{1,k}\mathbf{U}_k^T;\\
&= \mathbf{U}_k\mathbf{W}_{1,k}^T\left[\mathbf{I}_{n_h\times n_h}\ \mathbf{0}_{n_h\times(n_*-n_h)}\right](\mathbf{\Sigma}_k^*)^{\#}\begin{bmatrix}\mathbf{I}_{n_h\times n_h}\\ \mathbf{0}_{(n_*-n_h)\times n_h}\end{bmatrix}\mathbf{W}_{1,k}\mathbf{U}_k^T;\\
&= \mathbf{U}_k\mathbf{W}_{1,k}^T\left((\mathbf{W}_{1,k}^T)^{-1}\mathbf{\Sigma}_k^{-1}\mathbf{W}_{1,k}^{-1}\right)\mathbf{W}_{1,k}\mathbf{U}_k^T;\\
&= \mathbf{U}_k\mathbf{\Sigma}_k^{-1}\mathbf{U}_k^T = \mathbf{S}_k^{-1};
\end{aligned}
$$

which proves that any generalised inverse of $\mathbf{S}_k^*$ satisfies (D.5).

E

Partial Observation with the Kalman Filter

This appendix contains the proofs of Lemmas 7.1 and 7.2. These lemmas state that, for *static* systems, (i) a Kalman Filter can be run on only the *observable* state variables; and (ii) the full state estimate and covariance matrix (i.e., including the *unobservable* state variables) can be calculated at any time based on the full initial state estimate and covariance matrix and the new state estimate and covariance matrix of the observed part of the state.

The lemmas are proved using the KF equations for implicit measurement equations (7.4)–(7.8).

Problem statement

A measurement equation $\mathbf{h}_k(\mathbf{x}, \mathbf{z}_k) = \mathbf{0}_{n_h \times 1}$ is often function of only part of the state variables. These are called the *observable* state variables and are collected in the vector $\mathbf{x}_o$. The other are called the *unobservable* state variables and are collected in the vector $\mathbf{x}_u$. The dimensions of $\mathbf{x}_o$ and $\mathbf{x}_u$ are denoted as n_o and n_u respectively. The state estimate and its covariance matrix can be written as:

$$\hat{\mathbf{x}} = \begin{bmatrix} \hat{\mathbf{x}}_o \\ \hat{\mathbf{x}}_u \end{bmatrix}; \qquad \mathbf{P} = \begin{bmatrix} \mathbf{P}_{oo} & \mathbf{P}_{ou} \\ \mathbf{P}_{uo} & \mathbf{P}_{uu} \end{bmatrix}.$$

$\mathbf{P}_{oo}$ and $\mathbf{P}_{uu}$ are the covariance matrices on the estimates $\hat{\mathbf{x}}_o$ and $\hat{\mathbf{x}}_u$ respectively. $\mathbf{P}_{ou} = \mathbf{P}_{uo}^T$ describes the correlation between the estimates $\hat{\mathbf{x}}_o$ and $\hat{\mathbf{x}}_u$.

The Kalman Filter algorithm for the implicit measurement equation $\mathbf{h}_k(\mathbf{x}_o, \mathbf{z}_k) = \mathbf{0}_{n_h \times 1}$ is given in (7.4)–(7.8). Only the observable part $\mathbf{x}_o$ of the state is represented in the measurement equation, hence,

$$\mathbf{H}_k = \begin{bmatrix} \mathbf{H}_{o,k} & \mathbf{0}_{n_h \times n_u} \end{bmatrix}; \tag{E.1}$$

$$\mathbf{S}_k = \mathbf{D}_k \mathbf{R}_k \mathbf{D}_k^T + \mathbf{H}_{o,k} \mathbf{P}_{oo,k-1} \mathbf{H}_{o,k}^T; \tag{E.2}$$

$$\mathbf{K}_k = - \begin{bmatrix} \mathbf{P}_{oo} \\ \mathbf{P}_{uo} \end{bmatrix}_{k-1} \mathbf{H}_{o,k}^T \mathbf{S}_k^{-1}. \tag{E.3}$$

Lemma 7.1
The update for the estimate of the observable part of the state vector $\hat{\mathbf{x}}_o$ and its covariance matrix $\mathbf{P}_{oo}$ are independent *of the estimate of the unobservable part of the state vector $\hat{\mathbf{x}}_u$, its covariance matrix $\mathbf{P}_{uu}$ and the correlation $\mathbf{P}_{uo}$ between the estimates of the observable and unobservable part.*

Proof Lemma 7.1
Based on (7.4)–(7.8) and (E.1)–(E.3), we can calculate the update for the estimate $\hat{\mathbf{x}}_o$ and its covariance matrix $\mathbf{P}_{oo}$:

$$\begin{aligned}\hat{\mathbf{x}}_{o,k} &= \hat{\mathbf{x}}_{o,k-1} - \mathbf{P}_{oo,k-1}\mathbf{H}_{o,k}^T\mathbf{S}_k^{-1}\boldsymbol{\eta}_k;\\ \mathbf{P}_{oo,k} &= \left(\mathbf{I}_{n_o\times n_o} - \mathbf{P}_{oo,k-1}\mathbf{H}_{o,k}^T\mathbf{S}_k^{-1}\mathbf{H}_{o,k}\right)\mathbf{P}_{oo,k-1}.\end{aligned}$$

These equations are independent of the estimate $\hat{\mathbf{x}}_u$, its covariance matrix $\mathbf{P}_{uu}$, and the correlation $\mathbf{P}_{ou}$ between the estimates $\hat{\mathbf{x}}_o$ and $\hat{\mathbf{x}}_u$. This means that a reduced Kalman Filter can be run on the state vector $\mathbf{x}_o$, without considering $\mathbf{x}_u$.

Lemma 7.2
Suppose that from a time k until a time $k+l$ only the parameters $\mathbf{x}_o$ are observable. A reduced Kalman Filter has calculated $\hat{\mathbf{x}}_{o,k+l}$ and $\mathbf{P}_{oo,k+l}$. Then the updates for $\hat{\mathbf{x}}_{u,k+l}$, $\mathbf{P}_{uu,k+l}$ and $\mathbf{P}_{uo,k+l}$ are:

$$\hat{\mathbf{x}}_{u,k+l} = \hat{\mathbf{x}}_{u,k-1} + \mathbf{L}(\hat{\mathbf{x}}_{o,k+l} - \hat{\mathbf{x}}_{o,k-1}); \tag{E.4}$$

$$\mathbf{P}_{uo,k+l} = \mathbf{L}\mathbf{P}_{oo,k+l}; \tag{E.5}$$

$$\mathbf{P}_{uu,k+l} = \mathbf{P}_{uu,k-1} - \mathbf{L}(\mathbf{P}_{oo,k-1} - \mathbf{P}_{oo,k+l})\mathbf{L}^T; \tag{E.6}$$

where $\mathbf{L} = \mathbf{P}_{uo,k-1}\mathbf{P}_{oo,k-1}^{-1}$.

Proof Lemma 7.2
Based on (7.4)–(7.8) and (E.1)–(E.3), we can calculate the update of $\hat{\mathbf{x}}_o$, $\hat{\mathbf{x}}_u$, $\mathbf{P}_{oo}$, $\mathbf{P}_{uo}$ and $\mathbf{P}_{uu}$:

$$\hat{\mathbf{x}}_{o,k} = \hat{\mathbf{x}}_{o,k-1} - \mathbf{P}_{oo,k-1}\mathbf{H}_{o,k}^T\mathbf{S}_k^{-1}\boldsymbol{\eta}_k; \tag{E.7}$$

$$\hat{\mathbf{x}}_{u,k} = \hat{\mathbf{x}}_{u,k-1} - \mathbf{P}_{uo,k-1}\mathbf{H}_{o,k}^T\mathbf{S}_k^{-1}\boldsymbol{\eta}_k; \tag{E.8}$$

$$\mathbf{P}_{oo,k} = \mathbf{P}_{oo,k-1} - \mathbf{P}_{oo,k-1}\mathbf{H}_{o,k}^T\mathbf{S}_k^{-1}\mathbf{H}_{o,k}\mathbf{P}_{oo,k-1}; \tag{E.9}$$

$$\mathbf{P}_{uo,k} = \mathbf{P}_{uo,k-1} - \mathbf{P}_{uo,k-1}\mathbf{H}_{o,k}^T\mathbf{S}_k^{-1}\mathbf{H}_{o,k}\mathbf{P}_{oo,k-1}; \tag{E.10}$$

$$\mathbf{P}_{uu,k} = \mathbf{P}_{uu,k-1} - \mathbf{P}_{uo,k-1}\mathbf{H}_{o,k}^T\mathbf{S}_k^{-1}\mathbf{H}_{o,k}\mathbf{P}_{uo,k-1}^T. \tag{E.11}$$

Define the matrix $\mathbf{L}_{k-1} = \mathbf{P}_{uo,k-1}\mathbf{P}_{oo,k-1}^{-1}$. We prove that the $\mathbf{L}$ is time-independent, i.e., $\mathbf{L}_k = \mathbf{L}_{k-1}$ for all k:

$$\begin{aligned}
\mathbf{L}_k &= \mathbf{P}_{uo,k}\mathbf{P}_{oo,k}^{-1};\\
&= \left(\mathbf{P}_{uo,k-1}\left(\mathbf{I}_{n_o\times n_o} - \mathbf{H}_{o,k}^T\mathbf{S}_k^{-1}\mathbf{H}_{o,k}\mathbf{P}_{oo,k-1}\right)\right)\\
&\qquad \left(\mathbf{P}_{oo,k-1}\left(\mathbf{I}_{n_o\times n_o} - \mathbf{H}_{o,k}^T\mathbf{S}_k^{-1}\mathbf{H}_{o,k}\mathbf{P}_{oo,k-1}\right)\right)^{-1};\\
&= \mathbf{P}_{uo,k-1}\mathbf{P}_{oo,k-1}^{-1} = \mathbf{L}_{k-1}.
\end{aligned}$$

This proves (E.5).

Applying (E.7)–(E.11) l times and replacing the products $\mathbf{P}_{uo}\mathbf{P}_{oo}^{-1}$ by the symbol $\mathbf{L}$ gives:

$$\begin{aligned}
\hat{\mathbf{x}}_{o,k+l} &= \hat{\mathbf{x}}_{o,k-1} - \mathbf{P}_{oo,k-1}\mathbf{H}_{o,k}^T\mathbf{S}_k^{-1}\boldsymbol{\eta}_k - \dots\\
&\qquad - \mathbf{P}_{oo,k+l-1}\mathbf{H}_{o,k+l}^T\mathbf{S}_{k+l}^{-1}\boldsymbol{\eta}_{k+l};\\
\hat{\mathbf{x}}_{u,k+l} &= \hat{\mathbf{x}}_{u,k-1} + \mathbf{L}\left(-\mathbf{P}_{oo,k-1}\mathbf{H}_{o,k}^T\mathbf{S}_k^{-1}\boldsymbol{\eta}_k - \dots\right.\\
&\qquad \left.-\mathbf{P}_{oo,k+l-1}\mathbf{H}_{o,k+l}^T\mathbf{S}_{k+l}^{-1}\boldsymbol{\eta}_{k+l}\right);\\
\mathbf{P}_{oo,k+l} &= \mathbf{P}_{oo,k-1} - \mathbf{P}_{oo,k-1}\mathbf{H}_{o,k}^T\mathbf{S}_k^{-1}\mathbf{H}_{o,k}\mathbf{P}_{oo,k-1} - \dots\\
&\qquad - \mathbf{P}_{oo,k+l-1}\mathbf{H}_{o,k+l}^T\mathbf{S}_{k+l}^{-1}\mathbf{H}_{o,k+l}\mathbf{P}_{oo,k+l-1};\\
\mathbf{P}_{uu,k+l} &= \mathbf{P}_{uu,k-1} - \mathbf{L}\left(\mathbf{P}_{oo,k-1}\mathbf{H}_{o,k}^T\mathbf{S}_k^{-1}\mathbf{H}_{o,k}\mathbf{P}_{oo,k-1} + \dots\right.\\
&\qquad \left.+\mathbf{P}_{oo,k+l-1}\mathbf{H}_{o,k+l}^T\mathbf{S}_{k+l}^{-1}\mathbf{H}_{o,k+l}\mathbf{P}_{oo,k+l-1}\right)\mathbf{L}^T.
\end{aligned}$$

Hence, the updates for $\hat{\mathbf{x}}_{u,k+l}$ and $\mathbf{P}_{uu,k+l}$ are given by (E.4) and (E.6).

F

A NMSKF Linearizing State for the Geometrical Parameter Estimation

The measurement equations of a multiple-contact CF can be generated from the wrench spanning sets and closure equations of a number (l) of elementary CFs. This appendix describes how for each of these elementary contacts ($1 \leq j \leq l$) a state vector $\mathbf{x}'_j$ can be found which linearizes both the wrench spanning set $\mathbf{G}_{j,k}(\mathbf{x}'_j, \mathbf{t}_{d,k})$ and the closure equation $h_{dj,k}(\mathbf{x}'_j, \mathbf{t}_{d,k}) = 0$. Section 7.3.2 uses these $\mathbf{x}'_j$ to write quasi-linear contact models for the Non-minimal State Kalman Filter (NMSKF).

Section F.1 describes a method to find $\mathbf{x}'_j$. The method is at least applicable to all CFs in the elementary contact library of Chap. 6.

The measurement equations, and hence also the linearizing $\mathbf{x}'_j$, depend on the frame in which the measurements are expressed. Section F.2 describes the detailed formulas for the vertex-face CF when the measurements are expressed in the gripper frame. These formulas are used in the experiment of Chap. 8.

F.1 Choice of a Linearizing State x′

The elementary contact library of Chap. 6 describes for each of the elementary CFs a wrench spanning set $\mathbf{G}^c_{j,k}(\mathbf{x}, \mathbf{t}_{d,k})$ expressed in a frame {c}. This wrench spanning set expressed in any other frame {a} is

$$\mathbf{G}^a_{j,k}(\mathbf{x}, \mathbf{t}_{d,k}) = \, ^c_a\mathbf{S}_w(\mathbf{x}, \mathbf{t}_{d,k}) \; \mathbf{G}^c_{j,k}(\mathbf{x}, \mathbf{t}_{d,k}).$$

The wrench spanning sets are 6×1 matrices of which the 6 elements have the form $(\mathbf{g}_{ji}(\mathbf{x}))^T \, \mathbf{f}_{ji,k}(\mathbf{t}_{d,k})$, $1 \leq i \leq 6$. In other words, every element is a sum of terms which are functions of $\mathbf{x}$ and $\mathbf{t}_{d,k}$:

$$(\mathbf{g}_{ji}(\mathbf{x}))^T \; \mathbf{f}_{ji,k}(\mathbf{t}_{d,k}) = \sum_m g_{jim}(\mathbf{x}) \; f_{jim,k}(\mathbf{t}_{d,k}).$$

The elementary contact library of Chap. 6 describes a closure equation $h_{dj,k}(\mathbf{x}, \mathbf{t}_{d,k}) = 0$ for each of the elementary CFs, where $h_{dj,k}(\mathbf{x}, \mathbf{t}_{d,k})$ is of the form $(\mathbf{h}_j(\mathbf{x}))^T\ \mathbf{k}_{j,k}(\mathbf{t}_{d,k})$, i.e., the closure equation is a sum of terms which are functions of $\mathbf{x}$ and $\mathbf{t}_{d,k}$:

$$h_{dj,k}(\mathbf{x},\mathbf{t}_{d,k}) = (\mathbf{h}_j(\mathbf{x}))^T\ \mathbf{k}_{j,k}(\mathbf{t}_{d,k}) = \sum_m h_{jm}(\mathbf{x})\ k_{jm,k}(\mathbf{t}_{d,k}).$$

Define the vector $\mathbf{x}'_j$ as

$$\mathbf{x}'_j = \left[(\mathbf{g}_{j1}(\mathbf{x}))^T \dots (\mathbf{g}_{j6}(\mathbf{x}))^T\ (\mathbf{h}_j(\mathbf{x}))^T\right]^T.$$

The wrench spanning set $\mathbf{G}^a_{j,k}(\mathbf{x}'_j,\mathbf{t}_{d,k})$ and the closure function $h_{dj,k}(\mathbf{x}'_j,\mathbf{t}_{d,k})$ are *linear* functions of $\mathbf{x}'_j$.

F.2 Details for a Vertex-Face CF

For the experiment of Chap. 8, the elementary vertex-face CF is enough to generate all CF models. The measurements are expressed with respect to the gripper frame {g}.

This section describes the detailed formulas for the vertex-face $CF = CF_j$ between a vertex of the manipulated object and a face of the ith environment object (Sect. 6.5.1). A 12-dimensional state vector $\mathbf{x}'_j$ is found which linearizes the wrench spanning set expressed in the gripper frame {g} and the closure equation of the vertex-face CF.

Wrench Spanning Set

The wrench spanning set of the vertex-face CF expressed in the gripper frame {g} equals:

$$\begin{aligned}
\mathbf{G}^g_{j,k}(\mathbf{x},\mathbf{t}_{d,k}) &= {}^c_g\mathbf{S}_w(\mathbf{x},\mathbf{t}_{d,k})\ \mathbf{G}^c_j(\mathbf{x}^{di});\\
&= {}^m_g\mathbf{S}_w(\mathbf{x}^m)\ {}^c_m\mathbf{S}_w(\mathbf{x},\mathbf{t}_{d,k}) \begin{bmatrix}\mathbf{n}(\mathbf{x}^{di})\\ \mathbf{0}_{3\times 1}\end{bmatrix};\\
&= \begin{bmatrix} {}^m_g\mathbf{R}(\mathbf{x}^m) & \mathbf{0}_{3\times 3}\\ {}^m_g\mathbf{R}(\mathbf{x}^m)\ [{}_m\mathbf{p}^{g,m}(\mathbf{x}^m)\times]\ {}^m_g\mathbf{R}(\mathbf{x}^m)\end{bmatrix}\times\\
&\quad \begin{bmatrix} {}^g_m\mathbf{R}(\mathbf{x}^m)\ {}^w_g\mathbf{R}(\mathbf{t}_{d,k})\ {}^i_w\mathbf{R}(\mathbf{x}^i)\mathbf{n}(\mathbf{x}^{di})\\ [{}_m\mathbf{p}^{m,c}(\mathbf{x}^{dm})\times]\ {}^g_m\mathbf{R}(\mathbf{x}^m)\ {}^w_g\mathbf{R}(\mathbf{t}_{d,k})\ {}^i_w\mathbf{R}(\mathbf{x}^i)\mathbf{n}(\mathbf{x}^{di})\end{bmatrix};\\
&= \begin{bmatrix} {}^w_g\mathbf{R}(\mathbf{t}_{d,k})\ \mathbf{n}^w(\mathbf{x}^i,\mathbf{x}^{di})\\ [{}_g\mathbf{p}^{g,c}(\mathbf{x}^m,\mathbf{x}^{dm})\times]\ {}^w_g\mathbf{R}(\mathbf{t}_{d,k})\ \mathbf{n}^w(\mathbf{x}^i,\mathbf{x}^{di})\end{bmatrix};
\end{aligned}$$

where $\mathbf{n}^w(\mathbf{x}^i,\mathbf{x}^{di})$ is the normal to the face expressed in the world frame {w} and ${}_g\mathbf{p}^{g,c}(\mathbf{x}^m,\mathbf{x}^{dm})$ is the position of the contact point expressed in the gripper frame {g}:

$$\begin{aligned}
\mathbf{n}^w(\mathbf{x}^i,\mathbf{x}^{di}) &= {}^i_w\mathbf{R}(\mathbf{x}^i)\ \mathbf{n}(\mathbf{x}^{di});\\
{}_g\mathbf{p}^{g,c}(\mathbf{x}^m,\mathbf{x}^{dm}) &= {}^m_g\mathbf{R}(\mathbf{x}^m)\left({}_m\mathbf{p}^{m,c}(\mathbf{x}^{dm}) + {}_m\mathbf{p}^{g,m}(\mathbf{x}^m)\right).
\end{aligned}$$

We define an 11-dimensional state vector $\mathbf{x}'_j$:

1. the first two elements of the state vector $\mathbf{x}'_j$ define the vector $\mathbf{n}^w(\mathbf{x}^i, \mathbf{x}^{di})$ perpendicular to the contacting face, expressed in the world frame {w};

$$\alpha(\mathbf{x}^i, \mathbf{x}^{di}) \begin{bmatrix} \mathbf{x}'_j(1) \\ \mathbf{x}'_j(2) \\ 1 - \mathbf{x}'_j(1) - \mathbf{x}'_j(2) \end{bmatrix} = {}^i_w\mathbf{R}(\mathbf{x}^i)\, \mathbf{n}(\mathbf{x}^{di});$$

i.e.,

$$\begin{bmatrix} \mathbf{x}'_j(1) \\ \mathbf{x}'_j(2) \end{bmatrix} = \begin{bmatrix} 1 & 0 & 0 \\ 0 & 1 & 0 \end{bmatrix} \frac{{}^i_w\mathbf{R}(\mathbf{x}^i)\, \mathbf{n}(\mathbf{x}^{di})}{\alpha(\mathbf{x}^i, \mathbf{x}^{di})};$$

$$\alpha(\mathbf{x}^i, \mathbf{x}^{di}) = \begin{bmatrix} 1 & 1 & 1 \end{bmatrix} {}^i_w\mathbf{R}(\mathbf{x}^i)\, \mathbf{n}(\mathbf{x}^{di}).$$

2. the next three elements of the state vector $\mathbf{x}'_j$ define the position vector ${}_g\mathbf{p}^{g,c}(\mathbf{x}^m, \mathbf{x}^{dm})$ from the gripper origin to the contacting vertex, expressed in the gripper frame {g}.

$$\begin{bmatrix} \mathbf{x}'_j(3) \\ \mathbf{x}'_j(4) \\ \mathbf{x}'_j(5) \end{bmatrix} = {}^m_g\mathbf{R}(\mathbf{x}^m)({}_m\mathbf{p}^{m,c}(\mathbf{x}^{dm}) + {}_m\mathbf{p}^{g,m}(\mathbf{x}^m)).$$

3. the next six elements of $\mathbf{x}'_j$ are combinations of the first five elements.

$$\begin{aligned} \mathbf{x}'_j(6) &= \mathbf{x}'_j(1) \times \mathbf{x}'_j(3); \\ \mathbf{x}'_j(7) &= \mathbf{x}'_j(2) \times \mathbf{x}'_j(3); \\ \mathbf{x}'_j(8) &= \mathbf{x}'_j(1) \times \mathbf{x}'_j(4); \\ \mathbf{x}'_j(9) &= \mathbf{x}'_j(2) \times \mathbf{x}'_j(4); \\ \mathbf{x}'_j(10) &= \mathbf{x}'_j(1) \times \mathbf{x}'_j(5); \\ \mathbf{x}'_j(11) &= \mathbf{x}'_j(2) \times \mathbf{x}'_j(5). \end{aligned}$$

Expressed in $\mathbf{x}'_j$, the wrench spanning set $\frac{\mathbf{G}^g_{j,k}(\mathbf{x}, \mathbf{t}_{d,k})}{\alpha(\mathbf{x}^i, \mathbf{x}^{di})}$ is linear.

Closure Equation

The closure equation expresses that the vertex lies in the face, i.e.:

$$h_{dj,k}(\mathbf{x}, \mathbf{t}_{d,k}) = (\mathbf{n}(\mathbf{x}^{di}))^T\, {}_i\mathbf{p}^{i,c}(\mathbf{x}, \mathbf{t}_{d,k}) - d(\mathbf{x}^{di}) = 0;$$

where ${}_i\mathbf{p}^{i,c}(\mathbf{x}, \mathbf{t}_{d,k})$ is the vertex expressed in {i}:

$$\begin{aligned}
\begin{bmatrix} {}_i\mathbf{p}^{i,c}(\mathbf{x},\mathbf{t}_{d,k}) \\ 1 \end{bmatrix} &= {}^w_i\mathbf{T}(\mathbf{x}^i)\ {}^g_w\mathbf{T}(\mathbf{t}_{d,k})\ {}^m_g\mathbf{T}(\mathbf{x}^m) \begin{bmatrix} {}_m\mathbf{p}^{m,c}(\mathbf{x}^{dm}) \\ 1 \end{bmatrix}; \\
&= \begin{bmatrix} {}^w_i\mathbf{R}(\mathbf{x}^i) & {}_i\mathbf{p}^{i,w}(\mathbf{x}^i) \\ \mathbf{0}_{1\times 3} & 1 \end{bmatrix} \begin{bmatrix} {}^g_w\mathbf{R}(\mathbf{t}_{d,k}) & {}_w\mathbf{p}^{w,g}(\mathbf{t}_{d,k}) \\ \mathbf{0}_{1\times 3} & 1 \end{bmatrix} \times \\
&\quad \begin{bmatrix} {}^m_g\mathbf{R}(\mathbf{x}^m) & -{}^m_g\mathbf{R}(\mathbf{x}^m)\ {}_m\mathbf{p}^{m,g}(\mathbf{x}^m) \\ \mathbf{0}_{1\times 3} & 1 \end{bmatrix} \begin{bmatrix} {}_m\mathbf{p}^{m,c}(\mathbf{x}^{dm}) \\ 1 \end{bmatrix}; \\
{}_i\mathbf{p}^{i,c}(\mathbf{x},\mathbf{t}_{d,k}) &= {}^w_i\mathbf{R}(\mathbf{x}^i)\ {}^g_w\mathbf{R}(\mathbf{t}_{d,k})\ {}^m_g\mathbf{R}(\mathbf{x}^m)\ \left({}_m\mathbf{p}^{m,c}(\mathbf{x}^{dm}) - {}_m\mathbf{p}^{m,g}(\mathbf{x}^m)\right) \\
&\quad + {}^w_i\mathbf{R}(\mathbf{x}^i)\ {}_w\mathbf{p}^{w,g}(\mathbf{t}_{d,k}) + {}_i\mathbf{p}^{i,w}(\mathbf{x}^i); \\
&= {}^w_i\mathbf{R}(\mathbf{x}^i)\left({}^g_w\mathbf{R}(\mathbf{t}_{d,k})\ {}_g\mathbf{p}^{g,c}(\mathbf{x}^m,\mathbf{x}^{dm}) + {}_w\mathbf{p}^{w,g}(\mathbf{t}_{d,k})\right) \\
&\quad + {}_i\mathbf{p}^{i,w}(\mathbf{x}^i).
\end{aligned}$$

Hence, the closure equation equals

$$\begin{aligned}
h_{dj,k}(\mathbf{x},\mathbf{t}_{d,k}) &= \left(({}^w_i\mathbf{R}(\mathbf{x}^i))^T\mathbf{n}(\mathbf{x}^{di})\right)^T\ {}^g_w\mathbf{R}(\mathbf{t}_{d,k})\ {}_g\mathbf{p}^{g,c}(\mathbf{x}^m,\mathbf{x}^{dm}) + \\
&\quad \left(({}^w_i\mathbf{R}(\mathbf{x}^i))^T\mathbf{n}(\mathbf{x}^{di})\right)^T\ {}_w\mathbf{p}^{w,g}(\mathbf{t}_{d,k}) + \\
&\quad \left(\mathbf{n}(\mathbf{x}^{di})\right)^T\ {}_i\mathbf{p}^{i,w}(\mathbf{x}^i) - d(\mathbf{x}^{di}) = 0; \\
h_{dj,k}(\mathbf{x},\mathbf{t}_{d,k}) &= (\mathbf{n}^w(\mathbf{x}^i,\mathbf{x}^{di}))^T\ {}^g_w\mathbf{R}(\mathbf{t}_{d,k})\ {}_g\mathbf{p}^{g,c}(\mathbf{x}^m,\mathbf{x}^{dm}) + \\
&\quad (\mathbf{n}^w(\mathbf{x}^i,\mathbf{x}^{di}))^T\ {}_w\mathbf{p}^{w,g}(\mathbf{t}_{d,k}) + \\
&\quad \left(\mathbf{n}(\mathbf{x}^{di})\right)^T\ {}_i\mathbf{p}^{i,w}(\mathbf{x}^i) - d(\mathbf{x}^{di}) = 0.
\end{aligned}$$

We define a 12th variable $\mathbf{x}'_j$:

$$\mathbf{x}'_j(12) = \frac{d^w(\mathbf{x}^i,\mathbf{x}^{di})}{\alpha(\mathbf{x}^i,\mathbf{x}^{di})};$$

where $d^w(\mathbf{x}^i,\mathbf{x}^{di})$ is the (signed) distance from the origin of the world frame {w} to the face expressed in {w}:

$$d^w(\mathbf{x}^i,\mathbf{x}^{di}) = d(\mathbf{x}^{di}) - (\mathbf{n}(\mathbf{x}^{di}))^T\ {}_i\mathbf{p}^{i,w}(\mathbf{x}^i).$$

I.e., the face is defined in {w} by the normal vector $\mathbf{n}^w(\mathbf{x}^i,\mathbf{x}^{di})$ and the (signed) distance to the origin $d^w(\mathbf{x}^i,\mathbf{x}^{di})$. A point y lies in the contacting face if $(\mathbf{n}^w(\mathbf{x}^i,\mathbf{x}^{di}))^T\ {}_w\mathbf{p}^{w,y} = d^w(\mathbf{x}^i,\mathbf{x}^{di})$.

The closure equation $\frac{h_{dj,k}(\mathbf{x},\mathbf{t}_{d,k})}{\alpha(\mathbf{x}^i,\mathbf{x}^{di})} = 0$ is linear when expressed in the 12-dimensional state $\mathbf{x}'_j$.

G

CF-Observable Parameter Space for Twist and Pose Measurements

This appendix describes how to calculate a spanning set for the CF-observable parameter space for twist and pose measurements.

Measurement Information

Given the implicit nonlinear measurement equation $\mathbf{h}_k(\mathbf{x}, \mathbf{z}_k) = \mathbf{0}_{n_h \times 1}$ and the measurement covariance matrix $\mathbf{R}_k$; the set of *eigenvectors*[1] *of* $\mathbf{I}_{VT,meas,k}$ for which the corresponding eigenvalues are non-zero (see Sect. 9.2.3) is a spanning set of the *parameter directions observed* by a specific measurement $\hat{\mathbf{z}}_k$. The calculation of $\mathbf{I}_{VT,meas,k}$ is approximated by (9.4), i.e.,

$$\mathbf{I}_{VT,meas,k} = \mathbf{H}_k^T (\mathbf{D}_k \mathbf{R}_k \mathbf{D}_k^T)^{-1} \mathbf{H}_k;$$
$$\mathbf{H}_k = \left. \frac{\partial \mathbf{h}_k(\mathbf{x}, \mathbf{z}_k)}{\partial \mathbf{x}} \right|_{\mathbf{x}=\hat{\mathbf{x}}, \mathbf{z}_k=\hat{\mathbf{z}}_k}; \qquad \mathbf{D}_k = \left. \frac{\partial \mathbf{h}_k(\mathbf{x}, \mathbf{z}_k)}{\partial \mathbf{z}_k} \right|_{\mathbf{x}=\hat{\mathbf{x}}, \mathbf{z}_k=\hat{\mathbf{z}}_k}.$$

This appendix shows that the set of *right singular directions of* $\mathbf{H}_k$ corresponding to non-zero singular values is also a spanning set of the observed parameter directions.

CF-Observable Parameter Space for Twists and Poses

We calculate the space of parameter directions observed by one single pose measurement and by any possible twist measurement. The resulting CF-observable parameter space is equally the (i) CF-observable parameter space for pose measurements and (ii) the CF-observable parameter space for twist and pose measurements (see Sect. 7.2.2).

[1] Because the matrix $\mathbf{I}_{VT,meas,k}$ is a symmetric positive semi-definite matrix, its eigenvalue decomposition equals its singular value decomposition.

Overview of this Appendix

Section G.1 contains the proof that the problem of calculating a spanning set for the observable directions for a specific twist or pose measurement does not require the calculation of the information matrix $\mathbf{I}_{VT,meas,k} = \mathbf{H}_k^T(\mathbf{D}_k\mathbf{R}_k\mathbf{D}_k^T)^{-1}\mathbf{H}_k$, the analysis of only the matrix $\mathbf{H}_k$ is sufficient.

The CF-observable parameter space for twist measurements contains the directions observable by *any possible twist measurement at any possible pose.* The problem of calculating a spanning set for this space is reduced to the analysis of a limited number of twist measurements at a limited number of poses:

1. Section G.2 contains the proof that at a given pose, a spanning set of the observable parameter directions can be derived from the analysis of *a limited number of twists.*
2. The analysis needs to be done at only a limited number of poses, see Sect. G.3.

Section G.4 presents a short discussion on the relation between the CF-observable parameter space of a multiple-contact CF and the CF-observable parameter spaces of its elementary contacts.

G.1 The Matrix $\mathbf{H}_k$ Defines the Space of Observed Parameter Directions

The problem of calculating a spanning set for the parameter directions observed by a specific twist or pose measurement, is based on the eigenvalue decomposition[1, p. 235] $\mathbf{U}_{i,k}\mathbf{S}_{i,k}\mathbf{U}_{i,k}^T$ of the matrix

$$\mathbf{I}_{VT,meas,k} = \mathbf{H}_k^T(\mathbf{D}_k\mathbf{R}_k\mathbf{D}_k^T)^{-1}\mathbf{H}_k.$$

The calculation of $\mathbf{I}_{VT,meas,k}$, however, is not necessary: a spanning set can be calculated based on $\mathbf{H}_k$ only.

Proof

The result is derived in four steps:

1. $\mathbf{D}_k\mathbf{R}_k\mathbf{D}_k^T$ *is a positive-definite, regular* $n_h \times n_h$ *matrix.*
This is true for both the closure and the twist equation if (i) the pose and twist covariance matrices $\mathbf{R}_k$ are regular 6×6 matrices; and (ii) the contact description is minimal ($n_h \leq 6$ is then the number of directions constrained by the contact). Because a non-minimal contact description leads to the same estimation results (and hence the same observed parameter directions) as the corresponding minimal description (see Sect. 7.3.3) and because a minimal description is always possible; it is not a limitation to restrict the proof to minimal contact descriptions.

2. $\mathbf{I}_{VT,meas,k}$ *can be written as* $(\mathbf{A}_k\mathbf{H}_k)^T(\mathbf{A}_k\mathbf{H}_k)$ *where* $\mathbf{A}_k$ *is an* $n_h \times n_h$ *regular matrix.*
   The proof is trivial because the $n_h \times n_h$ regular, positive-definite, symmetric matrix $(\mathbf{D}_k\mathbf{R}_k\mathbf{D}_k^T)^{-1}$ can be written as $\mathbf{A}_k^T\mathbf{A}_k$ where $\mathbf{A}_k$ is an $n_h \times n_h$ regular matrix.
   Further on we use the singular value decomposition
$$\mathbf{A}_k\mathbf{H}_k = \mathbf{U}_{ah,k}\mathbf{S}_{ah,k}\mathbf{V}_{ah,k}^T.$$
3. $\mathbf{U}_{i,k}$ *equals* $\mathbf{V}_{ah,k}$ *and the same columns of these matrices correspond to non-zero elements in* $\mathbf{S}_{i,k}$ *and* $\mathbf{S}_{ah,k}$.
   Expanding both sides of the equation $\mathbf{I}_{VT,meas,k} = (\mathbf{A}_k\mathbf{H}_k)^T(\mathbf{A}_k\mathbf{H}_k)$ yields
$$\mathbf{U}_{i,k}\mathbf{S}_{i,k}\mathbf{U}_{i,k}^T = (\mathbf{U}_{ah,k}\mathbf{S}_{ah,k}\mathbf{V}_{ah,k}^T)^T(\mathbf{U}_{ah,k}\mathbf{S}_{ah,k}\mathbf{V}_{ah,k}^T); \tag{G.1}$$
$$= \mathbf{V}_{ah,k}\mathbf{S}_{ah,k}\mathbf{S}_{ah,k}\mathbf{V}_{ah,k}^T. \tag{G.2}$$
   Hence, $\mathbf{U}_{i,k} = \mathbf{V}_{ah,k}$ and $\mathbf{S}_{i,k} = \mathbf{S}_{ah,k}\mathbf{S}_{ah,k}$.
4. *The spaces spanned by the right singular directions* $\mathbf{V}_{h,k}$ *of* $\mathbf{H}_k$ *and* $\mathbf{V}_{ah,k}$ *of* $\mathbf{A}_k\mathbf{H}_k$*, corresponding to zero and non-zeros singular values are the same.*
   $\mathbf{A}_k$ is an $n_h \times n_h$ regular matrix, hence $\mathbf{A}_k\mathbf{H}_k\mathbf{x} = \mathbf{0}_{n_h\times 1} \iff \mathbf{H}_k\mathbf{x} = \mathbf{0}_{n_h\times 1}$.

These four steps prove that a spanning set of the observed parameter space can be found in the columns of $\mathbf{V}_{h,k}$ for which the corresponding singular values in $\mathbf{S}_{h,k}$ are non-zero.

G.2 CF-Observable Parameter Space for All Twists at One Pose

The CF-observable parameter space for twists contains the parameter directions observed by *any possible twist at any possible pose.* This section presents a way to find the directions observed by *any possible twist at a given pose* $\mathbf{t}_{d,k}$. A spanning set of these directions is found by means of the singular value decomposition of $\mathbf{H}_k$ *for a spanning set for the allowable twists, e.g., the columns of* $\mathbf{J}_k(\mathbf{x}, \mathbf{t}_{d,k})$.

Proof

Consider two admissible twists $\mathbf{t}_{k,1}$ and $\mathbf{t}_{k,2}$ at the pose $\mathbf{t}_{d,k}$. The matrix $\mathbf{H}_k$ for the twist $\mathbf{t}_k = a\mathbf{t}_{k,1} + b\mathbf{t}_{k,2}$ is

$$\mathbf{H}_k = \left.\frac{\partial \mathbf{G}_k^T(\mathbf{x}, \mathbf{t}_{d,k})\mathbf{t}_k}{\partial \mathbf{x}}\right|_{\hat{\mathbf{x}}}; \tag{G.3}$$
$$= a\left.\frac{\partial \mathbf{G}_k^T(\mathbf{x}, \mathbf{t}_{d,k})\mathbf{t}_{k,1}}{\partial \mathbf{x}}\right|_{\hat{\mathbf{x}}} + b\left.\frac{\partial \mathbf{G}_k^T(\mathbf{x}, \mathbf{t}_{d,k})\mathbf{t}_{k,2}}{\partial \mathbf{x}}\right|_{\hat{\mathbf{x}}}; \tag{G.4}$$
$$= a\mathbf{H}_{k,1} + b\mathbf{H}_{k,2}. \tag{G.5}$$

The directions observed by $\mathbf{t}_k = a\mathbf{t}_{k,1} + b\mathbf{t}_{k,2}$ are directions $\mathbf{x}$ for which $\mathbf{H}_k \mathbf{x} = a\mathbf{H}_{k,1} \mathbf{x} + b\mathbf{H}_{k,2} \mathbf{x} \neq \mathbf{0}_{n_h \times 1}$. I.e., the twist $\mathbf{t}_k$ only observes parameter directions which are also observed by $\mathbf{t}_{k,1}$ and/or by $\mathbf{t}_{k,2}$. Hence, the search for observable parameter directions can be limited to the analysis of a *spanning set for the allowable twists, e.g., the columns of* $\mathbf{J}_k(\mathbf{x}, \mathbf{t}_{d,k})$.

G.3 CF-Observable Parameter Space for All Twists at All Poses

The number of poses at which the observable parameter directions for twists needs to be calculated, depends on the CF.

Vertex-Face CF

At a given pose $\mathbf{t}_{d,1}$, the full orientation of the face in the world frame is observed by motions of the contacting vertex in the two allowed translational velocity degrees of freedom. The rotational velocities allow to identify two of the three position coordinates of the position of the vertex with respect to the gripper frame. The unobservable position coordinate is the position of the vertex along a normal to the face (see Fig. 9.6).

Hence, the CF-observable parameter directions are obtained by the analysis at two poses with different orientation of the face with respect to the gripper frame. This means that the analysis at two poses which are rotated about an axis which is not the face normal are sufficient to determine the CF-observable parameter directions of a vertex-face CF.

Other CFs

For other CFs, a similar analysis can be performed. *A CF-independent strategy,* on the other hand, is to calculate the observable parameter directions at a large number of poses. The computational cost of this offline calculation is of lower importance.

G.4 Multiple-Contact CFs

The theory of this appendix applies to general multiple-contact CFs. This section presents a small discussion on the relation between the CF-observable parameter space of a multiple-contact CF and the CF-observable parameter spaces of its elementary contacts.

A wrench spanning set for multiple-contact CFs is given by the concatenation of the wrench spanning sets of the elementary contacts:

$$\mathbf{G}_k(\mathbf{x}, \mathbf{t}_{d,k}) = \left[\mathbf{G}_{1,k}(\mathbf{x}, \mathbf{t}_{d,k}) \ \ldots \ \mathbf{G}_{m,k}(\mathbf{x}, \mathbf{t}_{d,k})\right] .$$

Hence,

$$\mathbf{H}_k = \frac{\partial \mathbf{G}_k^T(\mathbf{x}, \mathbf{t}_{d,k})\mathbf{t}_k}{\partial \mathbf{x}} = \begin{bmatrix} \mathbf{H}_{1,k} \\ \cdots \\ \mathbf{H}_{m,k} \end{bmatrix}.$$

This means that a direction $\mathbf{x}$ is unobserved under a motion $\mathbf{t}_k$ at a pose $\mathbf{t}_{d,k}$ in the multiple contact ($\mathbf{H}_k \mathbf{x} = \mathbf{0}_{n_h \times 1}$), if and only if it is unobserved under *all* of the elementary contacts, i.e., $\mathbf{H}_{i,k}\,\mathbf{x} = 0$, $\forall i : 1 \leq i \leq m$. If any of the elementary contacts observes a parameter direction under the twist $\mathbf{t}_k$ at the pose $\mathbf{t}_{d,k}$, the multiple-contact CF will also observe it.

However, the CF-observable parameter space of a multiple-contact CF is *not* the union of the CF-observable parameter spaces of its elementary contacts. The reason is that the space of admissible twists $\mathbf{t}_k$ and poses $\mathbf{t}_{d,k}$ for a multiple-contact CF is only the *intersection* of these admissible spaces for the elementary contacts. The CF-observable parameter space of a multiple-contact CF needs to be calculated with the *restricted* admissible twist and pose spaces.

References

1. A. Abbati-Marescotti, C. Bonivento, and C. Melchiorri. On the invariance of the hybrid position/force control. *Journal of Intelligent and Robotic Systems*, 3:233–250, 1990.
2. S. Ahmad and C.N. Lee. Shape recovery from robot contour-tracking with force feedback. In *Proceedings of the IEEE International Conference on Robotics and Automation*, pages 447–452, Cincinnati, OH, 1990.
3. D.S. Ahn, H.S. Cho, K. Ide, F. Miyazaki, and S. Arimoto. Learning task strategies in robotic assembly systems. *Robotica*, 10(5):409–418, 1992.
4. H. Akaike. Information theory and an extension of the maximum likelihood principle. In B.N. Petrov and F. Csaki, editors, *Proceedings of the Second International Symposium in Information Theory*, pages 267–281. Akadémiai Kiadó, Budapest, Hungary, 1973.
5. H. Akaike. A new look at the statistical model identification. *IEEE Transactions on Automatic Control*, 19:716–723, 1974.
6. D.L. Alspach and H.W. Sorenson. Nonlinear Bayesian estimation using Gaussian sum approximations. *IEEE Transactions on Automatic Control*, 17(4):439–448, August 1972.
7. R.J. Anderson and M.W. Spong. Hybrid impedance control of robotic manipulators. *IEEE Journal of Robotics and Automation*, 4(5):549–556, 1988.
8. T.W. Anderson. *An Introduction to Multivariate Statistical Analysis*. John Wiley & Sons, 1984.
9. V.I. Arnold. *Mathematical methods of classical mechanics*. Springer-Verlag, New York, Heidelberg, Berlin, 1980.
10. H. Asada. Teaching and learning of compliance using neural nets: Representation and generation of nonlinear compliance. In *Proceedings of the IEEE International Conference on Robotics and Automation*, pages 1237–1244, Cincinnati, OH, 1990.
11. H. Asada. Representation and learning of nonlinear compliance using neural nets. *IEEE Transactions on Robotics and Automation*, 9(6):863–867, December 1993.
12. H. Asada and H. Izumi. Direct teaching and automatic program generation for the hybrid control of robot manipulators. In *Proceedings of the IEEE International Conference on Robotics and Automation*, volume 3, pages 1401–1406, 1987.

13. H. Asada and H. Izumi. Automatic program generation from teaching data for the hybrid control of robots. *IEEE Transactions on Robotics and Automation*, 5(2):166–173, April 1989.
14. K.J. Astrom. Optimal control of Markov decision processes with incomplete state estimation. *Journal of Mathematical Analysis and Applications*, 10:174–205, 1965.
15. David J. Austin and Brenan J. McCarragher. Force control command synthesis for assembly using a discrete event framework. In *Proceedings of the IEEE International Conference on Robotics and Automation*, pages 933–938, Albuquerque, NM, 1997.
16. David J. Austin and Brenan J. McCarragher. Hybrid force/velocity discrete event controller synthesis for assembly tasks with friction. In *Proceedings of the IEEE International Conference on Robotics and Automation*, pages 3028–3033, San Francisco, CA, April 2000.
17. David J. Austin and Brenan J. McCarragher. Force control command synthesis for constrained hybrid dynamic systems with friction. *The International Journal of Robotics Research*, 20(9):753–764, September 2001.
18. F. Avnaim and J.D. Boissonnat B. Faverjon. A practical exact motion planning algorithm for polygonal objects amidst polygonal obstacles. In *Proceedings of the IEEE International Conference on Robotics and Automation*, volume 3, pages 1656–1661, 1988.
19. Y. Bar-Shalom and X.R. Li. *Estimation and Tracking: Principles, Techniques and Software.* Artech House, Boston, London, 1993.
20. L. Basañez and R. Suárez. Illustrating an automatic planner for robotic assembly task. In *Proceedings of the International Symposium on Industrial Robots*, pages 645–651, Barcelona, Spain, 1992.
21. Thomas Bayes. An essay towards solving a problem in the doctrine of chances. *Philosophical Transactions of the Royal Society of London*, 53:370–418, 1763. Reprinted in Biometrika, 45, 293–315, 1958.
22. Richard Bellman. *Dynamic Programming.* Princeton University Press, Princeton, New Jersey, 1957.
23. Richard Bellman. A Markov decision process. *Journal of Mathematical Mechanics*, 6:679–684, 1957.
24. A. Ben-Israel and T.N.E. Greville. *Generalized Inverses: Theory and Applications.* John Wiley & Sons, New York, 1974.
25. V.E. Beneš. Exact finite-dimensional filters for certain diffusions with nonlinear drift. *Stochastics*, 5:65–92, 1981.
26. Niclas Bergman. On the Cramér-Rao bound for terrain-aided navigation. Technical Report LiTH-ISY-R-1970, Department of Electrical Engineering, Linköping University, Linköping, Sweden, August 1997.
27. Dimitri P. Bertsekas. *Dynamic Programming and Optimal Control, Volume I.* Athena Scientific, Belmont Massachusetts, 1995.
28. Dimitri P. Bertsekas. *Dynamic Programming and Optimal Control, Volume II.* Athena Scientific, Belmont Massachusetts, 1995.
29. A. Bicchi, J.K. Salisbury, and P. Dario. Augmentation of grasp robustness using tactile sensing. In *Proceedings of the IEEE International Conference on Robotics and Automation*, pages 302–307, 1989.
30. A. Bicchi, K. Salisbury, and D.L. Brock. Contact sensing from force measurements. *The International Journal of Robotics Research*, 12(3):249–262, 1993.
31. G.J. Bierman. A comparison of discrete linear filtering algorithms. *IEEE Transactions on Aerospace and Electronic Systems*, 9(1):28–37, January 1973.

32. G.J. Bierman and C.L. Thornton. Numerical comparison of Kalman filter algorithms: Orbit determination case study. *Automatica*, 13:23–35, 1977.
33. M. Blauer and P.R. Bélanger. State and parameter estimation for robotic manipulators using force measurements. *IEEE Transactions on Automatic Control*, 32(12):1055–1066, 1987.
34. Giuseppe Borghi and Vincenzo Caglioti. Minimum uncertainty explorations in the self-localization of mobile robots. Technical Report 97.15, Department of Electronics and Information Engineering, Politecnico di Milano, Italy, 1997.
35. Frédéric Bourgault, Alexei A. Makarenko, Stefan B. Williams, Ben Grocholsky, and Hugh F. Durrant-Whyte. Information based adaptive robotic exploration. In *Proceedings of the IEEE/RSJ International Conference on Intelligent Robots and Systems*, pages 540–545, Lausanne, Switzerland, October 2002.
36. Craig Boutilier, Thomas Dean, and Steven Hanks. Decision-theoretic planning: Structural assumptions and computational leverage. *Journal of Artificial Intelligence Research*, 11:1–94, 1999.
37. Craig Boutilier and David Poole. Computing optimal policies for partially observable decision processes using compact representations. In *Proceedings of the National Conference on Artificial Intelligence*, volume 2, pages 1168–1175, Portland, Oregon, USA, 1996. AAAI Press / The MIT Press.
38. Lorenzo M. Brignone and Martin Howarth. A geometrically validated approach to autonomous robotic assembly. In *Proceedings of the IEEE/RSJ International Conference on Intelligent Robots and Systems*, pages 1626–1631, Lausanne, Switzerland, October 2002.
39. R.A. Brooks. Symbolic error analysis and robot planning. *International Journal of Robotics Research*, 1(4):29–68, 1982.
40. R. Brost. Computing metric and topological properties of C-space obstacles. In *Proceedings of the IEEE International Conference on Robotics and Automation*, volume 1, pages 170–176, 1989.
41. H. Bruyninckx. *Kinematics Models for Robot Compliant Motion with Identification of Uncertainties.* PhD thesis, Katholieke Universiteit, Leuven, Belgium, 1995.
42. H. Bruyninckx and J. De Schutter. Specification of force-controlled actions in the "task frame formalism": A synthesis. *IEEE Transactions on Robotics and Automation*, 12(4):581–589, August 1996.
43. H. Bruyninckx and J. De Schutter. Modelling and specification of compliant motions with two and three contact points. In *Proceedings of the IEEE International Conference on Robotics and Automation*, Leuven, Belgium, May 1998.
44. H. Bruyninckx, S. Demey, S. Dutré, and J. De Schutter. Kinematic models for model-based compliant motion in the presence of uncertainty. *The International Journal of Robotics Research*, 14(5):465–482, October 1995.
45. Stephen J. Buckley. Planning compliant motion strategies. *The International Journal of Robotics Research*, 8(5):28–44, October 1989.
46. V. Caglioti. Uncertainty minimization in the localization of polyhedral objects. *IEEE Transactions on Pattern Analysis and Machine Intelligence*, 16(5):524–530, May 1994.
47. J.F. Canny and J. Reif. New lower-bound techniques for robot motion planning problems. In *Proceedings of the IEEE Symposium on the Foundations of Computer Science*, pages 49–60, Los Angeles, CA, 1987.
48. Anthony R. Cassandra. Optimal policies for partially observable Markov decision processes. Technical Report CS-94-14, Brown University, Department of Computer Science, Providence RI, 1994.

49. Anthony R. Cassandra. *Exact and Approximate Algorithms for Partially Observable Markov Decision Processes.* PhD thesis, U. Brown, 1998.
50. A.R. Cassandra, L.P. Kaelbling, and J.A. Kurien. Acting under uncertainty: Discrete Bayesian models for mobile-robot navigation. In *Proceedings of IEEE/RSJ International Conference on Intelligent Robots and Systems*, pages 963–972, 1996.
51. Cassandra, A. R. POMDPs for dummies. http://www.cs.brown.edu/research/ai/pomdp/tutorial/.
52. E. Cervera, A. Del Pobil, E. Marta, and M. Serna. Perception-based learning for motion in contact in task planning. *Journal of Intelligent and Robotic Systems*, 17(3):283–308, November 1996.
53. M. Charlebois, K. Gupta, and S. Payandeh. Shape description of curved surfaces from contact sensing using surface normals. *The International Journal of Robotics Research*, 18(8):779–787, August 1999.
54. J.R. Chen and B.J. McCarragher. Programming by demonstration — Constructing task level plans in a hybrid dynamic framework. In *Proceedings of the IEEE International Conference on Robotics and Automation*, pages 1402–1407, San Francisco, CA, April 2000.
55. Hsien-Te Cheng. *Algorithms for Partially Observable Markov Decision Processes.* PhD thesis, University of British Columbia, British Columbia, Canada, 1988.
56. P.M. Cohn. *Algebra, Volume 1.* John Wiley & Sons, 1974.
57. Thomas M. Cover and Joy A. Thomas. *Elements of Information Theory.* Wiley Series in Telecommunications. Wiley-Interscience, 1991.
58. Harald Cramér. *Mathematical methods of Statistics.* Princeton. Princeton University Press, New Jersey, 1946.
59. F. d' Epenoux. Sur un problème de production et de stockage dans l'aléatoire. *Revue Francaise Recherche Oprationelle*, 14:3–16, 1960.
60. Gordon Dakin and Robin Popplestone. Simplified fine-motion planning in generalized contact space. In *Proceedings of the International Symposium on Intelligent Control*, pages 281–286, 1992.
61. F. Daum. New exact nonlinear filters. *J. C. Spall, editor, Bayesian Analysis of Time Series and Dynamic Models*, pages 199–226, 1988.
62. Nicolaas Govert De Bruijn. *Asymptotic methods in analysis.* Bibliotheca mathematica 4. North-Holland, Amsterdam, 1958.
63. J. De Geeter, J. De Schutter, H. Bruyninckx, H. Van Brussel, and M. Decréton. Tolerance-weighted L-optimal experiment design: A new approach to task-directed sensing. *Advanced Robotics*, 13(4):401–416, 1999.
64. J. De Geeter, J. De Schutter, M. Decréton, and H. Van Brussel. Sensor action planning driven by uncertainty. Application to object location with robust local sensors in a nuclear environment. *International Journal of Computing Anticipatory Systems*, 2:115–131, 1998.
65. Jan De Geeter, H. Van Brussel, J. De Schutter, and M. Decréton. Recognizing and locating objects with local sensors. In *Proceedings of the IEEE International Conference on Robotics and Automation*, pages 3478–3483, Minneapolis, Minnesota, April 1996.
66. Jan De Geeter, H. Van Brussel, J. De Schutter, and M. Decréton. Local world modelling for teleoperation in a nuclear environment using a Bayesian multiple hypothesis tree. In *Proceedings of the IEEE/RSJ International Conference on Intelligent Robots and Systems*, pages 1658–1663, Grenoble, France, September 1997.

67. J. De Schutter. A study of active compliant motion control methods for rigid manipulators using a generic scheme. In *Proceedings of the IEEE International Conference on Robotics and Automation*, pages 1060–1065, Raleigh, NC, 1987.
68. J. De Schutter, H. Bruyninckx, S. Dutré, J. De Geeter, J. Katupitiya, S. Demey, and T. Lefebvre. Estimating first-order geometric parameters and monitoring contact transitions during force-controlled compliant motions. *The International Journal of Robotics Research*, 18(12):1161–1184, 1999.
69. J. De Schutter and H. Van Brussel. Compliant robot motion I, A formalism for specifying compliant motion tasks. *The International Journal of Robotics Research*, 7(4):3–17, 1988.
70. J. De Schutter and H. Van Brussel. Compliant robot motion II, A control approach based on external control loops. *The International Journal of Robotics Research*, 7(4):18–33, 1988.
71. T. Debus, P. Dupont, and R. Howe. Contact state estimation using multiple model estimation and hidden Markov models. In B. Siciliano and P. Dario, editors, *Proceedings of the 8th International Symposium on Experimental Robotics*, Advanced Robotics Series, pages 517–526. Springer Verlag, 2002.
72. Nathan Delson and Harry West. Robot programming by human demonstration: Subtask compliance controller identification. In *Proceedings of the IEEE/RSJ International Conference on Intelligent Robots and Systems*, pages 33–41, Yokohama, Japan, July 1993.
73. S. Demey, H. Bruyninckx, and J. De Schutter. Model-based planar contour following in the presence of pose and model errors. *The International Journal of Robotics Research*, 16(6):840–858, 1997.
74. R.S. Desai. *On Fine Motion in Mechanical Assembly in Presence of Uncertainty.* PhD thesis, Department of Mechanical Engineering, University of Michigan, 1989.
75. R.S. Desai and R. Volz. Identification and verification of termination conditions in fine motion in presence of sensor errors and geometric uncertainties. In *Proceedings of the IEEE International Conference on Robotics and Automation*, pages 800–807, 1989.
76. B. Donald. A search algorithm for motion planning with six degrees of freedom. *Artificial Intelligence*, 31:295–353, 1987.
77. B.R. Donald. Robot motion planning with uncertainty in the geometric models of the robot and environment: A formal framework for error detection and recovery. In *Proceedings of the IEEE International Conference on Robotics and Automation*, pages 1588–1593, 1986.
78. B.R. Donald. A geometric approach to error detection and recovery for robot motion planning with uncertainty. *Artificial Intelligence*, 37(1–3):223–271, 1988.
79. A.W. Drake. *Observation of Markov Processes Through a Noisy Chanel.* PhD thesis, Massachusetts Institute of Technology, Cambridge, Massachusetts, 1962.
80. S.H. Drake. Using compliance in lieu of sensing feedback for automatic assembly. Technical Report Report T -657, Charles Stark Draper Laboratory, 1977.
81. B. Dufay and J.C. Latombe. An approach to automatic robot programming based on inductive learning. In *Proceedings of the International Symposium on Robotics Research*, August 1983.
82. Bruno Dufay and Jean-Claude Latombe. An approach to automatic robot programming based on inductive learning. *International Journal of Robotics Research*, 3(4):3–20, 1984.

83. J. Duffy. The fallacy of modern hybrid control theory that is based on "orthogonal complements" of twist and wrench spaces. *Journal of Robotic Systems*, 7(2):139–144, 1990.
84. D. Dugué. Applications des propriétés de la limite au sens du calcul des probabilités à l'étude des diverses questions d'estimation. *Journal de l'Ecole polytechnique*, 3(4):305–372, 1937.
85. S. Dutré. *Identification and Monitoring of Contacts in Force Controlled Robotic Manipulation*. PhD thesis, Katholieke Universiteit, Leuven, Belgium, 1998.
86. S. Dutré, H. Bruyninckx, S. Demey, and J. De Schutter. Solving contact and grasp uncertainties. In *Proceedings of the International Conference on Intelligent Robots and Systems*, pages 114–119, Grenoble, France, September 1997.
87. James N. Eagle. The optimal search for a moving target when the search path is constrained. *Operations research*, 32(5):1107–1115, 1984.
88. Brian Eberman. A model-based approach to Cartesian manipulation contact sensing. *International Journal of Robotic Research*, 16(4):508–528, August 1997.
89. Hoda A. ElMaraghy and S. Payandeh. Contact prediction and reasoning for compliant robot motions. *IEEE Transactions on Robotics and Automation*, 5(4):533–538, August 1989.
90. M. Erdmann. On motion planning with uncertainty. Technical Report report AI-TR-810 (Master Thesis), Massachusetts Institute of Technology, Artificial Intelligence Laboratory, 1984.
91. M. Erdmann. Using backprojections for fine motion planning with uncertainty. *International Journal of Robotic Research*, 5(1):19–45, 1986.
92. M. Erdmann. Randomization of robot tasks. *International Journal of Robotic Research*, 11(2):399–436, 1992.
93. Louis J. Everett, Rakesh Ravari, Richard A. Volz, and Marjorie Skubic. Generalized recognition of single-ended contact formations. *IEEE Transactions of Robotics and Automation*, 15(5):829–836, October 1999.
94. A.O. Farahat, B.S. Graves, and J.C. Trinkle. Identifying contact formations in the presence of uncertainty. In *Proceedings of the IEEE International Conference on Intelligent Robots and Systems*, volume 3, pages 59–64, Pittsburg, PA, August 1995.
95. Roy Featherstone, Stef Sonck-Thiebaut, and Oussama Khatib. A general contact model for dynamically-decoupled force/motion control. In *Proceedings of the IEEE International Conference on Robotics and Automation*, pages 3281–3286, Detroit, Michigan, May 1999.
96. A. Fedele, A. Fioretti, C. Manes, and G. Ulivi. On-line processing of position and force measures for contour identification and robot control. In *Proceedings of the IEEE International Conference on Robotics and Automation*, pages 369–374, Atlanta, GA, 1993.
97. Hans Jacob S. Feder, John J. Leonard, and Christopher M. Smith. Adaptive mobile robot navigation and mapping. *International Journal of Robotics Research*, 18(7):650–668, July 1999.
98. V.V. Fedorov. *Theory of optimal experiments*. Academic press, New York, 1972.
99. R.A. Fisher. On the mathematical foundations of theoretical statistics. *Philosophical Transactions of the Royal Society, A,*, 222:309–368, 1922.
100. R.A. Fisher. Theory of statistical estimation. *Proceedings of the Cambridge Philosophical Society*, 22:700–725, 1925.
101. D. Fox, W. Burgard, F. Dellaert, and S. Thrun. Monte Carlo localization: Efficient position estimation for mobile robots. In *Proceedings of the National Conference on Artificial Intelligence*, Orlando, FL, 1999.

102. D. Fox, W. Burgard, and S. Thrun. Active Markov localization for mobile robots. volume 25, pages 195–207, 1998.
103. D. Fox, W. Burgard, and S. Thrun. Markov localization for mobile robots in dynamic environments. *Journal of Artificial Intelligence Research*, 11:391–427, 1999.
104. Holger Friedrich and Rüdiger Dillmann. Robot programming based on a single demonstration and user intentions. In *Proceedings of the 3rd European Workshop on Learning Robots*, Heraklion, Crete, Greece, April 1995.
105. K. Gadeyne, T. Lefebvre, and H. Bruyninckx. Bayesian hybrid model-state estimation applied to simultaneous contact formation recognition and geometrical parameter estimation. *The International Journal of Robotics Research*, 2005.
106. A. Gelb, J.F. Kasper, R.A. Nash, C.F. Price, and A.A. Sutherland. *Applied Optimal Estimation*. MIT Press, Cambridge, MA, 1974.
107. A.A. Goldenberg and P. Song. Principles for design of position and force controllers for robot manipulators. *Robotics and Autonomous Systems, Special issue on Critical Issues on Robotics*, 21(3):263–277, September 1997.
108. V. Gullapalli, R.A. Grupen, and A.G. Barto. Learning reactive admittance control. In *Proceedings of the IEEE International Conference on Robotics and Automation*, volume 2, pages 1475–1480, Nice, France, May 1992.
109. Vijaykumar Gullapalli. Skillful control under uncertainty via direct reinforcement learning. *Robotics and Autonomous Systems*, 15(4):237–246, October 1995.
110. Blake Hannaford and Paul Lee. Hidden Markov model analysis of force/torque information in telemanipulation. *The International Journal of Robotics Research*, 10(5):528–539, October 1991.
111. M. Hauskrecht. Value-function approximations for partially observable Markov decision processes. *Journal of Artificial Intelligence Research*, 13:33–94, 2000.
112. S. Hirai and H. Asada. Kinematics and statics of manipulation using the theory of polyhedral convex cones. *The International Journal of Robotics Research*, 12(5):434–447, 1993.
113. S. Hirai and K. Iwata. Recognition of contact state based on geometric model. In *Proceedings of the IEEE International Conference on Robotics and Automation*, pages 1507–1512, Nice, France, 1992.
114. H. Hirukawa, T. Matsui, and K. Takase. Automatic determination of possible velocity and applicable force of frictionless objects in contact from a geometric model. *IEEE Transactions on Robotics and Automation*, 10(3):309–322, June 1994.
115. G. Hirzinger and K. Landzettel. Sensory feedback structures for robots with supervised learning. In *Proceedings of the IEEE International Conference on Robotics and Automation*, pages 627–635, 1985.
116. N. Hogan. Impedance control: An approach to manipulation. Parts I–III. *Transactions of ASME, Journal of Dynamic Systems, Measurement and Control*, 107:1–24, March 1985.
117. M.-O. Hongler, F. Bandano, M. Bétemps, and A. Jutard. A random exploration approach for automatic chamferless insertion. *International Journal of Robotics Research*, 14(2):161–173, 1995.
118. Geir E. Hovland and Brenan J. McCarragher. Frequency-domain force measurements for discrete event contact recognition. In *Proceedings of the IEEE International Conference on Robotics and Automation*, volume 2, pages 1166–1171, Minneapolis, MN, April 1996.

119. Geir E. Hovland and Brenan J. McCarragher. Hidden Markov models as a process monitor in robotic assembly. *International Journal of Robotics Research*, 17(2):153–168, February 1998.
120. Geir E. Hovland, P. Sikka, and Brenan J. McCarragher. Skill acquisition from human demonstration using a hidden Markov model. In *Proceedings of the IEEE International Conference on Robotics and Automation*, volume 3, pages 2706–2711, Minneapolis, MN, April 1996.
121. Ronald A. Howard. *Dynamic Programming and Markov Processes.* The MIT Press, Cambridge, Massachusetts, 1960.
122. S.L. Ipri and H. Asada. Force-guided robotic assembly using tuned dither and recursive parameter estimation. *ASME Dynamic Systems and Control Meeting*, pages 857–862, 1994.
123. K. Ito and K. Xiong. Gaussian filters for nonlinear filtering problems. *IEEE Transactions on Automatic Control*, 45(5):910–927, May 2000.
124. R.V. Ivanov. Scanning assembly. *The International Journal of Advanced Manufacturing Technology*, 4:95–102, 1989.
125. A.H. Jazwinski. *Stochastic Processes and Filtering Theory.* Academic Press, New York, 1970.
126. H. Jeffreys. Some tests of significance, treated by the theory of probability. *Proceedings of the Cambridge Philosophical Society*, 31:203–222, 1935.
127. K. Jeong and H. Cho. Development of a pneumatic vibratory wrist for robotic assembly. *Robotica*, 7:9–16, 1989.
128. Xuerong Ji and Jing Xiao. Planning motions compliant to complex contact states. *The International Journal of Robotics Research*, 20(6):446–465, June 2001.
129. L. Joskowicz and R.H. Taylor. Interference-free insertion of a solid body into a cavity: An algorithm and a medical application. *The International Journal of Robotics Research*, 15:211–229, 1996.
130. S. J. Julier and J. K. Uhlmann. Reduced sigma point filters for the propagation of means and covariances through nonlinear transformations. In *Proceedings of the IEEE American Control Conference*, pages 887–892, Anchorage, AK, May 2002.
131. S. J. Julier and J. K. Uhlmann. The scaled unscented transformation. In *Proceedings of the IEEE American Control Conference*, pages 4555–4559, Anchorage, AK, May 2002.
132. S.J. Julier, J.K. Uhlman, and H. Durrant-Whyte. A new method for the transformation of means and covariances in filters and estimators. *IEEE Transactions on Automatic Control*, 45(3):477–482, March 2000.
133. S.J. Julier and J.K. Uhlmann. A general method for approximating nonlinear transformations of probability distributions. Technical report, Robotics Research Group, Department of Engineering Science, University of Oxford, November 1996.
134. S.J. Julier and J.K. Uhlmann. Data fusion in non-linear systems. In D.L. Hall and J. Llinas, editors, *Handbook of Multisensor Data Fusion.* CRC Press, Chapter 13, 2001.
135. M. Kaiser and R. Dillmann. Building elementary robot skills from human demonstration. In *Proceedings of the IEEE International Conference on Robotics and Automation*, pages 2700–2705, Minneapolis, Minnesota, April 1996.
136. Michael Kaiser and Rüdiger Dillmann. Hierarchical learning of efficient skill application for autonomous robots. In *Proceedings of the International Symposium on Intelligent Robotic Systems*, Pisa, Italy, 1995.

137. R. E. Kalman. A new approach to linear filtering and prediction problems. *Transactions of the ASME, Journal of Basic Engineering*, 82:34–45, March 1960.
138. Paul G. Kaminski, Arthur E. Bryson, and Stanley F. Schmidt. Discrete square root filtering: A survey of current techniques. *IEEE Transactions on Automatic Control*, 16(6):727–736, December 1971.
139. Sungchul Kang, Yong K. Hwang, Munsang Kim, Chongwon Lee, and Kyo-Il Lee. A compliant controller dynamically updating the compliance center by contact localization. *Robotica*, 16:543–550, 1998.
140. R. W. Katz. On some criteria for estimating the order of a Markov chain. *Technometrics*, 23(3):243–249, August 1981.
141. P. Kazanzides, N.S. Bradley, and W.A. Wolovich. Dual-drive force/velocity control: Implementation and experimental results. In *Proceedings of the IEEE International Conference on Robotics and Automation*, pages 92–97, Scottsdale, Arizona, May 1989.
142. O. Khatib. A unified approach for motion and force control of robot manipulators: The operational space formulation. *IEEE Journal of Robotics and Automation*, 3:43–53, 1987.
143. K. Kitagaki, T. Ogasawara, and T. Suehiro. Methods to detect contact state by force sensing in an edge mating task. In *Proceedings of the IEEE International Conference on Robotics and Automation*, volume 2, pages 701–706, 1993.
144. K. Kitagaki, T. Suehiro, and T. Ogasawara. Monitoring of a pseudo contact point for fine manipulation. In *Proceedings of the IEEE/RSJ International Conference on Intelligent Robots and Systems*, pages 757–762, 1996.
145. S. Koenig and R.G. Simmons. Solving robot navigation problems with initial pose uncertainty using real-time heuristic search. In *Proceedings of the International Conference on Artificial Intelligence Planning Systems*, pages 145–153, 1998.
146. R. Koeppe, A. Breidenbach, and G. Hirzinger. Skill representation and acquisition of compliant motions using a teach device. In *Proceedings of the IEEE/RSJ International Conference on Intelligent Robots and Systems*, volume 2, pages 897–904, Osaka, Japan, November 1996.
147. R. Koeppe and G. Hirzinger. Learning compliant motions by task-demonstration in virtual environments. In *Proceedings of the 4th International Symposium on Experimental Robotics*, Stanford, CA, July 1995.
148. R. Koeppe and G. Hirzinger. A signal-based approach to localization and navigation of autonomous compliant motion. In *Proceedings of the IEEE/RSJ International Conference on Intelligent Robots and Systems*. Japan, 2000.
149. Steen Kristensen. Sensor planning with Bayesian decision theory. *Robotics and Autonomous Systems*, 19:273–286, 1997.
150. G.J.A. Kröse and R. Bunschoten. Probabilistic localization by appearance models and active vision. In *Proceedings of the IEEE conference on Robotics and Automation, Detroit*, May 1999.
151. S. Kullback and R.A. Leibler. On information and sufficiency. *Annals of mathematical Statistics*, 22:79–86, 1951.
152. Solomon Kullback. *Information theory and statistics.* Wiley publication in mathematical statistics, 1959.
153. Pierre-Simon Laplace. *Théorie Analytique des Probabilités.* Courcier Imprimeur, Paris, 1812.
154. J.-C. Latombe. *Robot Motion Planning.* Kluwer Academic Publisher, 1991.
155. C. Laugier. Planning fine motion strategies by reasoning in the contact space. In *Proceedings of the IEEE International Conference on Robotics and Automation*, pages 653–659, 1989.

156. C. Laugier and P. Theveneau. Planning sensor-based motions for part-mating using geometric reasoning techniques. In *Proceedings of the European Conference on Artificial Intelligence*, pages 617–629, Brighton, 1986.
157. S. Lee and H. Asada. Assembly of parts with irregular surfaces using active force sensing. *Proceedings of the IEEE International Conference on Robotics and Automation*, pages 2639–2644, 1994.
158. S. Lee and H.H. Hahn. A perturbation/correlation method for force guided robot assembly. *IEEE Transactions on Robotics and Automation*, 15(4):764–773, August 1999.
159. T. Lefebvre, H. Bruyninckx, and J. De Schutter. Comment on "A new method for the nonlinear transformation of means and covariances in filters and estimators". *IEEE Transactions on Automatic Control*, 47(8):1406–1408, August 2002.
160. T. Lefebvre, H. Bruyninckx, and J. De Schutter. Polyhedral contact formation modeling and identification for autonomous compliant motion. *IEEE Transactions On Robotics And Automation*, 19(1):26–41, February 2003.
161. T. Lefebvre, H. Bruyninckx, and J. De Schutter. Exact nonlinear Bayesian parameter estimation for autonomous compliant motion. *Advanced Robotics, Special issue on "Selected Papers from ICAR 2003"*, 18(8):787–800, 2004.
162. T. Lefebvre, H. Bruyninckx, and J. De Schutter. Kalman filters for nonlinear systems: A comparison of performance. *The International Journal of Control*, 77(7):639–653, May 2004.
163. T. Lefebvre, H. Bruyninckx, and J. De Schutter. Online statistical model recognition and state estimation for autonomous compliant motion. *IEEE Transactions on Systems, Man and Cybernetics, Part C, Special Issue on "Pattern Recognition for Autonomous Manipulation in Robotic Systems"*, 35(1):16–29, February 2005.
164. T. Lefebvre, H. Bruyninckx, and J. De Schutter. Polyhedral contact formation identification for autonomous compliant motion: Exact nonlinear Bayesian filtering. *IEEE Transactions On Robotics*, 21(1):124–129, February 2005.
165. T. Lefebvre, H. Bruyninckx, and J. De Schutter. Task planning with active sensing for autonomous compliant motion. *The International Journal of Robotics Research*, 24(1):61–81, January 2005.
166. T. Lefebvre, K. Gadeyne, H. Bruyninckx, and J. De Schutter. Exact Bayesian inference for a class of nonlinear systems with application to robotic assembly. In J. M. Bernardo, M. J. Bayarri, J. O. Berger, A. P. Dawid, D. Heckerman, A. F. M. Smith, and M. West, editors, *Bayesian Statistics 7*, pages 587–596. Oxford University Press, 2003.
167. T. Lefebvre, J. Xiao, H. Bruyninckx, and G. De Gersem. Active compliant motion: A survey. *Advanced Robotics, Special issue on "Compliant Motion: Modeling, Planning and Control"*, 2005.
168. H. Lipkin and J. Duffy. Hybrid twist and wrench control for a robotic manipulator. *Transactions of the ASME, Journal of Mechanisms, Transmissions and Automation in Design*, 110:138–144, 1988.
169. Michael L. Littman, Anthony R. Cassandra, and Leslie Pack Kaelbling. Efficient dynamic-programming updates in partially observable Markov decision processes. Technical Report CS-95-19, Brown University, Department of Computer Science, Providence RI, 1995.
170. Michael L. Littman, Thomas L. Dean, and Leslie Pack Kaelbling. On the complexity of solving Markov decision problems. In *Proceedings of the 11th International Conference on Uncertainty in Artificial Intelligence*, pages 394–402, Montreal, Québec, Canada, 1995.

171. M.-H. Liu. Force control for robotic deburring. In *Proceedings of the IMACS/SICE International Symposium on Robotics, Mechatronics and Manufacturing Systems*, pages 1373–1378, Kobe, Japan, September 1992.
172. I. López-Juárez and M. Howarth. Knowledge acquisition and learning in unstructured robotic assembly environments. *The International Journal of Information Sciences*, 145:89–111, 2002.
173. William S. Lovejoy. A survey of algorithmic methods for partially observed Markov decision processes. *Annals of Operations Research*, 18:47–65, 1991.
174. T. Lozano-Pérez. The design of a mechanical assembly system. Technical Report AI TR 397, AI. Lab, MIT, 1976.
175. T. Lozano-Pérez. A language for automatic mechanical assembly. In P.H.Winston and R.H. Brown, editors, *Artificial Intelligence: An MIT Perspective*, volume 2, pages 243–271. MIT Press, Cambridge, MA, 1979.
176. T. Lozano-Pérez. Automatic planning of manipulator transfer movements. *IEEE Transactions on Systems, Man, and Cybernetics*, 11(10):681–689, 1981.
177. T. Lozano-Pérez. Spacial planning: A configuration space approach. *IEEE Transactions on Computers*, 32(2):108–120, 1983.
178. T. Lozano-Pérez, M. Mason, and R. Taylor. Automatic synthesis of fine-motion strategies for robots. *The International Journal of Robotics Research*, 3(1):3–24, 1984.
179. David J. C. MacKay. Information theory, inference and learning algorithms. Textbook in preparation. http://wol.ra.phy.cam.ac.uk/mackay/itprnn/, 1999–2003.
180. Alexei A. Makarenko, Stefan B. Williams, Frédéric Bourgault, and Hugh F. Durrant-Whyte. An experiment in integrated exploration. In *Proceedings of the IEEE/RSJ International Conference on Intelligent Robots and Systems*, pages 534–539, Lausanne, Switzerland, October 2002.
181. M. T. Mason. Compliance and force control for computer controlled manipulators. *IEEE Transactions on Systems, Man, and Cybernetics*, SMC-11(6):418–432, 1981.
182. P.S. Maybeck. *Stochastic Models, Estimation, and Control, Volume 2*. Academic Press, 1982.
183. B.J. McCarragher. Petri net modeling for robotic assembly and trajectory planning. *IEEE Transactions on Industrial Electronics*, 41(6):631–640, December 1994.
184. B.J. McCarragher. Task primitives for the discrete event modeling and control of 6-DOF assembly tasks. *IEEE Transactions on Robotics and Automation*, 12(2):280–289, April 1996.
185. B.J. McCarragher and H. Asada. Qualitative template matching using dynamic process models for state transition recognition of robotic assembly. *Transactions of the ASME, Journal of Dynamic Systems, Measurement and Control*, 115:261–269, 1993.
186. Brenan J. McCarragher and David J. Austin. Model-adaptive hybrid dynamic control for constrained motion systems. *IEEE Transactions on Automatic Control, Special Issue on Hybrid Systems*, 43(4):560–563, April 1998.
187. R.K. Mehra. Approaches to adaptive filtering. *IEEE Transactions on Automatic Control*, 17(5):693–698, October 1972.
188. R.K. Mehra and J. Peschon. An innovations approach to fault detection and diagnosis. *Automatica*, 7:637–640, 1971.
189. J.-P. Merlet. C-surface applied to the design of an hybrid force-position robot controller. In *Proceedings of the IEEE International Conference on Robotics and Automation*, volume 2, pages 1055–1059, Raleigh, NC, April 1987.

190. L. Mihaylova, J. De Schutter, and H. Bruyninckx. A multisine approach for trajectory optimization based on information gain. In *Proceedings of the IEEE/RSJ International Conference on Intelligent Robots and Systems*, pages 661–666, Lausanne, Switzerland, October 2002.
191. L. Mihaylova, T. Lefebvre, E. Staffetti, H. Bruyninckx, and J. De Schutter. Tracking contact transitions during force-controlled compliant motion using an interacting multiple model estimator. In *Proceedings of the IEEE International Conference on Advanced Robotics*, pages 665–670, Budapest, Hungary, August 2001.
192. Lyudmila Mihaylova, Tine Lefebvre, Ernesto Staffetti, Herman Bruyninckx, and Joris De Schutter. Contact transitions tracking during force-controlled compliant motion using an interacting multiple model estimator. *Information and Security: An International Journal.*, 9:114–129, 2002.
193. N. Mimura and Y. Funahashi. Parameter identification in the grasp of an inner link mechanism. In *Proceedings of the IEEE International Conference on Robotics and Automation*, pages 209–215, Atlanta, Georgia, 1993.
194. N. Mimura and Y. Funahashi. Parameter identification of contact conditions by active force sensing. In *Proceedings of the IEEE International Conference on Robotics and Automation*, pages 2645–2650, San Diego, CA, 1994.
195. George E. Monahan. A survey of partially observable decision processes: Theory, models and algorithms. *Management Science*, 28(1):1–16, 1982.
196. H. Mosemann, A. Raue, and F. Wahl. Identification of assembly process states using polyhedral convex cones. In *Proceedings of the IEEE International Conference on Robotics and Automation*, pages 2756–2761, Detroit, MI, May 1999.
197. Kevin P. Murphy. A survey of POMDP solution techniques. Technical report, http://citeseer.nj.nec.com/murphy00survey.html, September 2000.
198. Shin-yo Muto and Ken-ichiro Shimokura. Accurate contact point detecting using force and velocity information complementarily. In *Proceedings of the IEEE International Conference on Robotics and Automation*, volume 1, pages 738–744, 1993.
199. Jorge Nocedal and Stephen J. Wright. *Numerical Optimization*. Springer Series in Operations Research. Springer, 1999.
200. M. Nørgaard, N.K. Poulsen, and O. Ravn. Advances in derivative-free state estimation for nonlinear systems. Technical Report IMM-REP-1998-15 (revised edition), Technical University of Denmark, Denmark, April 2000.
201. M. Nørgaard, N.K. Poulsen, and O. Ravn. New developments in state estimations for nonlinear systems. *Automatica*, 36(11):1627–1638, November 2000.
202. M. Nuttin, J. Rosell, R. Suárez, H. Van Brussel, L. Basañez, and J. Hao. Learning approaches to contact estimation in assembly tasks with robots. In *Proceedings of the 3rd European Workshop on Learning Robots*, Heraklion, Crete, Greece, April 1995.
203. M. Nuttin and H. Van Brussel. Learning assembly operations: A case study with real-world objects. *Studies in Informatics and Control*, 5(3):205–221, 1996.
204. M. Nuttin and H. Van Brussel. Learning the peg-into-hole assembly operation with a connectionist reinforcement technique. *Computers in Industry*, 33(1):101–109, Augustus 1997.
205. M. Nuttin and H. Van Brussel. Learning and re-calibration in flexible assembly. In K. Morik, M. Kaiser, and V. Klingspor, editors, *Making Robots Smarter — Combining Sensing and Action through Robot Learning*, chapter 4, pages 53–64. Kluwer Academic Publishers, 1999.

206. M. Oussalah. Fuzzy linear regression for contact identification. In *Proceedings of the IEEE International Conference on Robotics and Automation*, pages 3616–3621, San Francisco, CA, April 2000.
207. Y.K. Park and H.S. Cho. A self-learning rule-based control algorithm for chamferless part mating. *Control Engineering Practice*, 2(5):773–783, 1994.
208. J.A. Parkin. Co-ordinate transformations of screws with applications to screw systems and finite twists. *Mechanism and Machine Theory*, 25(6):689–699, 1990.
209. S. Patarinski and R. Botev. Robot force control: A review. *Mechatronics*, 3(4):377–398, 1993.
210. M.A. Peshkin. Programmed compliance for error-corrective assembly. *IEEE Transactions on Robotics and Automation*, 6(4):473–482, August 1990.
211. V. Peterka. Bayesian approach to system identification. In P. Eykhoff, editor, *Trends and progress in Systems Identification*, pages 239–304. Pergamon Press, Oxford, New York, 1981.
212. F. Pukelsheim. *Optimal Design of Experiments*. John Wiley, New York edition, 1993.
213. Martin L. Puterman. *Markov Decision Processes: Discrete Stochastic Dynamic Programming*. John Wiley & Sons, Wiley series in probability and mathematical statistics, New York, 1994.
214. M. Raibert and J. J. Craig. Hybrid position/force control of manipulators. *Transactions of ASME, Journal of Dynamic Systems, Measurement and Control*, 102:126–133, 1981.
215. C. Radhakrisuna Rao. Information and the accuracy attainable in the estimation of statistical parameters. *Bulletin of the Calcutta Mathematical Society*, 37:81–91, 1945.
216. J. Rissanen. Stochastic complexity and modeling. *The Annals of Statistics*, 14(3):1080–1100, 1986.
217. J. Rissanen. *Stochastic complexity in statistical inquiry*. Series in Computer Science, Volume 15. World Scientific, 1989.
218. Nicholas Roy, Wolfram Burgard, Dieter Fox, and Sebastian Thrun. Coastal navigation — Mobile robot navigation with uncertainty in dynamic environments. In *Proceedings of the IEEE International Conference on Robotics and Automation*, volume 1, pages 35–40, Detroit, MI, May 1999.
219. J. Rust. Numerical dynamic programming in economics. In H. Amman, D. Kendrick, and J. Rust, editors, *Handbook of Computational Economics*, pages 619–729. Elsevier, Amsterdam, 1996.
220. J.K. Salisbury. Active stiffness control of a manipulator in Cartesian coordinates. In *Proceedings of the IEEE Conference on Decision and Control*, pages 83–88, Albuquerque, NM, December 1980.
221. T.S. Schei. A finite-difference method for linearization in nonlinear estimation algorithms. *Automatica*, 33(11):2053–2058, November 1997.
222. J.M. Schimmels and M.A. Peshkin. Admittance matrix design for force-guided assembly. *IEEE Transactions on Robotics and Automation*, 8(2):213–227, 1992.
223. T.M. Schulteis, P.E. Dupon, P.A. Millman, and R.D. Howe. Automatic identification of remote environments. In *Proceedings of the ASME Dynamic Systems and Control Division*, pages 451–458, Atlanta, GA, November 1996.
224. Gideom Schwartz. Estimating the dimension of a model. *Annals of Statistics*, 6:461–5, 1978.
225. Uwe Schweigert. Precision assembly with industrial robots. *Industrial Robot*, 19(2):28–34, 1992.

226. P. Schweitzer and A. Seidmann. Generalized polynomial approximations in Markovian decision processes. *Journal of Mathematical Analysis and Applications*, 110:568–582, 1985.
227. C.E. Shannon. A mathematical theory of communication, I. *The Bell System Technical Journal*, 27:379–423, July 1948.
228. C.E. Shannon. A mathematical theory of communication, II. *The Bell System Technical Journal*, 27:623–656, October 1948.
229. R. Shibata. Selection of the order of an autoregressive model by Akaike's information criterion. *Biometrika*, 63:117–126, 1976.
230. David Siegmund. *Sequential analysis, Test and Confidence Intervals.* Springer-Verlag, 1985.
231. Pavan Sikka and Brenan J. McCarragher. Monitoring contact using clustering and discriminant functions. In *Proceedings of the IEEE International Conference on Robotics and Automation*, volume 2, pages 1351–1356, Minneapolis, MN, 1996.
232. Pavan Sikka and Brenan J. McCarragher. Rule-based contact monitoring using examples obtained by task demonstration. In *Proceedings of the 15th International Joint Conference on Artificial Intelligence*, pages 514–521, Nagoya, Japan, August 1997.
233. Reid Simmons and Sven Koenig. Probabilistic robot navigation in partially observable environments. In *Proceedings of the International Joint Conference on Artificial Intelligence*, pages 1080–1087, Montréal, Canada, 1995. Springer-Verlag, Berlin, Germany.
234. J. Simons, H. Van Brussel, J. De Schutter, and J. Verhaert. A self-learning automaton with variable resolution for high precision assembly by industrial robots. *IEEE Transactions on Automatic Control*, 27(5):1109–1113, October 1982.
235. M. Skubic and R. Volz. Learning force-based assembly skills from human demonstration for execution in unstructured environments. In *Proceedings of the IEEE International Conference on Robotics and Automation*, pages 1281–1288, Leuven, Belgium, May 1998.
236. Marjorie Skubic, Benjamin Forrester, and Brent M. Nowak. Clustering of qualitative contact states for a transmission assembly. In *Proceedings of the IEEE International Conference on Robotics and Automation*, pages 1281–1288, Leuven, Belgium, May 1998.
237. Marjorie Skubic and Richard A. Volz. Identifying single-ended contact formations from force sensor patterns. *IEEE Transactions of Robotics and Automation*, 16(5):597–603, October 2000.
238. S. Soatto, R. Frezza, and P. Perona. Motion estimation via dynamic vision. *IEEE Transactions on Automatic Control*, 41(3):393–414, March 1996.
239. Edward J. Sondik. *The Optimal Control of Partially Observable Markov Processes.* PhD thesis, Stanford University, Stanford, California, 1971.
240. Harold W. Sorenson. *Kalman filtering: Theory and application.* IEEE Press, New York, NY, 1985.
241. M. Spreng. Dealing with unexpected situations during the execution of robot motions. In *Proceedings of the IEEE International Conference on Robotics and Automation*, pages 64–69, April 1991.
242. M. Spreng. A probabilistic method to analyze ambiguous contact situations. In *Proceedings of the IEEE International Conference on Robotics and Automation*, pages 543–548, Atlanta, USA, 1993.
243. R. Suárez and L. Basañez. Assembly with robots in presence of uncertainty. In *Proceedings of the International Symposium on Industrial Robots*, pages 19/1–19/15, Detroit, USA, 1991.

244. R. Suárez, L. Basañez, M. Nuttin, H. Van Brussel, and J. Rosell. Learning versus analytical approach to contact estimation in assembly tasks with robots. In *Proceedings of the IEEE International Conference on Advanced Robotics*, pages 769–775, Sant Feliu de Guíxols, Spain, September 1995.
245. R. Suárez, L. Basañez, and J. Rosell. Assembly contact force domains in the presence of uncertainty. In *Proceedings of the 4th IFAC Symposium on Robot Control (SYROCO)*, pages 653–659, Capri, Italy, September 1994.
246. R.S. Sutton and A.G. Barto. *Reinforcement Learning, An introduction.* The MIT Press, 1998.
247. Tomoichi Takahashi, Hiroyuki Ogata, and Shin-yo Muto. A method for analyzing human assembly operations for use in automatically generating robot commands. In *Proceedings of the IEEE International Conference on Robotics and Automation*, volume 2, pages 695–700, 1993.
248. H. Tanizaki. *Nonlinear Filters: Estimation and Applications — Second, Revised and Enlarged Edition.* Springer-Verlag, Berlin-Heidelberg, 1996.
249. R.H. Taylor. *A synthesis of manipulator control programs from task-level specifications.* PhD thesis, Stanford University, AI. Lab., 1976.
250. C.L. Thornton and G.J. Bierman. Filtering and error analysis via the UDU covariance factorization. *IEEE Transactions on Automatic Control*, 23(5):901–907, October 1978.
251. S. Thrun. Monte Carlo POMDPs. In S. A. Solla, T. K. Leen, and K. R. Muller, editors, *Advances in Neural Processing Systems*, volume 12, pages 1064–1070. MIT Press, 2000.
252. S. Thrun. Probabilistic algorithms in robotics. *AI Magazine*, 21(4):93–109, 2000.
253. Petr Tichavský, Carlos H. Muravchik, and Arye Nehorai. Posterior Cramér-Rao bounds for discrete-time nonlinear filtering. *IEEE Transactions on Signal Processing*, 46(5):1386–1396, May 1998.
254. T. Tsujimura and M. Yabuta. Object detection by tactile sensing method employing force/torque information. *IEEE Transactions on Robotics and Automation*, 5(4):444–450, August 1989.
255. J.K. Tugnait. Detection and estimation for abruptly changing systems. *Automatica*, 18(5):607–615, 1982.
256. J.K. Tugnait and A.H. Haddad. A detection-estimation scheme for state estimation in switching environments. *Automatica*, 15:477–481, 1979.
257. Matthew A. Turk. A fine-motion planning algorithm. In David P. Casasent, editor, *Proceedings of the SPIE Conference on Intelligent Robots an Computer Vision*, volume 579, pages 113–120, Cambridge, Massachusetts, September 1985.
258. J. Uhlmann. *Dynamic Map Building and Localization: New Theoretical Foundations.* PhD thesis, University of Oxford, UK, 1995.
259. Harry L. Van Trees. *Detection, Estimation and Modulation Theory, Volume I.* Wiley and Sons, New York, 1968.
260. Abraham Wald. *Sequential analysis.* John Wiley & Sons, Inc., 1947.
261. Qi Wang and Joris De Schutter. Towards real-time robot programming by human demonstration for 6D force controlled actions. In *Proceedings of the IEEE International Conference on Robotics and Automation*, volume 3, pages 2256–2261, Leuven, Belgium, 1998.
262. Qi Wang, Joris De Schutter, Wim Witvrouw, and Sean Graves. Derivation of compliant motion programs based on human demonstration. In *Proceedings of the IEEE International Conference on Robotics and Automation*, pages 2616–2621, April 1996.

263. H. Warnecke, B. Frankenhauser, D. Gweon, and H. Cho. Fitting of crimp contacts to connectors using industrial robots supported by vibrating tools. *Robotica*, 6:123–129, 1988.
264. Jing Wei and Wyatt S. Newman. Improving robotic assembly performance through autonomous exploration. In *Proceedings of the IEEE International Conference on Robotics and Automation*, pages 3303–3308, Washington, DC, May 2002.
265. H. West and H. Asada. A method for the design of hybrid position/force controllers for manipulation constrained by contact with the environment. In *Proceedings of the IEEE International Conference on Robotics and Automation*, pages 251–259, 1985.
266. D.E. Whitney. Historic perspective and state of the art in robot force control. *The International Journal of Robotics Research*, 6(1):3–14, 1987.
267. D.E. Whitney and J.L. Nevins. What is the remote center compliance and what can it do ? In *Proceedings of the 9th International Symposium on Industrial Robots*, pages 135–152, Washington, D.C., 1979.
268. A.S. Willsky. A survey of design methods for failure detection in dynamic systems. *Automatica*, 12:601–611, 1976.
269. A.S. Willsky and H.L. Jones. A generalized likelihood ratio approach to the detection and estimation of jumps in linear systems. *IEEE Transactions on Automatic Control*, 21(6):108–112, February 1976.
270. J. Xiao. Replanning with compliant rotation in the presence of uncertainties. In *Proceedings of the International Symposium on Intelligent Control*, pages 102–108, Glasgow, Scotland, 1992.
271. J. Xiao. Automatic determination of topological contacts in the presence of sensing uncertainties. In *Proceedings of the IEEE International Conference on Robotics and Automation*, pages 65–70, Atlanta, Georgia, USA, 1993.
272. J. Xiao and X. Ji. A divide-and-merge approach to automatic generation of contact states and planning of contact motions. In *Proceedings of the IEEE International Conference on Robotics and Automation*, pages 750–756, April 2000.
273. J. Xiao and R. Volz. On replanning for assembly tasks using robots in the presence of uncertainties. In *Proceedings of the IEEE International Conference on Robotics and Automation*, pages 638–645, Scottsdale, Arizona, USA, 1989.
274. Jing Xiao and Lianzhong Liu. Contact states: Representation and recognizability in the presence of uncertainties. In *Proceedings of the IEEE/RSJ International Conference on Intelligent Robots and Systems*, Victoria, B.C., Canada, October 1998.
275. Jing Xiao and Lixin Zhang. Towards obtaining all possible contacts — Growing a polyhedron by its location uncertainty. *IEEE Transactions on Robotics and Automation*, 12(4):553–565, August 1996.
276. Jing Xiao and Lixin Zhang. Contact constraint analysis and determination of geometrically valid contact formations from possible contact primitives. *IEEE Transactions on Robotics and Automation*, 13(3):456–466, June 1997.
277. T. Yoshikawa and A. Sudou. Dynamic hybrid position/force control of robot manipulators — On-line estimation of unknown constraint. *IEEE Transactions on Robotics and Automation*, 9(2):220–226, 1993.
278. T. Yoshikawa, T. Surgie, and M. Tanaka. Dynamic hybrid position force control of robot manipulators. *IEEE Journal of Robotics and Automation*, 4(6):699–705, 1988.

279. T. Yoshikawa, Y. Yu, and M. Koike. Estimation of contact position between object and environment based on probing of robot manipulator. In *Proceedings of the IEEE/RSJ International Conference on Intelligent Robots and Systems*, pages 769–776, 1996.
280. G. Zeng and A. Hemami. Overview of robot force control. *Robotica*, 15(5):473–482, 1997.
281. N.L. Zhang and W. Liu. Planning in stochastic domains: Problem characteristics and approximation. Technical Report HKUST-CS96031, Hong Kong University of Science and Technology, 1996.

Index

Symbols, Definitions and Abbreviations

General Abbreviations

CC	: contact configuration
CF	: contact formation
KF	: Kalman Filter
EKF	: Extended Kalman Filter
IEKF	: Iterated Extended Kalman Filter
MNSKF	: Non-minimal State Kalman Filter
CKF	: Cascaded Kalman Filter
NIS	: Normalised Innovation Squared
SNIS	: Summed Normalised Innovation Squared
pdf	: probability density function
1D, 2D, 3D	: 1-, 2-, or 3-dimensional

General Symbols and Definitions

a	: scalar (unbold lower case)
$\mathbf{a}$	: vector (bold lower case)
$\mathbf{A}$	: matrix (bold upper case)
$\mathbf{A}_{m\times n}$	: m-by-n matrix $\mathbf{A}$
$\mathbf{a}(i)$	: ith element of the vector $\mathbf{a}$
$\mathbf{a}^T$ or $\mathbf{A}^T$	: transpose of the vector $\mathbf{a}$ or the matrix $\mathbf{A}$
$\mathbf{A}^{-1}$	: inverse of the matrix $\mathbf{A}$
$\mathbf{A}^{\#}$	: generalised inverse of the matrix $\mathbf{A}$
$\mathbf{A}^{\#}_W$	: generalised inverse of the matrix $\mathbf{A}$ with weighting matrix $\mathbf{W}$
$\lvert\mathbf{A}\rvert$	: determinant of the matrix $\mathbf{A}$
$\mathrm{diag}(\sigma_1, \ldots, \sigma_n)$	: n-by-n diagonal matrix with diagonal elements $\sigma_1, \ldots, \sigma_n$
$\mathbf{a} \times \mathbf{b}$	: vector product of the vectors $\mathbf{a}$ and $\mathbf{b}$
$[\mathbf{p}\times]$	: matrix representing the vector product with $\mathbf{p}$
$\mathbf{I}_{m\times m}$	: m-by-m identity matrix

$\mathbf{0}_{m\times n}$	: m-by-n zero matrix
$\mathbf{0}_{m\times 1}$	: m-dimensional zero vector
$\hat{a}$ or $\hat{\mathbf{a}}$	: value for the scalar a or the vector $\mathbf{a}$
$\cdot_k$	: time index
$\cdot_{k\|k-1}$	: at time k given the measurements up to time $k-1$
$\cdot_{k\|k}$	: at time k given the measurements up to time k
$\sum_a$ or $\sum_{\mathbf{a}}$	: sum over the domain of the variable a or the vector $\mathbf{a}$
$\int_a$ or $\int_{\mathbf{a}}$	: integral over the domain of the variable a or the vector $\mathbf{a}$

Bayesian Probability Theory

$p(a)$	: probability of a
$p(a\|b)$	: probability of a given b
$\mathbf{x}$	: n-dimensional state vector
$\mathbf{x}^*$	: true value of the state vector
$\hat{\mathbf{x}}$	: state estimate
$\mathbf{P}$	: covariance matrix of the state estimate
$\mathbf{I}$	: information matrix of the state estimate
$\mathbf{z}$	: m-dimensional measurement vector
$\hat{\mathbf{z}}$	: measured value
$\mathbf{u}$	: system input
$\mathbf{s}$	: sensor input
$\mathbf{X}$	: sequence of state vectors
$\mathbf{Z}$	: sequence of measurement vectors
$\mathbf{U}$	: sequence of system inputs
$\mathbf{S}$	: sequence of sensor inputs
$p(\mathbf{x})$	: prior pdf over the state $\mathbf{x}$
$p(\mathbf{x}\|\hat{\mathbf{z}})$	: posterior pdf over the state $\mathbf{x}$ given the measurement $\hat{\mathbf{z}}$
$p(\mathbf{x}\|\hat{\mathbf{Z}})$	: posterior pdf over the state $\mathbf{x}$ given the measurements $\hat{\mathbf{Z}}$
$p(\hat{\mathbf{z}}\|\mathbf{x})$	: likelihood of the measurement $\hat{\mathbf{z}}$ as a function of the state $\mathbf{x}$
$p(\hat{\mathbf{Z}}\|\mathbf{x})$	: likelihood of the measurements $\hat{\mathbf{Z}}$ as a function of the state $\mathbf{x}$
$\mathbf{f}(.)$	: process model
$\mathbf{h}(.)$	: measurement model
$\boldsymbol{\rho}_p$	: process model uncertainty
$\boldsymbol{\rho}_m$	: measurement model uncertainty
H	: depending on the context: hypothesis or entropy
M	: model
$D(.\|\|.)$	: Kullback-Leibler distance
$\mathbf{I}_{MI}$	: Mutual information
$\mathbf{I}_{FI}$	: Fisher information matrix
$\mathbf{I}_{VT}$	: Expected information in the posterior state estimate
$\mathbf{I}_{VT,prior}$	: Expected information in the prior state estimate
$\mathbf{I}_{VT,meas}$	: Expected information in the measurements
$\mathbf{a}$	: action to optimise
$\boldsymbol{\pi}$	: policy (sequence of actions) to optimise
V	: value function to optimise

MDP : Markov Decision Process
POMDP : Partially Observable Markov Decision Process

Kalman Filtering

$\mathbf{f}(.)$: process model
$\mathbf{h}(.)$: measurement model
$\mathbf{F}, \mathbf{b}, \mathbf{C}$: linear(ized) process model
$\mathbf{H}, \mathbf{d}, \mathbf{E}$: linear(ized) measurement model
$\boldsymbol{\rho}_p^*$: process uncertainty due to linearization
$\boldsymbol{\rho}_m^*$: measurement uncertainty due to linearization
$\mathbf{Q}$: covariance matrix of the process model uncertainty
$\mathbf{R}$: covariance matrix of the measurement model uncertainty
$\mathbf{Q}^*$: covariance matrix representing the extra uncertainty on the linearized process model due to linearization errors
$\mathbf{R}^*$: covariance matrix representing the extra uncertainty on the linearized measurement model due to linearization errors
$\mathbf{K}$: Kalman gain
$\boldsymbol{\eta}$: innovation vector
$\mathbf{S}$: covariance matrix of the innovation vector
$\boldsymbol{\mathcal{X}}, \boldsymbol{\mathcal{Z}}$: regression point
$\mathbf{x}'$: higher-dimensional state vector with dimension n'
$\mathbf{P}'$: covariance matrix of the higher-dimensional state vector

Contact Modelling

$\mathbf{t}$: twist
$\mathbf{v}$: translational velocity
$\boldsymbol{\omega}$: rotational velocity
$\mathbf{w}$: wrench
$\mathbf{f}$: force
$\mathbf{m}$: moment
$\mathbf{t}_d$: pose
$\mathbf{G}$: wrench spanning set
$\boldsymbol{\phi}$: wrench decomposition
$\mathbf{J}$: twist spanning set
$\boldsymbol{\chi}$: twist decomposition
$\mathbf{h}_d(.) = \mathbf{0}$: closure equation
{w} : world frame
{e} : environment frame
{i} : environment frame for the ith environment object
{g} : gripper frame
{m} : manipulated object frame
{c} : frame in a contact point
$\mathbf{x}^m$: grasping uncertainties $\mathbf{x}^m = [x^m \; y^m \; z^m \; \theta_x^m \; \theta_y^m \; \theta_z^m]^T$

x^m, y^m, z^m : position of the manipulated object
$\theta_x^m, \theta_y^m, \theta_z^m$: orientation of the manipulated object
$\mathbf{x}^{dm}$: uncertain dimensions of the manipulated object
$\mathbf{x}^e$: environment uncertainties $\mathbf{x}^e = [x^e\ y^e\ z^e\ \theta_x^e\ \theta_y^e\ \theta_z^e]^T$
x^e, y^e, z^e : position of the environment object
$\theta_x^e, \theta_y^e, \theta_z^e$: orientation of the environment object
$\mathbf{x}^{de}$: uncertain dimensions of the environment object
$\mathbf{x}^i$: environment uncertainties for the ith environment object $\mathbf{x}^i = [x^i\ y^i\ z^i\ \theta_x^i\ \theta_y^i\ \theta_z^i]^T$
x^i, y^i, z^i : position of the ith environment object
$\theta_x^i, \theta_y^i, \theta_z^i$: orientation of the ith environment object
$\mathbf{x}^{di}$: uncertain dimensions of the ith environment object
${}_b^a\mathbf{S}_w$: wrench transformation matrix to transform a wrench expressed in {a} to a wrench expressed in {b}
${}_b^a\mathbf{S}_t$: twist transformation matrix to transform a twist expressed in {a} to a twist expressed in {b}
${}_b^a\mathbf{R}$: rotation matrix to rotate a vector from {a} to {b}
${}_c\mathbf{p}^{b,a}$: vector from the origin of {b} to the origin of {a} expressed in the frame {c}
${}_b^a\mathbf{T}$: homogeneous transformation matrix to transform a point expressed in {a} to a point expressed in {b}

Geometrical Parameter Estimation

$\mathbf{H}, \mathbf{d}, \mathbf{E}, \mathbf{D}$: linearization of an implicit, nonlinear measurement model
$\mathbf{R}_c, n_c$: parameters determining the test on transient behaviour
$\mathbf{x}_o$: observable part of the state vector
$\mathbf{x}_u$: unobservable part of the state vector
$\mathbf{P}_{oo}$: covariance matrix of the estimates of $\mathbf{x}_o$
$\mathbf{P}_{uu}$: covariance matrix of the estimates of $\mathbf{x}_u$
$\mathbf{P}_{uo}, \mathbf{P}_{ou}$: correlation between the estimates of $\mathbf{x}_o$ and $\mathbf{x}_u$

Task Planning with Active Sensing

$\mathbf{O}$: spanning set of the observable parameter space
$\mathbf{P}_{init}$: initial covariance matrix of the state estimates
$\mathbf{P}_{end}$: final covariance matrix of the state estimates
$\mathbf{P}_{toleranced}$: toleranced uncertainty matrix
DIRECTIONSTOOBSERVE($\mathbf{P}_1, \mathbf{P}_2$) :
parameter directions for which the uncertainty is larger under covariance matrix $\mathbf{P}_1$ than under covariance matrix $\mathbf{P}_2$
OBSERVABLESPACE($\mathbf{O}$) :
space of the observable parameter directions given its spanning set $\mathbf{O}$
$\mathbf{U}$: CF-compliant path

$\mathbf{U}_a$: active sensing motion
$\mathbf{U}_t$: transition motion
$\boldsymbol{\pi}$: CF sequence
n_{Seq} : number of CFs in the CF sequence
$\mathbf{f}_{obj}$: objective function to minimise
$\mathbf{c}_{ineq}$: inequality constraint
$\mathbf{c}_{eq}$: equality constraint
E_{kin} : kinetic energy
t_s : stabilisation time
t_t : time of the transition motion
t_a : extra execution time due to active sensing
t_{Seq} : time of the total task execution
$\mathbf{h}_{d,next}$: distance vector of contacts to close
$\mathbf{h}_{d,avoid}$: distance vector of contacts to avoid
$h_{avoid,desired}$: desired minimum distance to contacts to avoid
$\Delta\mathbf{h}_d$: vector containing $\mathbf{h}_{d,next}$ and the difference between $h_{avoid,desired}$ and $\mathbf{h}_{d,avoid}$
$\mathbf{A}_{next}$: relation between the twist decomposition $\boldsymbol{\chi}$ of the transition twist and $\mathbf{h}_{d,next}$
$\mathbf{A}_{avoid}$: relation between the twist decomposition $\boldsymbol{\chi}$ of the transition twist and $\mathbf{h}_{d,avoid}$
$\mathbf{A}$: relation between the twist decomposition $\boldsymbol{\chi}$ of the transition twist and $\Delta\mathbf{h}_d$
$\mathbf{J}_{actsens}$: twist spanning set for the active sensing motion

MIX
Papier aus verantwortungsvollen Quellen
Paper from responsible sources
FSC® C105338

If you have any concerns about our products,
you can contact us on
ProductSafety@springernature.com

In case Publisher is established outside the EU,
the EU authorized representative is:
Springer Nature Customer Service Center GmbH
Europaplatz 3, 69115 Heidelberg, Germany

Printed by Libri Plureos GmbH
in Hamburg, Germany